ADVANCES IN BIOMEDICAL POLYMERS

POLYMER SCIENCE AND TECHNOLOGY

Recent volumes in the series:

Volume 23 POLYMERS IN MEDICINE: Biomedical and Pharmacological Applications
Edited by Emo Chiellini and Paolo Giusti

Volume 24 CROWN ETHERS AND PHASE TRANSFER CATALYSIS IN POLYMER SCIENCE
Edited by Lon J. Mathias and Charles E. Carraher, Jr.

Volume 25 NEW MONOMERS AND POLYMERS
Edited by Bill M. Culbertson and Charles U. Pittman, Jr.

Volume 26 POLYMER ADDITIVES
Edited by Jiri E. Kresta

Volume 27 MOLECULAR CHARACTERIZATION OF COMPOSITE INTERFACES
Edited by Hatsuo Ishida and Ganesh Kumar

Volume 28 POLYMERIC LIQUID CRYSTALS
Edited by Alexandre Blumstein

Volume 29 ADHESIVE CHEMISTRY
Edited by Lieng-Huang Lee

Volume 30 MICRODOMAINS IN POLYMER SOLUTIONS
Edited by Paul Dubin

Volume 31 ADVANCES IN POLYMER SYNTHESIS
Edited by Bill M. Culbertson and James E. McGrath

Volume 32 POLYMERIC MATERIALS IN MEDICATION
Edited by Charles G. Gebelein and Charles E. Carraher, Jr.

Volume 33 RENEWABLE-RESOURCE MATERIALS: New Polymer Sources
Edited by Charles E. Carraher, Jr., and L. H. Sperling

Volume 34 POLYMERS IN MEDICINE: Biomedical and Pharmacological Applications II
Edited by E. Chiellini, P. Giusti, C. Migliaresi, and L. Nicolais

Volume 35 ADVANCES IN BIOMEDICAL POLYMERS
Edited by Charles G. Gebelein

ADVANCES IN BIOMEDICAL POLYMERS

Edited by
Charles G. Gebelein
Youngstown State University
Youngstown, Ohio

PLENUM PRESS • NEW YORK AND LONDON

Library of Congress Cataloging in Publication Data

Advances in biomedical polymers.

(Polymer science and technology; v. 35)
"Proceedings of a symposium on advances in biomedical polymers, held September 8–15, 1985 at the 190th National American Chemistry Society Meeting in Chicago, Illinois"—Verso of t.p.
Sponsored by the Division of Polymeric Materials: Science & Engineering.
Includes bibliographical references and index.
1. Polymers in medicine—Congresses. 2. Prosthesis—Congresses. I. Gebelein, Charles G. II. American Chemistry Society. Meeting (190th: 1985: Chicago, Ill.) III. American Chemistry Society. Division of Polymeric Materials: Science and Engineering. IV. Series.
R857.P6A38 1986 610′.28 86-25415
ISBN 0-306-42467-3

Proceedings of a symposium on Advances in Biomedical Polymers, held September 8–15, 1985, at the 190th National American Chemistry Society meeting in Chicago, Illinois

A Division of Plenum Publishing Corporation
233 Spring Street, New York, N.Y. 10013

Printed in the United States of America

PREFACE

This book is derived from a Symposium held at the 190th National American Chemical Society Meeting, which was held in the Fall of 1985, in Chicago, and was sponsored by the Division of Polymeric Materials: Science & Engineering. This Symposium was, in turn, a follow-up on an earlier one held in Houston, TX, in the Spring of 1980, which was published as the book "Biomedical and Dental Applications of Polymers" [Plenum Press, New York, 1981]. In that earlier book, our opening Preface passage quoted King David, "I will praise Thee; for I am fearfully and wonderfully made..." (Psalm 139:14). As we noted five years ago, sickness of many types does occur in our wonderfully made bodies, but much human suffering can be aided by biomedical polymers. That earlier book considered much of the fantastic progress that had been made in biomedical polymers during the previous quarter century and brought many of these topics up to date. That Symposium, and book, noted that much help was available for the varied afflictions and problems that sometimes beset, and upset, our God-given bodies, and the promise of new and important advances was held out as a shining ray of hope amidst the gloom of sickness and affliction.

The present volume is an update on the advances that have occurred since the 1981 book and sets the stage for even greater advances in the future. The field of biomedical polymers has become increasingly important since 1980 and many symposia are now held each year. This is an indication of the partial fulfillment of the promises in this field. Unfortunately, not all problems have been solved in the five years between our Symposia and some new, difficult problems have arisen during this time. Such is the nature of scientific research. As we strive to solve these problems and alleviate human suffering we would do well to note the words of two important sages of the past - Moses and Jesus. Moses stated, while talking to the children of Israel, "I [God] will put none of these diseases upon thee, ... for I am the LORD that healeth thee" (Exodus 15:26). Centuries later Jesus remarked, to his followers, "... they shall lay hands on the sick, and they shall recover" (Mark 16:18). The problems of disease and afflictions still linger with us today, but much is being done in scientific research to solve these problems. Ways are being found to "lay hands" on the sick and help heal them with polymeric prosthetic devices. Controlled release medication, and related concepts, are enabling our bodies to recover more effectively from diseases that would have taken our lives only a few decades ago. Progress is being made, although slower than we might desire. This book is a report on part of this progress.

Most of the papers in this book are highly research oriented. This is proper because our earlier book covered much of the important background information on biomedical polymers. Like its predecessor, this present book spans a broad range of activity, but, in our opinion, these widely diverse areas can profitably learn from each other. One of our goals has always been to attempt to bring these apparently disassociated concepts into closer proximity so that they could cross-fertilize and produce better fruit than either concept alone. We did, however, resist the temptation to place the papers in a random order in the book, even though this might have aided in the cross-over process.

The thirty-three chapters follow a loose arrangement of Surgical Applications (5 papers), Dental Applications (2 papers), Orthopedic Uses (6 papers), Cardiovascular Applications (6 papers), New Cardiovascular Biomaterials (8 papers), and Other Biomedical Polymer Applications (6 papers). There is much overlap between these divisions, however. For example, polyurethanes are discussed in several sections and natural polymers, or modifications thereof, also are present in several "parts" of this book. For this reason alone, no clear distinction has been drawn between these "sections" which are more arbitrary than might be supposed. Nonetheless, some sense of order is necessary in order to develop the theme of this book, which is current advances in a rapidly moving field. These thirty-three papers consider such varied topics as nerve regeneration, facial surgery, denture liners, bone plates, replacement joints, pacemaker coatings, artificial blood vessels, artificial red blood cells, many kinds of polymeric biomaterials, and a potential diabetes treatment. The continuing thread in this book is the fact that these diverse endpoints all utilize polymeric materials, usually as the major component of these systems. Much, but not all, of the progress in biomaterials is occurring with polymers which are now being made specifically for this important application. This is in itself an important advance over what was being done only a few decades ago.

We would like to thank the officers of the Division of Polymeric Materials: Science and Engineering of the American Chemical Society for sponsoring the Symposium from which this book derives. Obviously I would like to thank each author for their papers describing their research work. The Session Chairpersons deserve a special note of thanks; these were: Dr. Roberto Benson (U. Utah), Dr. Eugene Goldberg (U. Florida), Dr. Samuel Huang (U. Connecticut), Dr. Smarajit Mitra (3M) and Dr. Danute Nitecki (Cetus). All of the manuscripts were typeset by CG Enterprises. Finally, I would like to thank my wife and the members of my family for their special help during the time in which this book was being formed.

Charles G. Gebelein,
Department of Chemistry
Youngstown State University
Youngstown, OH 44555

CONTENTS

REGENERATION OF SCIATIC NERVE ACROSS 15mm GAP BY USE OF A POLYMERIC TEMPLATE

I. V. Yannas[a], D. P. Orgill[a], J. Silver[b], T. V. Norregaard[c], N. T. Zervas[c], and W. C. Schoene[d]

(a) Fibers and Polymers Laboratories, Massachusetts Institute of Technology, Cambridge, MA 02139
(b) Department of Medicine, Case Western Reserve University, Cleveland, OH 44106
(c) Massachusetts General Hospital, Boston, MA 02114
(d) Department of Pathology (Neuropathology), Brigham and Women's Hospital, Boston, MA 02114

A porous biodegradable collagen-glycosaminoglycan bridge induced regeneration of myelinated and unmyelinated axons over large distances (15 mm) between the severed ends of the adult rat sciatic nerve. The newly formed strand of nerve tissue was well vascularized and sheathed in connective tissue 6 weeks following implantation of the polymeric guide. Nerve tissue did not grow in the absence of the polymeric guide.

INTRODUCTION

Previously we have shown that certain porous and biodegradable collagen-glycosaminoglycan (CG) polymers containing no growth factors or cells induce regeneration of a physiologic dermis at the site of full thickness skin wounds in guinea pigs (1,2) and humans (3,4). If seeded with epidermal cells CG polymers induce regeneration both of a new dermis and a new epidermis in full thickness skin wounds in guinea pigs (5).

We now present preliminary evidence that CG polymers containing no cells or growth factors induce regeneration of well vascularized nerve tissue when implanted as bridges over a 15 mm gap in the rat sciatic nerve. In this preliminary study we did not attempt to optimize the structural features of the CG bridge. Nor was restitution of sciatic nerve function studied in this exploratory phase. A detailed study is underway in our laboratories. However, the results obtained to date support the hypothesis (5) that highly porous, biodegradable CG polymers (2,6) can elicit and utilize the apparently underestimated regeneration potential of mammalian tissues other than skin.

MATERIALS AND METHODS

Test implants were fabricated by filling a 25 mm length of transparent silicone tubing, internal diameter 1.5 mm (Dow Corning, Midland MI), with an aqueous suspension of CG polymer, prepared as previously described (7). The filled tube was then placed on the shelf of a freeze dryer (Freezemobile, Virtis Corporation, Gardener, NY) and a highly porous CG plug was prepared thereby (8) (Figure 1). With microforceps, 5 mm of the CG plug was removed from each end of the tube to allow space for insertion of the severed nerve ends. The polymeric plugs were studied by use of scanning electron microscopy (mean pore size, about 120 μm), measurement of the average weight between crosslinks (about 44,000), and confirmation of the fully native content of collagen using infrared spectroscopy, as described elsewhere in detail (7).

Adult male rats (Wistar, approx. 300 g) were anesthetized with sodium pentobarbital. A 2 cm skin incision was made parallel and just posterior to the right femur. The avascular intermuscular plane between vastus lateralis and hamstring muscles was entered and followed to the loose fibroareolar tissue surrounding the sciatic nerve. The loose tissue was divided longitudinally thereby freeing the sciatic nerve over its full extent without devascularizing any portion. The sciatic nerve was sharply divided with microscissors. The proximal and distal nerve ends were abutted against a CG polymer bridge placed within a cylindrical silicone rubber tube. Three 10-0 nylon sutures through the epineurium and silicone rubber tubing secured the nerve ends in place. The length of the CG bridge, and thus the distance between the nerve ends, was 15 mm. To provide a 15 mm sciatic nerve gap, a cross anastomosis was performed between the cut ends of the proximal left sciatic nerve and the distal right sciatic nerve. A furrow was created in the spinal extensor muscles and the adjacent spinous process was removed. A 25 mm length of transparent silicone tube was placed in this furrow and the left proximal and right distal sciatic nerve was inserted 5 mm into the ends of the tube and fixed to it with three 10-0 nylon sutures as above. In control animals, the proximal and distal nerves were placed within a silicone rubber tube which contained no CG bridge, as described by Lundborg et al. [9]. After hemostasis was achieved the wound was closed with a single layer of 4-0 skin sutures. No antibiotics were administered. The proce-

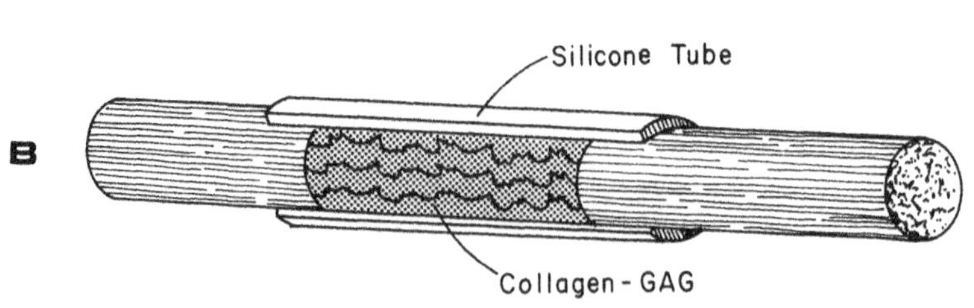

Figure 1. Schematic representation of the implant geometry. (A) No CG bridge (control); (B) CG bridge was used (test).

dure was performed with microsurgical technique using a surgical microscope.

Animals were perfused with mixed aldehydes at six weeks following implantation of the silicone tube implant. After perfusion, the implants were removed along the proximal and distal nerve ends. The silicone tube was slit longitudinally, removed, and the tissue contained therein was postfixed with the same mixed aldehydes, processed through ethanol and embedded in Spurr's plastic. Serial thick, one-micrometer or ultrathin sections were cut with an MT5000 DuPont Microtome. Sections were taken from the midpoint and distal three-quarter position of the tissue bridge. One micrometer sections were stained with toluidine blue and the ultrathin sections were stained with uranyl acetate and lead citrate. Electron microscope sections of the distal three quarter point of the explanted strand were used to prepare montages at 5500x in order to resolve the unmyelinated fibers of the smallest diameter. A total of 1/20 of the cross section area of the strand was carefully inspected for uniformity of distribution (excluding the perineurium); two small areas were then picked at random and the axons were counted.

A total of 10 animals were grafted; 5 animals were implanted with test grafts and 5 were implanted with controls. At six weeks the surviving four animals from the test group and five animals from the control group were sacrificed. Inspection of the implantation site showed that migration of the stumps had occurred in one of the test sites and in one of the control sites leading to a gap length of significantly less than 15 mm, while one of the control sites was infected; accordingly, the latter three implants were not included in subsequent studies. In the remaining three test implants and three control implants the distance between the two cut ends was measured following explantation and was found to be 15 $\pm$ 1 mm.

RESULTS

In spite of substantial variability in the data significant differences were observed at 6 weeks between test grafts and their controls (Table I; Figures 2-4). In all cases, the results were observed at the three quarter distal point of each regenerated strand.

Inspection of the morphometric data in Table I shows that essentially no axons were observed over the cross section of control strands. The experimental gap was not bridged in one of the control sites (Table I; animal no. 1). In the presence of CG polymer the average number of myelinated plus unmyelinated axons over the cross section of the regenerated strand was about 3000 and 8000 respectively. The variability observed from one test site to the other was large.

The average cross section area of regenerated strands formed in the presence of the CG polymer was about 0.8 mm^2 and was only 0.008 mm^2 when no CG polymer was used (Table I). Vascularization was extensive in two of three test strands; the capillaries were well formed, were endothelialized and basal lamina was present. Vascularization was present but poorly developed in the third test strand. No evidence of vascularization was observed in strands which formed in the absence of CG polymer. Test strands contained a perineural area which ensheathed an inner core and constituted about 38% of the total area. In contrast, control strands had no inner core but comprised only perineural tissue.

A gross examination of a test and a control strand are shown in Figure 2. Light microscopic views appear in Figures 3A and 3B. Figure 3A

Table I. Morphometric Data At the Three Quarter Distal Point of 15 mm Strands

(6 weeks following implantation)

Animal Number	Total Area Cross section $mm^2 \pm 1\%$	Inner Core Area $mm^2 \pm 5\%$	No. Fibers Unmyelinated $\pm 10\%$	No Fibers Myelinated $\pm 1\%$	Vascularization
A. Controls (no CG polymer in gap)					
9	0	0	0	0	none
3	0.01	0	4	2	none
8	0.014	0	0	0	none
B. Test Animals (gap filled with CG polymer)					
4	0.31	0.21	4,575	1,530	excellent
5	1.4	0.91	18,840	8,560	excellent
6	0.6	0.32	450	0	poor

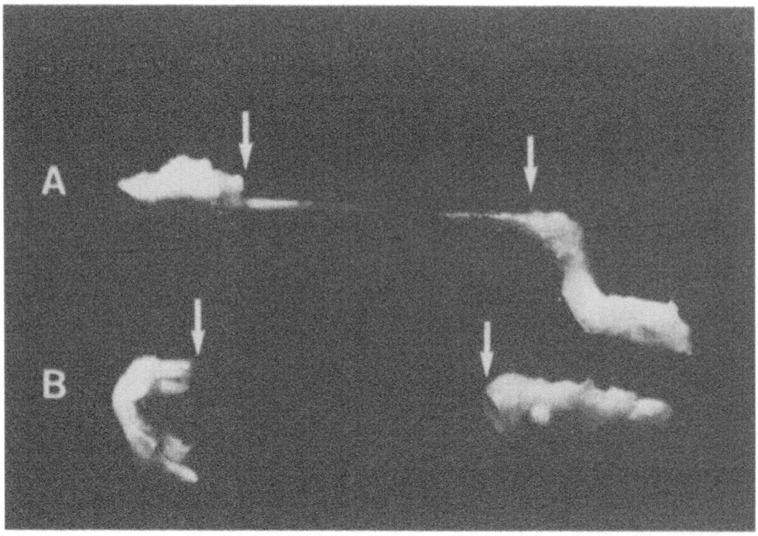

Figure 2. Gross view (3.3x) of sciatic nerve regenerating across a 15 mm gap at 6 weeks. Vascularized nerve tissue has almost totally replaced the biodegradable CG cylinder that originally bridged the experimental gap (Figure 2A; animal No. 6; Table I). In a control (Figure 2B; animal No. 8; Table I) where no CG cylinder was used, a very thin strand of nonvascularized tissue spanned the gap. Silicone tubing (see Figure 1) was removed from explants prior to photography.

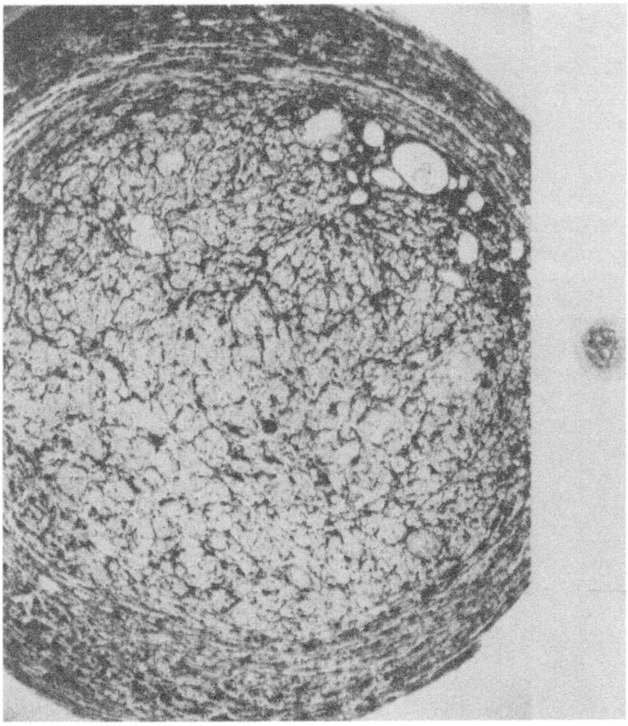

Figure 3A. Light micrograph (100x) of a toluidine blue stained, transverse section through a CG nerve bridge ("test"). The nerve guide was fixed with mixed aldehydes, processed through ethanol and embedded in Spurr's plastic. 1 m or ultrathin sections were cut with an MT5000 Dupont Microtome. The micron section was taken from the three quarter distal end of the regenerated strand. The CG guides were largely biodegraded by six weeks; however, very small residual strands of the CG polymer appeared as dark wavy lines or small lobes within the section. In test grafts large numbers of myelinated and even greater numbers of unmyelinated axons were seen thoughout the section by light microscopy and were also observed with the electron microscope (Figure 4A). In addition to Schwann cells and axons, a number of cell types, newly synthesized collagen and other extracellular matrices were observed. Rich vascular networks were seen interspersed throughout the cross section of test grafts. Dense, multiple wrappings of elongated perineural cells occupied the periphery of test grafts. This photograph is from animal No. 5 (Table I).

Figure 3B. Light micrograph (100x) of a control (no CG polymer used). Specimen preparation and magnification was as with Figure 3A. The cross section area of the test specimen (Figure 3A) is 100 times larger than that of the control (Figure 3B). This photograph is from animal No. 8 (Table I).

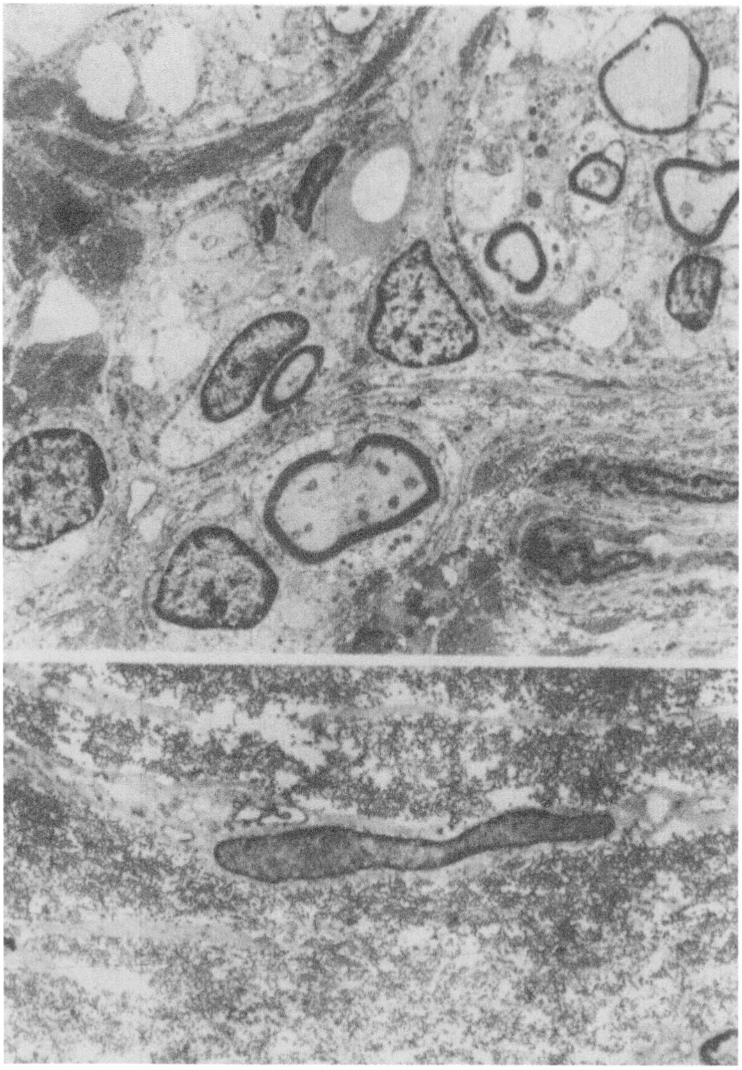

A

B

Figure 4A. An electron micrograph of the regenerating nerve in test animal No. 5 (Table I) obtained from the center of a cross section cut at the distal three quarter position of the 15 mm gap 6 weeks following grafting. CG polymer was used to bridge the gap at the time of grafting. Both myelinated and unmyelinated axons are present (4400x).

Figure 4B. In contrast, the control, also viewed at the center of the cross section, shows only newly synthesized collagens and an occasional fibroblast 6 weeks following grafting. In controls, no CG polymer was present in the gap at the time of grafting. This figure is from animal No. 3 (Table I) (7000x).

is a toluidine blue stained, transverse section of a test regenerated strand from animal No. 5 (Table I; CG polymer was present) and Figure 3B is a control explanted from animal No. 8 (Table I; no CG polymer was present). The interior of the test strand is shown to be densely populated with unmyelinated and myelinated axons, as well as Schwann cells. The inner core of the test strand contained a rich vascular network. There was no evidence of residual CG polymer. In contrast, the control strand shown in Figure 3B did not show axons or Schwann cells nor was vasculature observed.

Transmission electron microscopy of several regions in the explanted test strands showed large numbers of myelinated and even larger numbers of unmyelinated fibers throughout the cross section (Figure 4A; animal No. 5; Table I). In addition to Schwann cells and axons, a number of cell types, newly synthesized collagen and other extracellular matrices were observed. Rich vascular networks were seen. Dense, multiple wrappings of elongated perineural cells occupied the periphery of the test strands. Inspection at a comparable location within a control showed a very small number of axons in only one of three animals (animal No. 3; Table I). Blood vessels were not observed in any of the strands recovered from the three control animals. Controls consisted almost exclusively of newly synthesized collagen fibers and fibroblasts (Figure 4B; animal No. 3; Table I).

DISCUSSION

The large variablity in the data obtained with test implants is possibly due to lack of adequate control of certain aspects of the pore structure in individual implants. The freeze drying procedure used in the preparation of tubular implants yields CG polymers with apparently similar values of the average pore width as measured by scanning electron microscopy. However, microscopic examination of the implants also showed substantial channelling, with areas free of CG polymer adjacent to areas containing polymer with relatively small average pore size. We hypothesize that the differences in morphology among implants (Table I) is largely due to inability to control adequately the pore structure, and in particular the distribution of pore size, in this preliminary study. This problem is under study in our laboratory.

In a previous study, Lundborg, Dahlin, Danielsen, Gelberman, Longo, Powell and Varon (10) used empty silicone tubes and observed axonal regeneration across a 6 mm and a 10 mm gap but not across a 15 mm gap. In the series of explants which contained a 15 mm gap the authors (10) noted that, in three of five specimens, the proximal and distal stumps had small outgrowths extending less than 1 mm into the silicone chamber but bridging of the gap did not occur. In the remaining two specimens, however, bridging did occur. The authors studied these latter two specimens by light microscopy and reported "only amorphous material without axons or vessels" (10). Our observations with the controls (Table I) confirm the results of Lundborg, et al. (10) with the exception of a very small number of axons which was observed in one of our controls (animal No. 3; Table I). By direct comparison of our findings with those of Lundborg, et al., (9,10) we conclude that the copious synthesis of axons observed generally with test strands would not have taken place if the tubes did not contain CG polymer.

The density of myelinated and unmyelinated axons in the normal rat sciatic nerve has been recently reported as 8000 and 16000, respectively, per cross section, with 33% of the fibers myelinated (11). Comparison with the data in Table I shows that at least one regenerated test strand (animal No. 5; Table I) contained an approximately normal number of myelinated and unmyelinated axons.

A 15 mm gap has been previously bridged by regenerating axons using an autologous sural implant (12). In addition, several investigators have recently ensheathed lesions in the peripheral and central nervous systems using autologous tissues and polymeric devices (13-18), including collagen in various states (16,17). In the studies cited above (13-18), as well as in our own study, the effect of structure of the polymer on specific morphological events accompanying regeneration was not studied. It is, therefore, not yet possible to assign the observed regeneration events to specific features of the polymeric graft. Studies are currently under way in our laboratories.

The results obtained support the hypothesis (5) that highly porous, biodegradable CG polymers, free of exogenous growth factors, can be used to induce regeneration of tissues other than dermis and epidermis. We now hypothesize that CG polymers provide a spatial template which after being rapidly infiltrated by host cells and growth factors from a fresh wound, maintains these cells in specifically correlated "addresses" while the template is being degraded by host enzymes and new tissue is being synthesized. To explain the regeneration of peripheral nerve over a large gap we hypothesize that maintenance of infiltrating Schwann cells and endothelial cells at specific locations in an extracellular matrix over several days following wounding is a necessary, though probably not sufficient, condition for the observed regeneration. Future studies will address these unresolved issues. Nevertheless, our observations that skin (1-5) and peripheral nerve (19) can each be regenerated by use of polymeric templates following trauma suggest that the regenerative potential of adult mammalian tissues is substantial.

ACKNOWLEDGEMENT

We thank W. Nauta and Alan C. Nelson for the loan of a surgical microscope. We also thank F. O. Schmitt, V. Ingram, F. Solomon, R. D. G. McKay and E. Bizzi for useful discussions.

REFERENCES

1. I. V. Yannas, in: "The Surgical Wound", P. Dineen, Ed., Lea and Febiger, Philadelphia, 1981, Chapter 15, p. 171.
2. I. V. Yannas, J. F. Burke, P. L. Gordon and C. Hauang, U.S. Patent 4,060,081 Nov. 29, 1977; filed July 15, 1975.
3. I. V. Yannas, J. F. Burke, M. Warpehoski, P. Stasikelis, E. M. Skrabut, D. Orgill and D. J. Giard, Trans. Am. Soc. Artif. Organs, 27, 19 (1981).
4. J. F. Burke, I. V. Yannas, W. C. Quinby. Jr., C. C. Bondoc and W. K. Jung, Ann Surg., 194, 413 (1981).
5. I. V. Yannas, J. F. Burke, D. P. Orgill, and E. M. Skrabut, Science, 215, 174 (1982).
6. I. V. Yannas, J. F. Burke, C. Huang and P. L. Gordon, Poly. Preprints. Am. Chem. Soc. Polymer Chem. Div. 16(2), 209 (1975).
7. I. V. Yannas, J. F. Burke, P. L. Gordon, C. Huang and R. H. Rubenstein, J. Biomed. Mater. Res. 14, 107 (1980).
8. N. Dagalakis, J. Flink, P. Stasikelis, J. F. Burke, and I. V. Yannas, J. Biomed. Mater. Res. 14, 511 (1980).
9. G. Lundborg, R. H. Gelberman, F. M. Longo, H. C. Powell and S. F. Varon, J. Neuropath. Exp. Neurol. 41, 412 (1982).
10. G. Lundborg, L. B. Dahlin, N. Danielsen, R. H. Gelberman, F. M.

Longo, H. C. Powell and S. Varon, Exp. Neurol., 76, 361 (1982).
11. C-B. Jenq and R. E. Coggeshall, Soc. Neurosci. Abs., 15, 589 (1985).
12. H. Millesi, in: "Management of Peripheral Nerve Problems", G. E. Omer, Jr., and M. Spinner, Eds., W. B. Saunders Co., Philadelphia, 1980, Chapter 24.
13. L. R. Williams, H. C. Powell, G. Lundborg and S. Varon, Brain Res., 293, 210 (1984).
14. R. J. Keynes, W. G. Hopkins and C. L-H. Huang, Brain Res., 295, 275 (1984).
15. B. G. Uzman and G. M. Villegas, J. Neurosci. Res., 9, 325 (1983).
16. W. Colin and R. B. Donoff, J. Dental Res., 63, 987 (1984).
17. J. C. DeLaTorre, P. K. Hill, M. Gonzales-Carvajal and J. C. Parker, Jr., Exp. Neurol., 84, 188 (1984).
18. R. Madison, R. L. Sidman, E. Nyilas, T-C. Chiu and D. Greatorex, Exp. Neurol., 86, 448 (1984).
19. I. V. Yannas, D. P. Orgill, J. Silver, T. V. Norregard, N. T. Zervas and W. C. Schoene, Trans. Soc. Biomat., 8, 146 (1985).

TEMPORARY WOUND COVERING: EFFECT OF CHEMICAL MODIFICATION

Annie Y. C. Ma and Paul Y. Wang

Institute of Biomedical Engineering
University of Toronto
Toronto, Ontario, Canada M5S 1A8

Extensive skin loss resulting from thermal injuries poses traumatic and life threatening problems. The major causes of death are from bacterial infection and excessive fluid loss developed post-burn. Victims who manage to survive these early threats are, nonetheless, permanently damaged psychologically and physically by the disfiguring scars which arise from wound contraction. The absence of a template for cell propagation retards epithelialization and subsequent wound closure. Previous studies on hydrogels prepared from non-immunogenic dextran (mol. wt. 151,000) and impregnated with bacteriostatic agents have shown them to be effective in controlling infection. Therefore, it is desirable to modify these hydrogels such that they will also support the growth of cells. Since positively charged groups such as 2-diethyl-aminoethyl (DEAE) have been shown to facilitate cell attachment and growth in commercial microcarriers, we investigated the binding of DEAE groups onto preformed gels. A preliminary evaluation of the DEAD-treated hydrogels indicated that DEAE-binding significantly facilitates the propagation and growth of cells on the hydrogel surface.

INTRODUCTION

One of the primary objectives of a temporary wound covering is to reduce evaporative loss and absorb exudate, both of which delay healing as well as present life threatening problems. With these problems under control, bacterial infection developed postburn becomes the major cause of death. Dressings or the more expensive xenograft have only been partially satisfactory in alleviating these problems because the plastic sheets often fail to seal the wound properly against bacterial ingress, while the porcine xenograft suffers from possible antigenic reactions (1,2). Recent use of hydrogels as a wound covering has been promising. Preliminary evaluation of the dextran hydrogel has shown that it functions effectively as a barrier against fluid loss and that gels impregnated with silver nitrate are effective against bacterial infection (3).

Another objective in burn treatment is to minimize the degree of wound contraction during healing and hence eliminate the physically and psychologically disfiguring scars of the burn victims. Wound contraction upon healing is not as acute in second degree as in third degree burns. Nevertheless, it poses a major obstacle that the victims have to overcome in rehabilitation. The wound contraction results from the inability of cells to migrate and propagate across the wound surface. Therefore, it is desirable to construct a wound covering that would also function as a temporary anchorage for the propagation of cells during the healing process. The proper growth of cells over the wound surface will eliminate the problem of wound contaction.

One of the major factors which promotes the anchorage and spreading of cells is the presence of charged groups on the surface on which the cells can propagate, because this mimics the charge and surface interactions that the cells experience in vivo. Recent work on cell culture technology, namely the microcarriers, has demonstrated the effectiveness of positively charged groups associated with the beads, in the improvement of cell growth (4). One example is the commercially available microcarrier, cytodex 1, which contains 2-diethylaminoethyl (DEAE) groups attached to cross-linked dextran beads. Therefore we decided to investigate the binding of DEAE onto preformed dextran sheets and the effect of these charged groups on the growth and proliferation of cells on the gel surface.

MATERIALS AND METHODS

Preparation of dextran gel

Dextran was obtained from Sigma Chemical Co. (industrial grade, average molecular weight 151,000). The hydrogel sheet was prepared using epiclorohydrin (Aldrich Chemicals) as the cross-linking agent according to a procedure described previously (5). The preformed gels were then soaked in excess distilled water for 5-12 hours before the reaction with DEAE.

The DEAE reaction

The gels were treated with DEAE (BDH Chemical) in an aqueous solution containing NaOH. This allows the incorporation of the DEAE groups onto the gel. The reaction time was 60 minutes at 60 $\pm$ 3° C. After the reaction, the gels were washed thoroughly with distilled water, followed by 0.1 N HCL. The amount of DEAE groups present on the gel was determined using the silver nitrate titration procedure described by Lavine et al (6). In a typical run, the gel was soaked in 50 mL of a 10% w/w Na_2SO_4 solution for one hour before the solution was titrated with 0.1 N $AgNO_3$ using 5% potassium chromate solution as the indicator (Fisher). It takes about three volumes of Na_2SO_4 to complete the displacement of all Cl-ions associated with the DEAE groups. The gels were then washed thoroughly with water and sterilized using UV radiation before they were used in the cell culture experiments.

Cell culture

Normal mouse fibroblasts BALB/3T3, clone A31 (ATCC NO. CCL 163) were obtained frozen at passage 64 from the American type Culture Collection (Rockville, MD). The cells were grown to confluence in 25 mm^2 tissue culture flasks (Falcon). prior to inoculation confluent cultures were incubated in a 0.25% trypsin solution prepared in Hanks balanced salt solution and resuspended in the medium. The medium used was Dulbeccos-modified Eagles medium supplemented with 10% bovine calf serum (Flow Lab, Virginia). All cell cultures were maintained at 37° C, in a 95% air: 5%

CO_2 atmosphere. Cell number was determined by counting the cells released by 0.25% trypsin. Both the control and DEAE-treated dextran gels containing different degrees of DEAE substitution were seeded with 1.5 x 10^5 cells and incubated for 5-10 days before the cells were leased and enumerated. The size of the gel was 7 cm^2.

RESULTS

In order to establish the optimum conditions for the reaction of dextran hydrogel with DEAE, the following experiments were carried out to correlate the reactant concentrations and reaction time to the extent of DEAE binding. Table I shows the effect of DEAE concentrations used on the extent of DEAE binding. The dextran sheet was cast into a polypropylene jar and therefore only the surface of the gel was exposed to the alkaline DEAE solution. However, titration results indicate that the extent of reaction was not limited to the surface of the gel. When five pieces of the reacted gel were equilibrated in a large excess of 10% w/w Na_2SO_4 solution for different time periods (Table II), subsequent titrations showed that this exchange reaction is time dependent. The results clearly show that about 60 minutes equilibration time is necessary to achieve equilibrium. For a typical run, three volumes of solution with equilibration time of 1 hour each are adequate for the complete displacement of the Cl-ions. Additional experiments indicate that when a short equilibration time (5-10 mins) is used then 7-10 volumes of the Na_2SO_4 solution are necessary for complete displacement.

The results from Table I show that as the reactant concentration is increased, more DEAE groups are bound to the gel. The amount of DEAE bound is observed to level off quite rapidly as the DEAE concentration increases. A similar trend is observed for the effect of the reaction time on DEAE binding. The results are summarized in Table III.

Studies have shown that with the microcarriers, the extent of DEAE substitution is a critical parameter which affects the propagation of cells in the culture (6). This effect was also observed in our study. When a series of DEAE-treated gels were inoculated with cells, initial attachment and spreading of cells was observed on the gel surface. Within 5 days, the cells were found to withdraw from the surface, with the formation of large clumps. Successful spreading and growth of cells was observed for one set of DEAE gels which showed greater than 80% coverage of the gel surface, 5 days after inoculation (Table IV).

Table I. The effect of soaking time on the displacement of Cl^- ions associated with the bound DEAE groups.

Soaking time/mins	mmol Cl^-/g of gel
30	0.018
60	0.045
120	0.042
180	0.042
240	0.042

Table II. The effect of DEAE concentration on the number of DEAE groups covalently bonded to the dextran gel.

DEAE (mmol/g gel)	DEAE bound (mmol/g gel)
0.28	0.070
0.64	0.12
1.1	0.16
1.8	0.18

DISCUSSION

The dressing method has been used extensively for burn treatment to prevent excessive evaporation loss, wound contraction and infection. The healing process takes place through the epithelialization and regeneration of dermal tissues. However, in the full-thickness burn, re-epithelialization is hindered by the absence of substrate for the growth and spreading of cells across the wound surface. Therefore, a wound covering which also facilitates and directs epithelial migration and growth of cells is desirable.

In cell culture technology, factors which affect the cell attachment, spreading and growth have been well documented (7,8). The most important of all is the state of the microcarrier surface, namely the charge distribution and the presence of absorbed macromolecules on the surface. Cell adhesion studies with a number of hydrogel materials where the surfaces were modified by protein adsorption have been reported (9) and wound covering design based on this approach has been actively pursued by others (10). Our objective is to study the effectiveness of charged groups attached to the preformed hydrogel on the propagation of cells.

Results from Table II showed that DEAE groups can be incorporated into the hydrogel through either linkages onto the -D-Glucose backbone. Under the reaction conditions used in these experiments, the maximum concentration of DEAE that could be introduced was about 0.18 mmol/g gel. The concentration could be further increased in a linear fashion by an increase in the reaction time.

Early studies with DEAE-Sephadex have shown that the degree of sub-

Table III. The effect of the reaction time on the number of DEAE groups that bind to the dextran gel (16 mmol of DEAE was used in the reaction)

Reaction time (min.)	mmol DEAE
30	1.05
60	1.63
90	1.82
120	2.41

Table IV. The effect of DEAE substitution on the growth of cells.

Gel	cells/cm^2
DEAE-treated gel (0.014 mmol/cm^2)	1.8×10^5
Control gel	8.7×10^3

stitution greatly influences cell toxicity effects (6). Our results also showed the clumping of cells that were initially growing on the gel surface after 5 days. This suggested that the growth of cells was sensitive to the degree of DEAE substitution. The limited spreading of cells observed in the majority of the DEAE-treated gels may be related to the distribution of the DEAE groups on the gel surface. Of special significance is the fact that there was a large difference in the number of cells that attached and grew on the DEAE-treated gel and the control gel. This clearly demonstrated that, while the hydrogel itself does not have good cell adhesion properties, the presence of DEAE groups on the gel does result in a better substrate for supporting cell growth. Therefore, a wound covering containing DEAE-treated hydrogel should facilitate the wound healing process by functioning as a substrate and providing a directional influence for cell propagation.

REFERENCES

1. G. B. Park, J. M. Courtney, A. McNair and J. D. Gaylor, Engineering in Medicine, 7, 11 (1978).
2. H. M. Bruch in: "Basic Problems in Burns", R. Vrabec, Z. Konickova and J. Moserova, eds., Springer-Verlag, New York, 1975, p 17.
3. P. Y. Wang in: "Polymeric Materials and Artificial Organs", C. G. Gebelein, ed., American Chem. Soc. Symp. Series No. 256, American Chemical Society, Washington, 1984, p 181.
4. M. Hirtenstein, J. Clark, G. Lindgren and P. Vretblad, Develop. Biol. Standard, 46, 109 (1980).
5. P. Y. Wang and N. Samji, Poly. Sci. Technol., 14, 29 (1980).
6. D. W. Levine, D. I. C. Wang and W. G. Thilly, Biotechnol. Bioeng., 21, 821 (1979).
7. D. W. Levine, W. G. Thilly and D. I. C. Wang, Develop. Biol. Standard, 42, 159 (1979).
8. S. Reuveny, A. Mizrahi, M. Kotler and A. Freeman, Biotech. Bioeng., 25, 469 (1983).
9. R. D. Ratner, H. Horbett, A. S. Hoffman and S. D. Hanschka, J. Biomed. Mater. Res., 9, 407 (1975).
10. I. V. Yannis and J. F. Burke, J. Biomed. Mater. Res., 14, 65 (1980).

CUSTOM MADE SILICONE IMPLANTS IN FACIAL CONTOUR RECONSTRUCTION

Mutaz B. Habal

Tampa Bay Craniofacial Center, Institute of Human Resource
University of South Florida, Tampa, Florida 33624

The poly(dimethylsiloxane) (silicones) are one of the most widely used implantable biomaterial in the eighties. It's use as an implantable polymer with wide clinical application was achieved after success in implantation to correct breast deformities. In the preceding years over 100,000 women had silicone implants to improve their breast contour, or to correct the deformity created by radical resection for malignant disease. The silicone can be also polymerized to create harder biomaterial that can be implanted in the facial region. These polymerized implants are best used when they are custom fabricated to fit the defects created by the surgical resection. Two areas of such contour abnormalities are present in the face. The non-dynamic area (Figure 1), such as the forehead and maxilla, is either conserved and/or contoured, while the dynamic region (Figure 2), such as the mandible, requires complete fixation and immobilization with the adjacent structures for a perfect fit and a good final result.

The polysiloxanes (silicones) are widely used today as inorganic implants or tissue substitutes in biological systems as the best biocompatible material. The present applications of these biomaterials developed from the search for an ideal biocompatible implant. These inorganic materials are used externally (explants) or, more frequently, placed subcutaneously to replace parts of the body (implants). These implants are used to substitute for a missing part, to reposition parts, to augment an existing unit of the human body, or to serve a functional purpose. During the past decade, experience with this biomaterial has advanced to higher levels of sophistication.

The inert polysiloxanes can be prepared in many different forms which vary from hard rubber to the liquid polymer that can be injected through a small-caliber hypodermic needle. Use of the polysiloxanes is widely accepted in almost every medical and dental specialty, perhaps most commonly in reconstructive surgery. This wide use of the silicones has spawned the development of special industries that focus their entire

Address correspondence to: Dr. Habal, 4211 Carrollwood Village Drive, Tampa, FL 33624

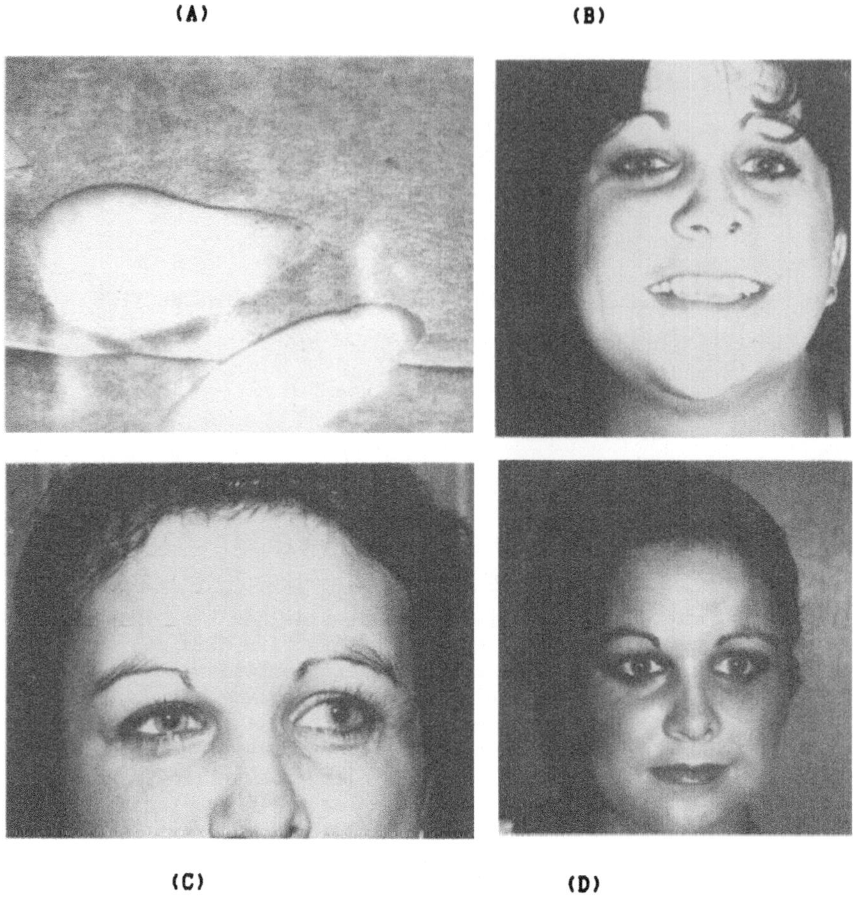

Figure 1. Patient with contour deformity treated with prefabricated silicone medical-grade implant.

research and production on the reactions of the body to the silicone implants, as well as the solubility, quality controls, and physical properties of the silicones themselves. The biological applications of the silicones and the interaction between the body as a whole and this inorganic material is the main interest of the surgeon and the bioengineer. The polysiloxanes that are called medical-grade silicones, are high grade materials that are produced pure (without any additives), and are recommended for implant use. To understand the biological basis for the silicones clinical application requires examination of their biophysical and biochemical properties (1,2).

From a historical background, F. S. Kippling, who lived in the late 1800's and early 1900's, is considered the founding father of silicone chemistry. Later, around the time of World War II, two chemists working independently investigated silicone as a possible tissue substitute for further development in the plastics industry. At the end of the war, a

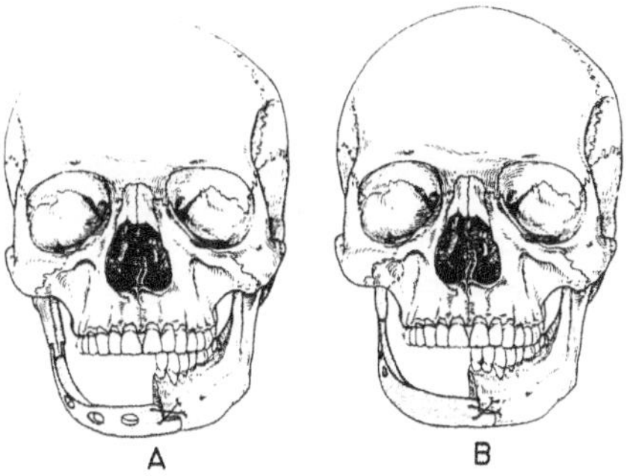

Figure 2. Patient with tumor of the mandible, after major resection with immediate reconstruction. The patient was able to be rehabilitated with minimal contour deformity, complete preservation of the oral splinter, and absence of drooling - a major step when further surgery and/or radiation therapy is anticipated.

classic publication on the toxicology of silicone in industry described its effect when accidentally injected under the skin. DeNicola was credited with the first implantation in 1959; he used silicone rubber tubing to replace a urethra. Subsequently, other authors reported their experience with the use of silicone fluid in medicine. In 1955, Holter developed the first successful shunt made of silicone for drainage of excess fluid in children with hydrocephalic condition. Work done by Marzoni and his group further studied the biological reaction of tissues to silicone fluid injections. A major advance was made in 1961, when Cronin and Gerow developed a mammary prosthesis that used the new material in a baglike fashion, now one of the most frequent and popular uses of silicone in clinical medicine (1,4).

The chemistry of silicone is very interesting, the basic element silicone, under certain circumstances, unites with oxygen to make silica. Silicone can unite with other elements to create compounds that are formed in nature and are used by man. Such compounds do not appear to be stable in spite of the simplicity of their reactions. In the formation of polysiloxanes, silica rock is heated, causing it to give up silicone. The silicone is treated with methyl chloride, producing dimethyl dichlorosilane, which forms dimethylsilanediol when treated with water. This last compound is unstable and it condenses with neighboring molecules to form long chains of silicone-oxygen molecules with two methyl groups on each silicon atom. The length of the polymer chain determines the viscosity; an increase in viscosity is paralleled by the number of units in the chain.

Most medical-grade silicones are of the poly(dimethyl silicone) group. The viscosity of the polysiloxanes with two units is 0.65 centistokes (cS), and the viscosity of a 200 unit polysiloxane chain is 1,000 cS. The usual medical-grade polysiloxane is one with very high viscosity to which silica has been added for strength. There are two forms of the medical-grade polysiloxane: the heat-vulcanizing (HV) forms and the room temperature-vulcanizing (RTV) form. The latter is used to reconstitute certain delicate structures and to form a quick mold, usually in the operating room. Stannous octoate is used as the catalyst for RTV silicone. It is essential to wait until the catalyst is completely consumed by chemical reaction before implanting this material under the skin, because the catalyst is noxious to the tissues and could produce a slight internal chemical burn.

For HV silicone, dichlorobenzoyl peroxide is used as the catalyst during vulcanization when the polymer chains crosslink, changing the liquid to a solid. The heat used in the HV silicone causes the catalyst decompose leaving only inert material. This is particularly important in subdermal implantation and characterizes the superiority of the HV silicone over the RTV silicone.

The structure of the silicone thus depends on the number of units of dimethylsiloxane group in the polymer, which in turn determines the biophysical density of the materials to be used.

Silicones for most medical uses are fabricated by one manufacturer (Dow Corning. Midland. Mich.). Other small companies fabricate silicone for medical and nonmedical uses around the world. The basic chemical composition and manufacturing techniques are the same. Silicones are among the very few bioimplantable materials that are presterilized. Sterilization packages were perfected during the middle 1960's to satisfy bacteriologists and microbiologists in the medical and industrial fields.

The surface of silicone is smooth, therefore, when implanted under the skin, it does not allow the ingrowth of blood vessels or fibrous tissue. The silicones are also nonwelding agents and are used on the surface of the glass to prevent the adherence of the liquid material to the surface of the glassware.

Silicone is one of the most inert materials available for use in the medical field today. By definition, inert means that there is minimal biological reactivity toward the material when it is implanted in the human body. When silicone is injected subdermally as a fluid, however, there is a giant-cell reaction with macrophage vaculization of the injected material. Such implants are called inorganic to differentiate them from organic or absorbable implants that are placed in the biological systems.

The toxicology and properties described herein relate primarily to medical-grade silicone. Other silicones, thousands of which have been developed and are used in industry, do not have all these properties. In particular, many are not inert. Additives that are used to make these silicones heat stable decrease their physiological acceptance by the biological system. Under no circumstances should any polysiloxane not identified as medical-grade be implanted surgically. The polysiloxane produced for medical use meets precise specifications, is manufactured under strict condition, and is prepared in a bacteria-free environment.

The rubber, variety of silicone has low tensile strength, so it cannot be used to replace regions bearing major stress or needing the structure of hard tissue (Figure 3). Where strength is needed, reinforcement of silicone with stainless steel netting or solid wires can produce the proper form with "shock-absorbing" capacity. The degree of polymerization is important in the final formulation of the silicone rubbers, which are the hardest of the polysiloxanes in use. Polysiloxanes should not be used in areas of high tensions, since they cannot withstand the major applied physical forces. However, this lack of proper strength is compensated by their mechanical durability. Silicone rubber can be flexed in stress forms for changing position over one million cycles without fragmentation. Other mechanical properties of the silicones are their

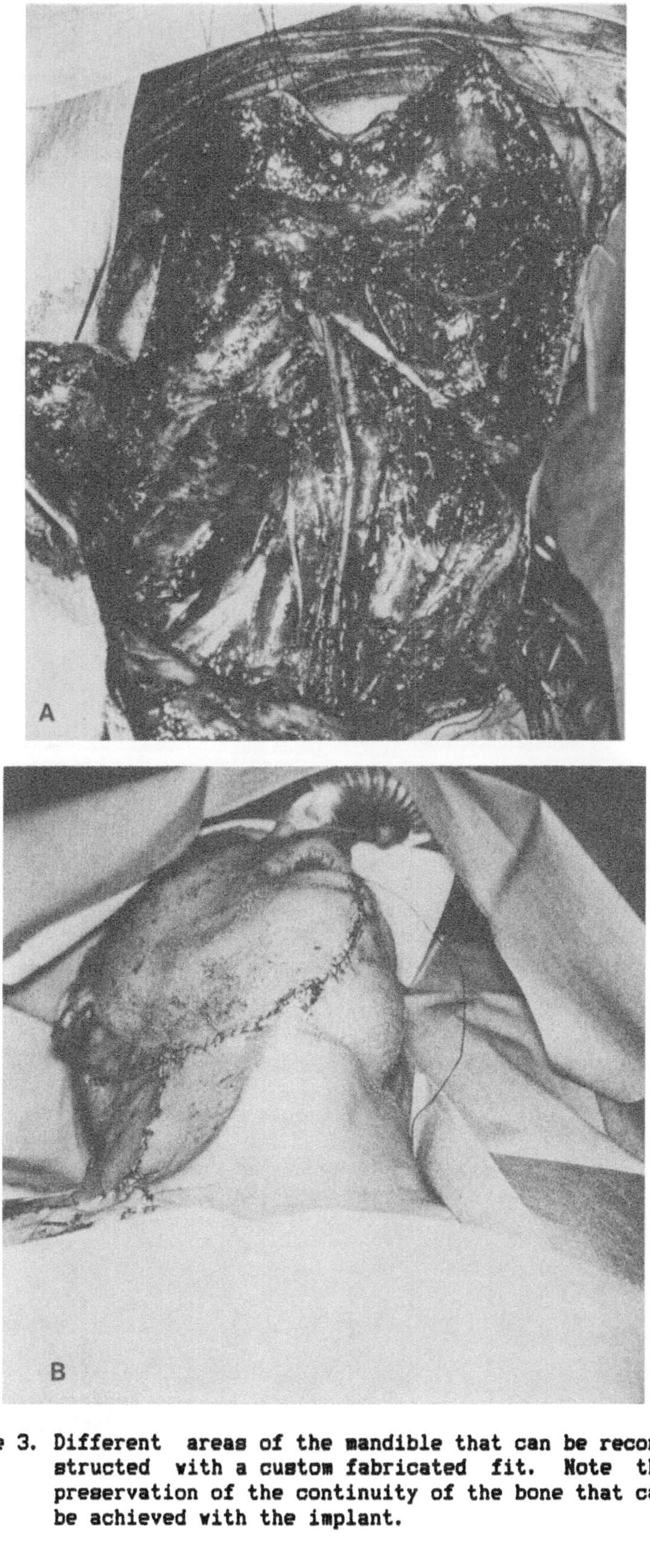

Figure 3. Different areas of the mandible that can be reconstructed with a custom fabricated fit. Note the preservation of the continuity of the bone that can be achieved with the implant.

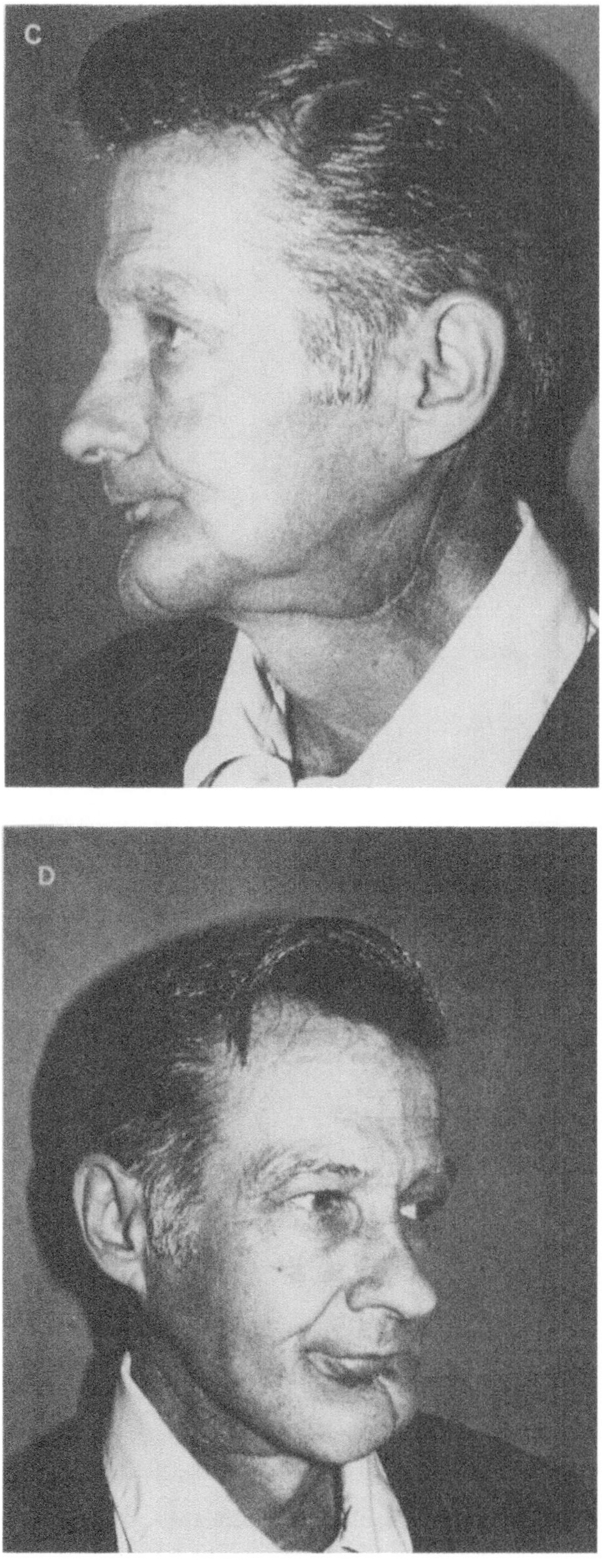

Figure 3. Continued.

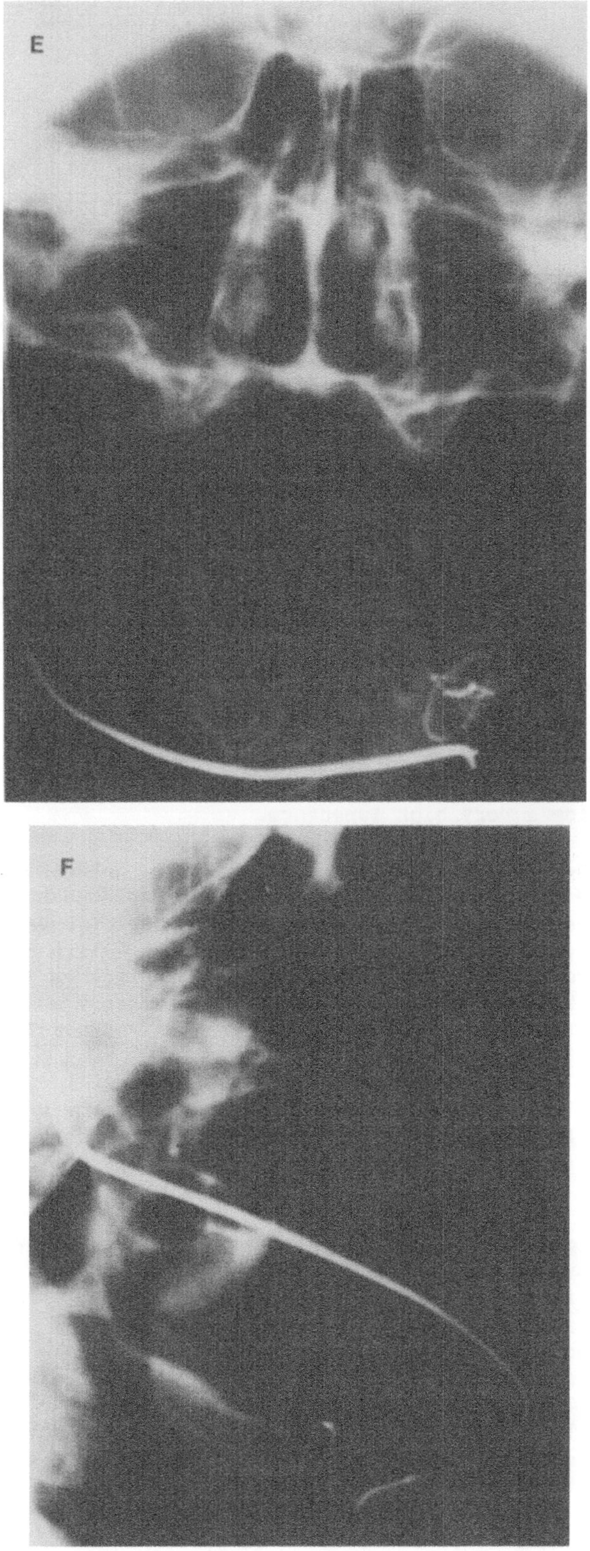

Figure 3. Concluded.

softness, texture, and conformity, which makes them ideal substitutes for soft-tissue replacement.

The chemical properties of silicone are primarily characterized by their lack of reactivity when placed under the soft tissue in the biological system. The silicones do not produce an acute foreign body reaction in the tissue with immediate destruction. They are not chemically toxic or abrasive to the adjacent biologic structures. There is not an immunological reactivity to their presence, and they do not act as a hapten to combine with other proteins in the body to produce an antigen-antibody complex with intense immunological reaction. While their presence goes mainly unnoticed, there is a chronic form of reaction around them, with minimal inflammatory reaction. Silicone's surface chemistry allows certain chemical to be absorbed on it so they can be used while the silicone is implanted to increase their physical and chemical surface properties. Heparin sulfate decreases fibrous tissue reaction, and povidone (polyvinyl pyrrolidone) reduces the desiccation at the silicone-tissue interface and reduces fibrous capsule formation. All these compounds are used with a slight change in surface properties, which decrease the fibrous pseudocapsule formed around them. The silicones are not carcinogens.

The electrical properties of the silicones are not well understood. The surface provides a major electronegative static charge. Therefore, they can be used as a nonadherent surface for all particles that are electronegative. This is particularly important when used in blood collecting and transferring systems. Silicones are poor conductors of electricity and there is no electrolysis noted during their implantation. Silicone is an excellent dielectric, developing a static charge that attracts other objects such as lint. Their electrical resistance does not change after long term immersion in saline solution or other fluids. The silicones provide a very high impedance.

The surface properties of the silicones are interesting. The body reacts to the implant by producing a fibrous pseudocapsule to surround it. This is basically an internal scar; a group of specialized fibroblasts, myofibroblasts, forms the pseudocapsule, which is composed of a layer of fibrous tissue. Such capsules are advantageous when silicone is used as a hard tissue substitute, but disadvantageous when the silicone is used as a soft tissue substitute, as spherical contracture will negate the purpose of their use.

When the silicones are used as a unit to transfer visual images, such as the intraocular lens, another polar material is coated on them. This substance minimizes the fibrous tissue formation by acting as a wetting agent on the hydrophobic surface of the silicone, which is known to enhance scar formation around the implant and is implicated as the basic causal factor of the fibrous tissue reaction. In heavy implants that do not need to move in the tissue, fixation patches made of material with fenestrations larger than 50 um in diameter are used, resulting in the ingrowth of a strong net of fibrous tissue through the patches, which completely immobilize the implant. Such is called an "anchoring patch". On the other hand, capsules of fibrous tissue are not desirable in areas where free mobility is required around the implant. Toxic side effects of the steroids will be maximized in the soft tissue around these implants, however, and these effects are dose related, which in part limits the use of the corticosteroids in such applications as scar dissolvers.

The biocompatibility of silicone is primarily evidenced by the inability of the body to reject such implants. Neither long or short term rejection has been noted. No major cause for any immunologic rejection has been reported.

Some toxic effects may be related to the method of sterilization. The silicones are sterilized with conventional materials and autoclaves. Chemicals that may adsorb on the surface should not be used, because such chemicals will produce a toxic reaction in the body when the silicones are implanted. Thus, gas sterilization may not be advantageous in certain instances.

The major questions remaining concern long term biocompatibility and the development of cancer. To date, there is not reported observation or evidence to indicate that a tumorigenic response has developed in the presence of these implants available today. Since the early 1960's when silicones were first implanted in human beings, various groups have studied the carcinogenicity of these substances and found no correlation with tumor formation. Studies also focused on the tumorigenic response to silicone polymers. The silicones are neither promoters nor inducers of malignant disease. To date, the silicones are not known to have any carcinogenic effect in human beings, nor does a tumor near the implant receive any enhancing biophysical or biochemical effect from being adjacent to a silicone implant (3).

The polysiloxanes do not induce an immunological reaction in human beings or in experimental work. They also do not act as haptens with other chemicals in human beings. However, there may be a change in the immunological system in patients who have silicone implants. Changes in the efferent or the afferent immune limbs or alterations in lymphocyte functions, may lead to some manifestation later on. Some changes were noted in experimental tumor design; however, further experimentation is required.

CLINICAL APPLICATIONS

The use of silicones is limited to necessary replacement of soft tissue or tubular structures, and to coating other materials that do not have the inertness of medical grade silicones. However, silicone implants have also been custom fabricated in individual laboratories for specific fits in patients with a contour deformity; for example, as maxillofacial prosthesics or as replacements for parts of skeletal structures after ablative procedures for cancer or after extensive trauma (Figure 1).

Outside the surgical field, silicones are used externally as explants, which are fixed or glued on the patient's skin as a substitute for a missing part. Explants can be custom made to a specific shape, and color is added for the perfect match to the surrounding skin. A tissue adhesive is used to affix the explant to the skin. The most common use of such silicones is as external ear prosthesics. Other uses include replacing parts of the face lost due to trauma or resection for cancer. Because of the low degree of reactivity, such prosthesics can be used internally in areas of bony resection to rebuild a lost structure (4). The most common application is an obdurator for palatal lift or for the maxilla after maxillectomy. Silicone tubes are used as a lacrimal drainage system in patients with severe obstruction of the nasolacrimal duct and for ventilation of the middle ear. In all of these applications, the silicones are used externally, and the advantage of their use is the lack of any external tissue reactivity. Under no circumstances does the patient become allergic to their presence on the skin, since the silicones are not haptens (Figures 4 and 5).

Silicone is used in two ways in facial contour surgery. First, custom made parts are used for onlays; these components can be used in the area of the malar bone to augment the chin in microgenia, or over the fore-

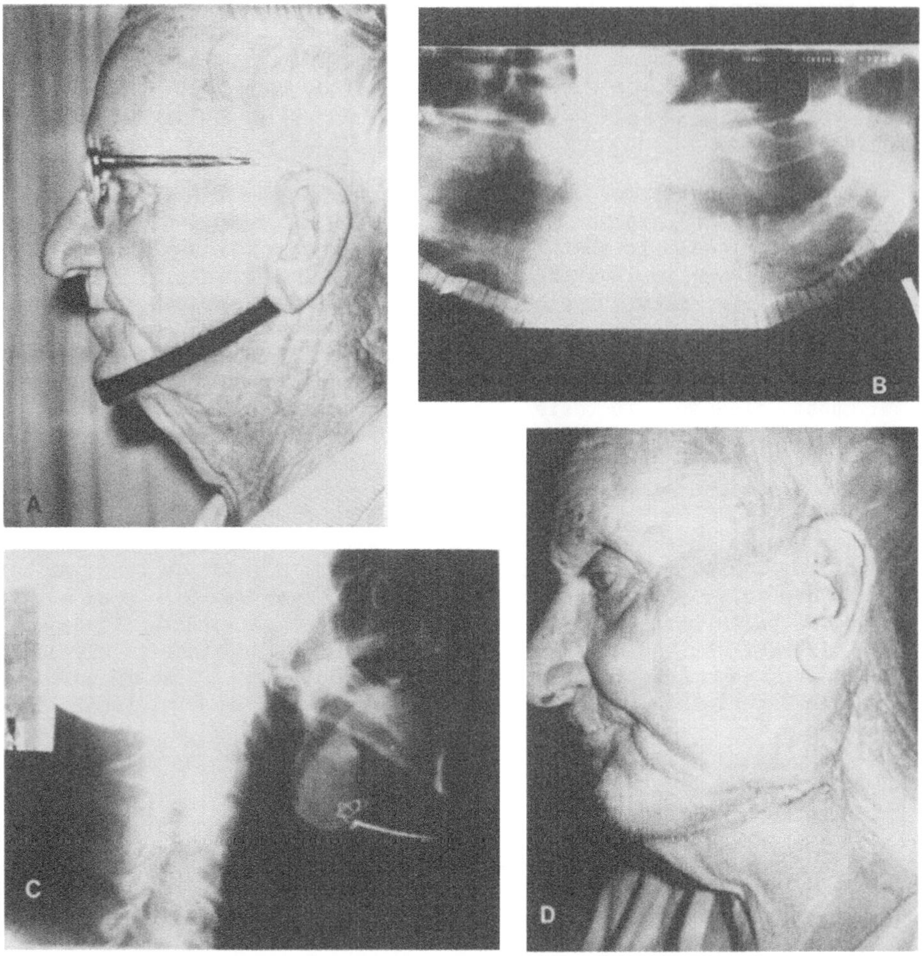

Figure 4. Patient with tumor of the mandible. Preoperative planning included a radiograph of the mandible and proper fit was prefabricated. The tumor was resected and complete immobilization was achieved with interosseous wiring.

head to smooth it (5). Second, missing parts are replaced for discontinuity defects; for example, to replace the mandible, pieces are interposed between the facial bones. Reinforcement of the silicone with stainless steel Kirschner wires is necessary. The soft silicone can be used as a soft frame to reconstruct the missing ear in patients with microtia. Some of the parts most commonly used for reconstruction are produced commercially for use when custom fabrication is not available.

Injectable silicone has been used for many years by certain qualified investigators, but it is not yet available for general use. It is injected into the dermis in the treatment of certain diseases and for certain deformities, thus having very limited indication. Restriction of its use will continue until its pathophysiology, tissue reactivity, and fate are better understood (Figure 6).

Polysiloxane lenses coated with wetting agent such as povidone have

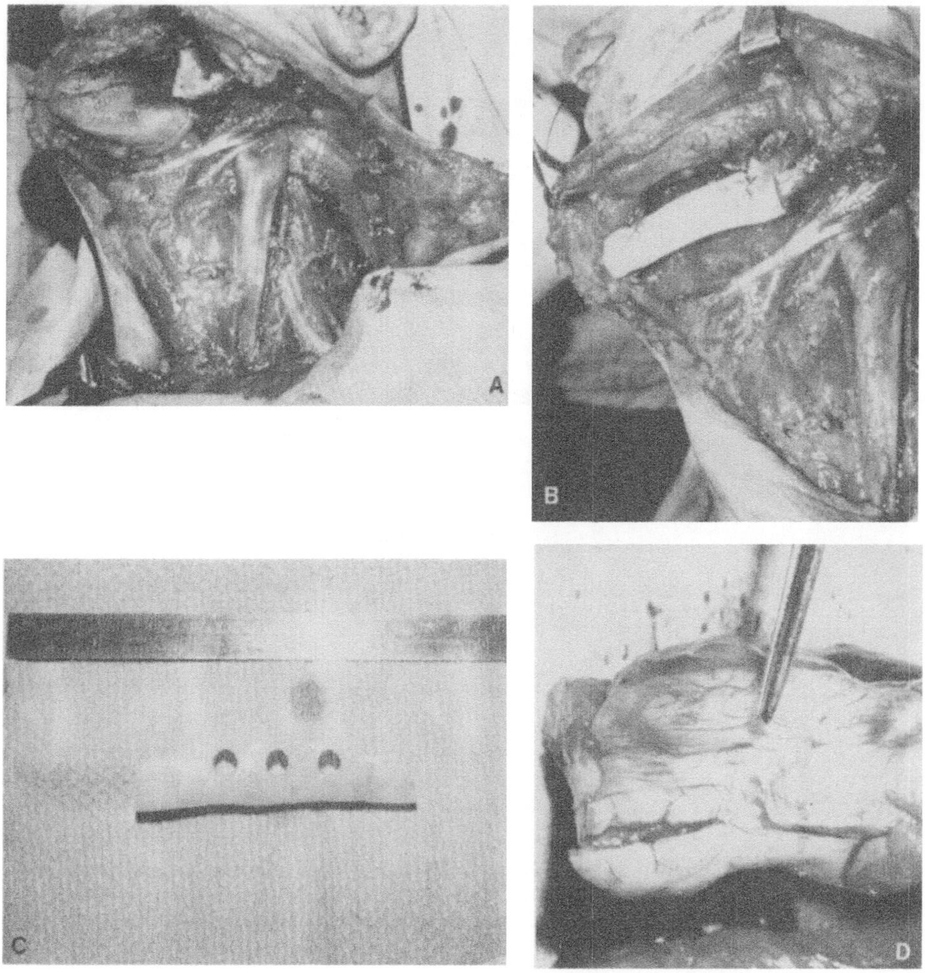

Figure 5. The intraoperative photographs after resection and after reconstruction with a silicone implant in a patient with malignant disease of the mandible.

been used for patients after cataract surgery. The results have been encouraging. The wetting agent has decreased the scar around the implant and the destruction of the arterior chamber endothelium. This scar produces opacification as a thin film in front of the implant, thus decreasing the transparency of the silicone implanted. Povidone prevents such interference. The injectable silicone is used for relief of various intraocular problems.

The future research on polysiloxanes will focus on minimizing the body's fibrous capsule formation around silicone implants so they can be better used for soft tissue replacement. The last decade has witnessed a tremendous improvement in the use and understanding of the silicones. We still have a long path to follow, however, before we reach the ideal state of the art in biological inorganic implantation, where we can take from the shelf and replace every component of the human body. We need to understand more about its chemistry.

The use of wetting agents such as povidone on the surface of the

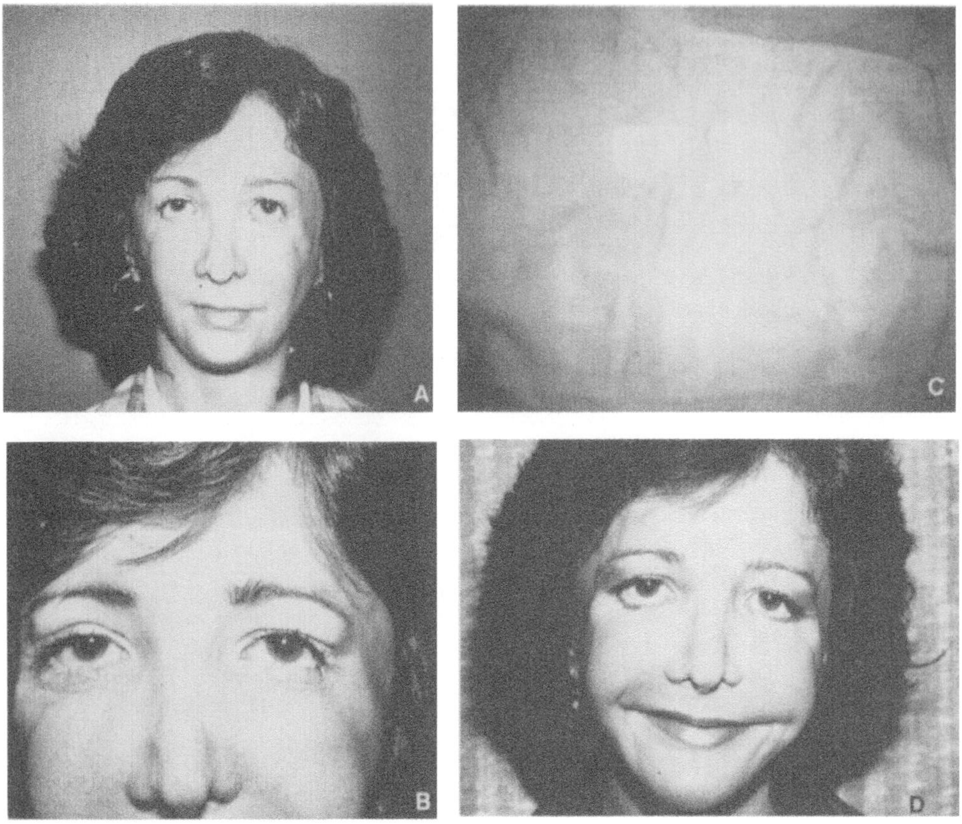

Figure 6. Patient with contour deformity which was congenital in nature. The orbits were contoured with custom fabricated silicone implants.

silicone has a promising future. Povidone is very easily adsorbed in the surface of silicone by using x-rays. The technology is well perfected. Povidone is a local chemical that adsorbs on the surface of the silicone and will alter the physical properties of the implanted units. Collagen and fibrous tissue formation seem to be decreased when such wetting agents are used. At present, povidone adsorption on silicone lenses is perfected in intraocular lenses only. It is hoped that this application can be adapted to other uses of silicone in other parts of the body in the future, such as for breast implants.

One of the most promising uses of the silicone in recent years, and most prominent advance made in reconstructive surgery, is the use of soft tissue expanders made of silicone implants. The expander is basically an empty silicone gel bag with a valve placed at a distance from the expander. After the expander is placed surgically in a subcutaneous pocket, the valve is used to inflate the expander to almost 3 sizes the distance of the soft tissue to be used. Once the soft tissue and skin is expanded maximally, the expander is removed and the defect is covered with a local expanded flap. Such new methodology allows the reconstructive surgeon to depend more on soft tissue in the adjacent region to cover defects that are large and may necessitate local or distant flaps for coverage. More indications for the use of the tissue expanders are on the rise as their use is becoming more acceptable and popular by the reconstructive surgeons, to treat burn wounds, and non malignant cutaneous conditions.

Certainly, a study of the immunological systems after silicone implants should be pursued in the future. Any noted alterations in the experimental animal models should be investigated for long term effects in patients who received silicone implants.

The future use of the silicones, especially in the fields of ophthalmology and reconstructive plastic surgery, will rely on further understanding of the fibrous tissue (pseudocapsule) reaction formation following implantation and the control of scar formation in internal wounds. Changing the surface properties of silicone implants by wetting agents certainly appears to be most promising.

REFERENCES

1. M. B. Habal and V. A. Chalian, "Prefabricated Silicone Subdermal Mandibular Implants", Observations and problems related to their application, J. Biomed. Mater. Res., 7, 529-536 (1976).
2. M. B. Habal, M. L. Powell and R. D. Schimpff, "Immunological Evaluation of the Tumorigenic Response to Implanted Polymers", J. Biomed. Mater. Res., 14, 455-466 (1980).
3. M. B. Habal and R. D. Powell, "Biophysical Evaluation of Tumorigenic Responses to Implanted Polymers", J. Biomed. Mater. Res., 14, 447-454 (1980).
4. M. B. Habal, "Current Status of Biomaterials: Clinical Applications in Plastic and Reconstructive Surgery", Biomaterials Medical Devises, Artifical Organs, 7, 229-241 (1979).
5. D. L. Leake, J. E. Murray and M. B. Habal, "Custom Fabrication for Mandible Reconstruction", Oral Surg., 33, 879-883 (1972).

THEMOPLASTIC CHLORINATED POLYETHYLENE FOR MAXILLOFACIAL PROSTHESES

Lawrence Gettleman, Janet M. Vargo, Paul H. Gebert
and H. Ralph Rawls

Gulf South Research Institute, P.O. Box 26518
New Orleans, LA 70186-6518

Portions of the face such as ears, noses, eyes and chins are most often lost due to skin cancers, with trauma and birth defects secondary in incidence. Such devastating losses can be replaced by the use of elastomeric materials by specially trained prosthodontists to make removable prostheses for rehabilitation. Because conventional silicone rubbers often fray, tear, or absorb coloring agents, multiple prostheses are usually made. Such thermosetting materials allow for few corrections after molding; chlorinated polyethylene has been chosen for further development because of its excellent mechanical properties, coloring ability, and the possibility of repeated molding until perfection is achieved. Gypsum molds have been found to be useful when heated in a steam autoclave for 30 minutes. Coloring has been achieved by applying pigment or fibers to the CPE surface and laminating beneath a thin sheet of CPE, followed by cycling in the autoclave once more. Clinical trials are now underway, alternating CPE prostheses with identically-shaped ones made from medical-grade silicone rubbers.

INTRODUCTION

Facial and oral cancer is increasingly treated vigorously by surgical procedures which leave devastating defects in parts of the face. Such disease is more often seen in the sunbelt because of greater sunlight exposure. The facial defect is most tragic because the face, head and neck are very important to adequate functioning in society. Patients who are missing ears (see Figure 1), noses, eyes (orbits), or chins are often the victorious winners of their fights with cancer, yet are usually left without financial resources or insurance at the end of their fight. Even with health insurance, carriers often refuse payment for rehabilitation with extraoral maxillofacial prostheses because they consider them "cosmetic". In fact, these patients often withdraw from society if not presentable in public, and prostheses can often make the difference between a useful citizen and a recluse. Other facial defects are also caused by traumatic injuries or birth defects, but the cancers are the most widespread. Rehabilitation of such patients is a true "orphan" in health care.

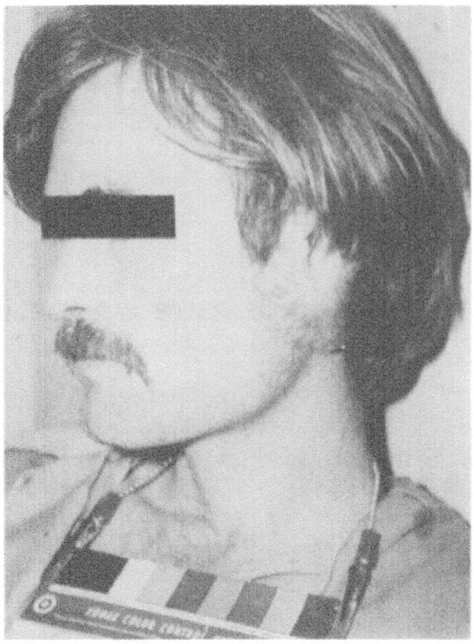

Figure 1. Patient presenting to the dental clinic after loss of left ear due to cancer surgery. Such cases are treated by prosthodontists trained in maxillofacial prosthetics because of the skills and techniques involved.

Although plastic surgery can close open wounds with keratinized tissue, the proper rehabilitation of the intricate anatomy of ears, noses, and orbits is difficult to achieve. To solve these medical problems, extraoral prostheses are usually constructed by maxillofacial prosthodontists, dentists specially trained in this form of prosthetics. They are made by obtaining a positive model of the patient's face, carving the prosthesis in clay, and molding the device in two or three-part gypsum, epoxy, or metal molds. Coloring is applied intrinsically to establish a base color and extrinsically to match the patient's skin at that time of year.

The most common prosthetic materials are the medical grade silicone rubbers. They are soft and stable, but suffer from inability to make corrections and repairs during fabrication, are subject to fraying or tearing in use, discoloration, and difficulty using skin adhesive.

EXPERIMENTAL METHODS

Chlorinated polyethylene has been chosen for development because of its strength and tear resistance, repeated moldability because of its thermoplasticity, and neutral color. The material is used for such rugged duty as chemical waste pond liners and as roofing membranes. Dow Chemical USA supplies its CPE 3614A in granular form which is then compounded as shown in Table I. Inorganic pigments are added to achieve proper coloring to human skin. Compounding is performed on a heated rubber mill at 135°C (275°F). Representative properties of the maxillofacial CPE formulation are 3.9 MPa (657 lbf/in^2) tensile strength, 1081% elongation at break, 0.55 MPa (80 lbf/in^2) stress at 100% strain, 9.2 kN/m (53 lbf/in) tear resistance, and hardness of 45 on the Durometer A scale. These measures were made after 34 days at 100% relative humidity at 25°C; the mechanical

Table I. GSRI-CPE 726/19-15 for Extraoral Maxillofacial Prostheses

Chemical Name	Trade Name	Manufacturer
Chlorinated polyethylene elastomer	CPE-3614A	Dow Chemical USA
Chlorinated paraffin plasticizer	Chlorowax 40	Diamond Shamrock
Low density polyethylene	Dow 991	Dow Chemical USA
Calcium stearate lubricant	-----	Mallinckrodt
Oxidation inhibitor	Irganox 1010	Ciba-Geigy
Ultraviolet absorber	Tinuvin	Ciba-Geigy
Pigments	as needed	Artskin

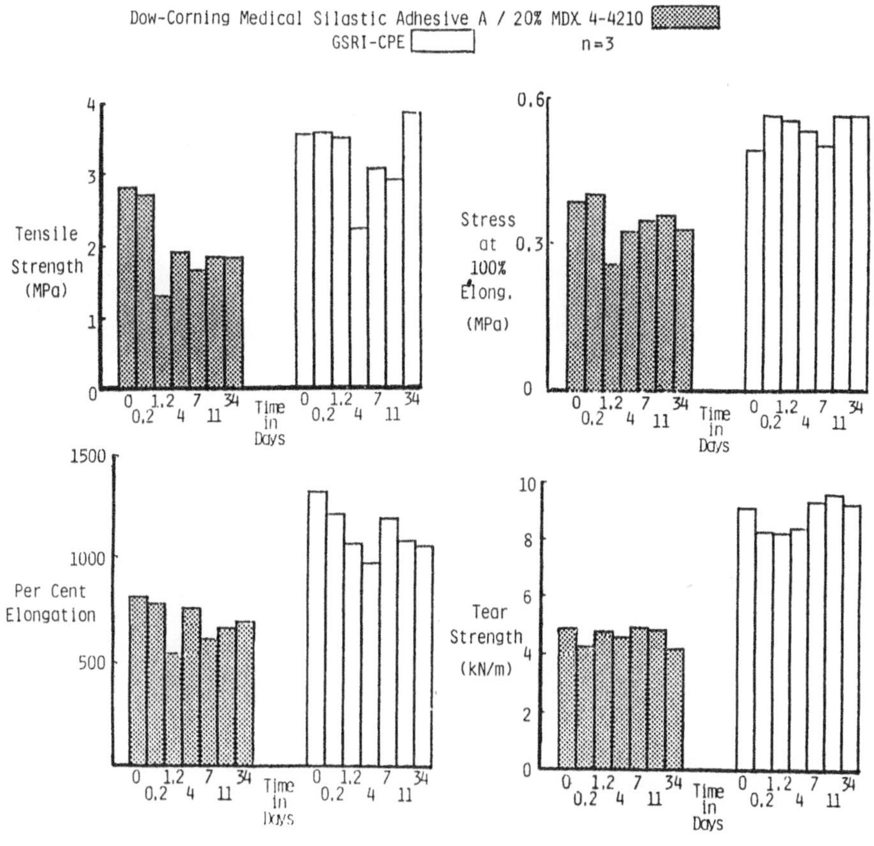

Figure 2. Physical properties of extraoral maxillofacial prosthetic materials. 100% relative humidity, 25°C. Dow-Corning Medical Silastic Adhesive A/20% MDX 4-4210

properties are approximately twice those of Dow-Corning medical Silastic Adhesive A with 20% MDX 4-4210 (see Figure 2).

The GSRI-CPE can be molded in two-part hard gypsum molds which are heated in a steam autoclave or pressure pot for up to 30 minutes. The heat at 115°C at a pressure of 69 kPa (10 lbf/in^{2}) is sufficient to soften the CPE and yet maintain the integrity of the gypsum for repeated molding until the prosthesis is free of voids (see Figures 3a and 3b). Customized coloring is then achieved by applying pigments and fibers to the prosthesis and overlaying it with a thin sheet of the CPE, followed by another cycle in the autoclave. This fuses the sheet to the body of the prosthesis and laminates the pigment in permanently (see Figures 4a through 4g).

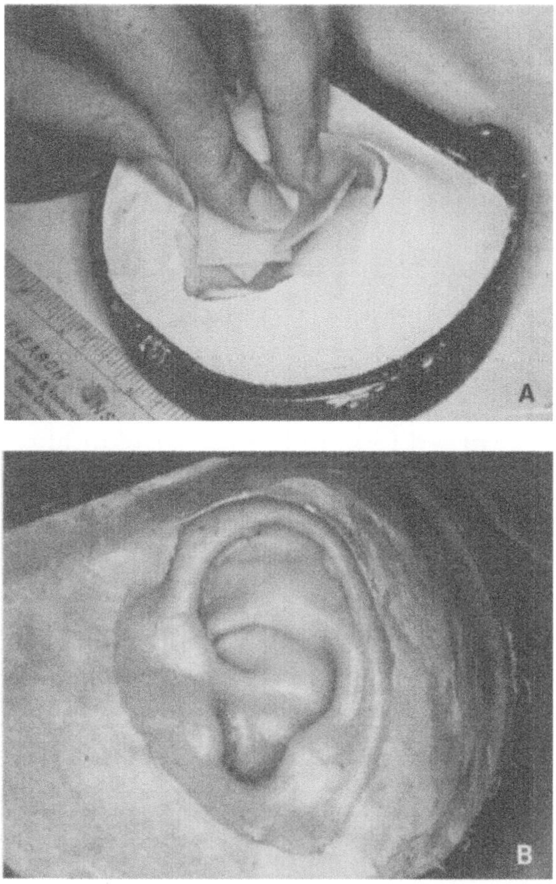

Figure 3a. Chips of GSRI-CPE are packed into hard gypsum mold of ear prosthesis. Other half of mold is then closed and clamped. Heating is by steam autoclave or pressure pot at 69 kPa (10 lbf/in^{2}) for up to 30 minutes, achieving a temperature of 115°C.

3b. Since GSRI-CPE is thermoplastic, repeated moldings is possible until a void-free prosthesis is produced, colored in the base shade for that patient.

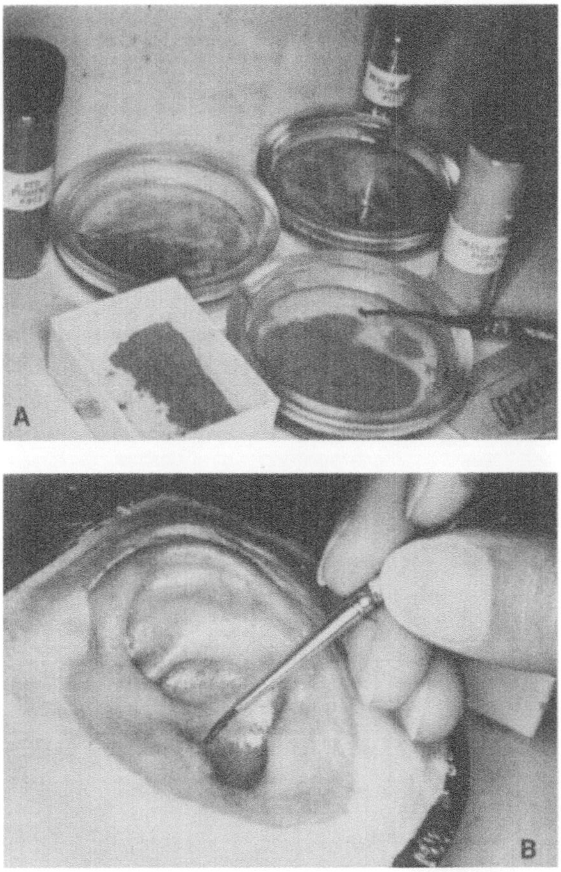

Figure 4a. Inorganic pigments ana resin fibers are selected for color and texture. They can be applied to the prosthesis with a brush dipped in a solvent such as cyclohexane. Features to be mimicked include capillaries, freckles, moles, striations, etc.

4b. Coloring should be applied in some excess due to translucient optical effect of the CPE overlay to be done next. This step is performed with the patient present.

4c. Overlay of unpigmented GSRI-CPE is placed over hand-colored prosthesis. Thickness is 100-300 micrometers.

4d. Mold is again closed over the overlay, clamped, and heat cycled.

4e. Completed prosthesis. Additional changes may be performed using a hot spatula to add warts, craters, etc., or to correct errors.

4f. Prosthesis is then separated from the mold and the area of flash trimmed with a scissors, ground, and polished. If necessary, the other side of the prosthesis also can be colored at the same time as the first lamination.

4g. Same patient as in Figure 4a with prosthesis attached using skin adhesive.

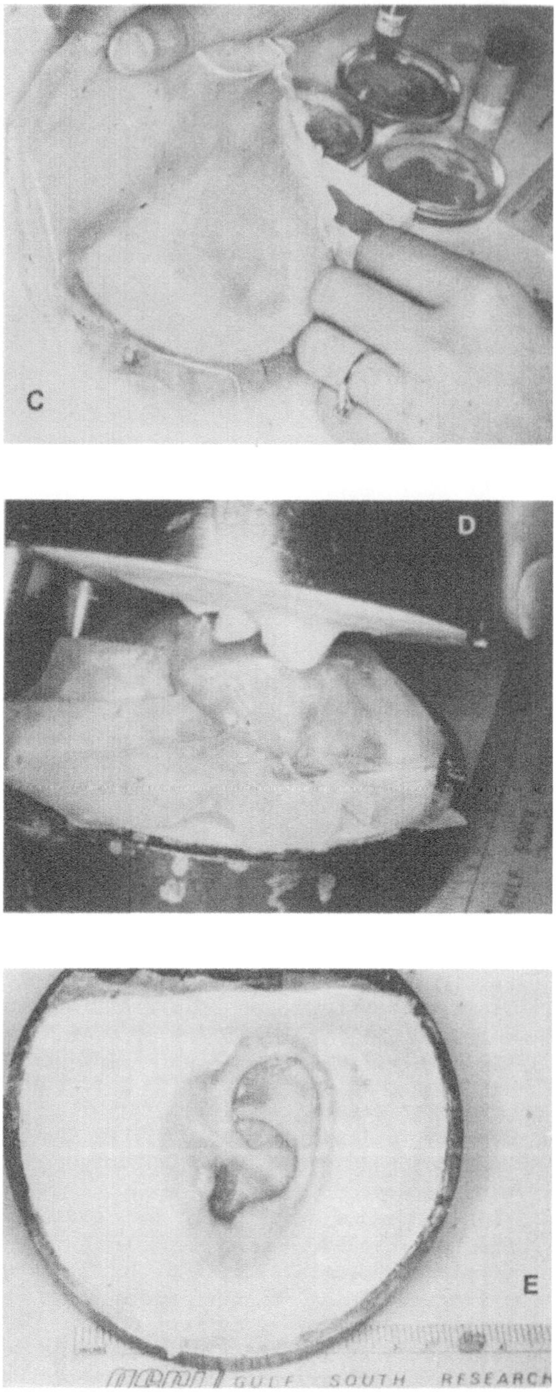

Figure 4 continued.

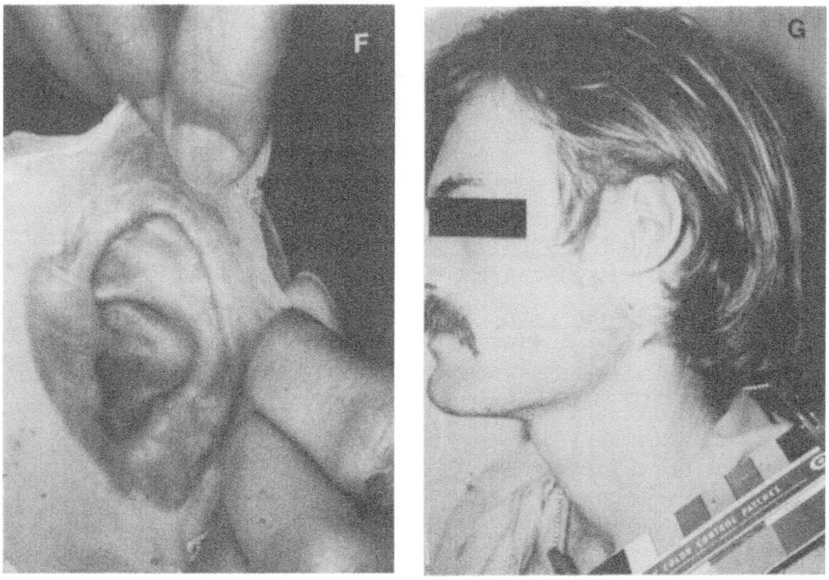

Figure 4 continued.

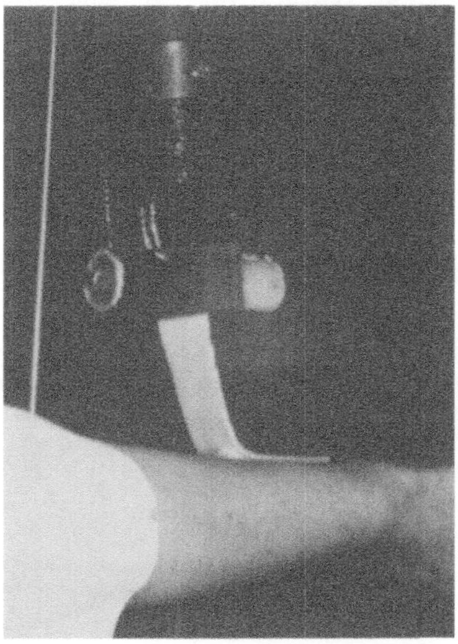

Figure 5. Adhesion of GSRI-CPE and Silastic was tested against human skin by placing skin-adhesive-prosthesis sandwich in Instron mechanical tester and peeling at 10 cm per minute.

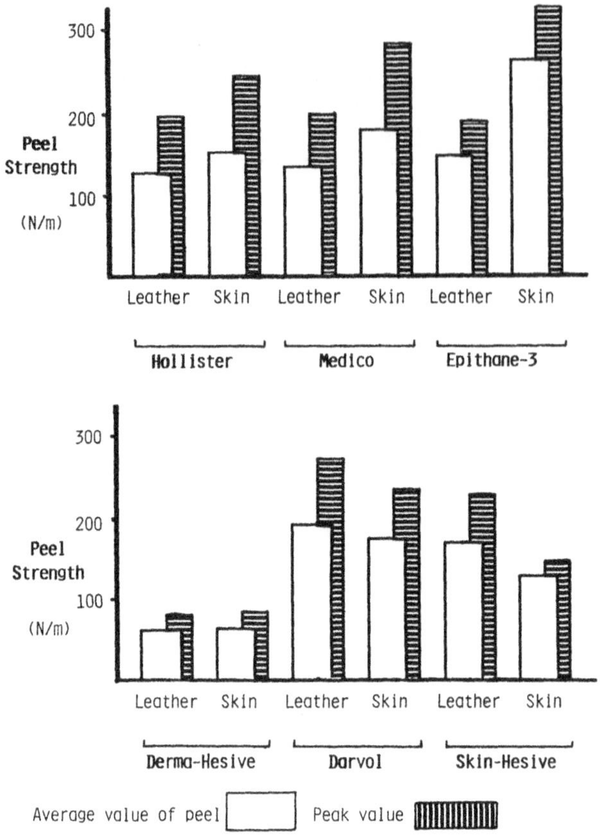

Figure 6. Peel strength of CPE using six skin adhesives against leather (n = 3) and human skin (n = 1). (Tested at 1 hour at 100% relative humidity).

Tests of adhesion after one hour at 100% relative humidity were run comparing the ability of the two prosthetic materials to adhere to human skin (see Figure 5) and leather using six commonly used medical adhesives. Results indicated that GSRI-CPE bonds well when using any of the six adhesives, but Silastic bonds well only when using Hollister adhesive (see Figures 6 and 7).

DISCUSSION

The conventional silicone rubber and the experimental GSRI-CPE are undergoing clinical trials at Charity Hospital of New Orleans (LSU Medical Center). Patients will wear prostheses from each material for two months and then exchange them. Experience up to two years is anticipated. A detailed clinical protocol is in use to determine whether one material or the other is superior at the end of the trial.

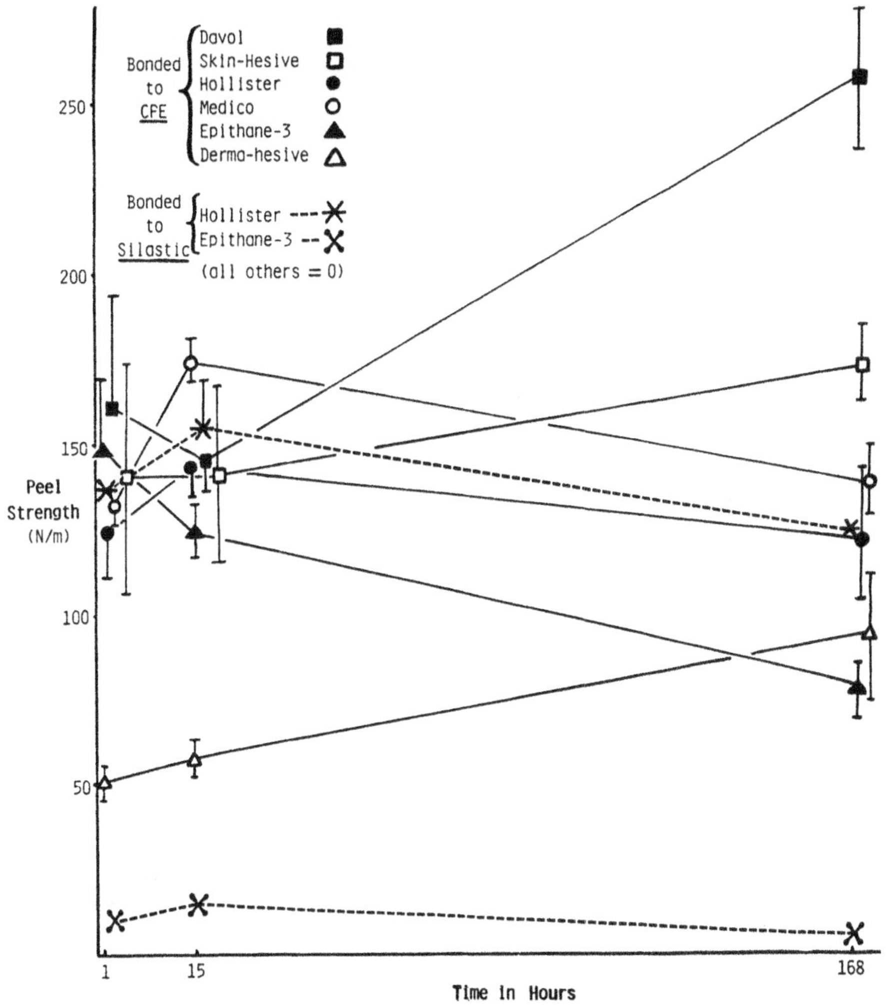

Figure 7. Adhesive bonding of skin adhesives to two extra-oral maxillofacial materials. Leather substrate, 100% relative humidity, n = 5.

ACKNOWLEDGEMENTS

Supported by research grant No. R01-DE06409 from the National Institute of Dental Research, National Institutes of Health, Bethesda, Maryland.

REFERENCES

1. P. D. May and L. R. Guerra, Maxillofacial Prostheses of Chlorinated Polyethylene, J. Biomed. Mater. Res., 12, 421-431 (1978).
2. L. Gettleman, P. H. Gebert, L. M. Ross, T. J. Huber and L. R.

Guerra, Extraoral Maxillofacial Prostheses Molded from Chlorinated Polyethylene, J. Dent. Res., 64, 368, No. 1745 (1985).
3. L. Gettleman, L. Ross-Bertrand, P. H. Gebert and L. R. Guerra . Novel Elastomers for Denture and Maxillofacial Prostheses. Biomed. Engin. IV, Recent Devel., B. Sauer, ed., Pergamon, pp 141-144 (1985).

EXPERIENCE OVER 10 YEARS WITH A NEW IMPLANTING POLYMER "OSTEOMESH" USED WITH BONE GRAFTS IN CRANIOMAXILLOFACIAL REHABILITATION

Mutaz B. Habal[a,b,c] and Donald L. Leake[b]

(a) Tampa Bay Craniofacial Center
Institute of Human Resource
University of South Florida
Tampa, FL

(b) Dental Research Institute
UCLA
Los Angeles, CA

The best possible implantable biomaterial for bony contour abnormalities is autogenous bone grafts. However, some conditions may be present that alloplastic materials have to be utilized; such alloplastic material had to be inert and there should be no reaction to it in the body. The composite graft is a combination of the two different materials described. To have a better take, the bone graft used is grinded with a craniotome, and then mixed with blood. This produces an inductive matrix that promotes osteogenesis. The alloplastic material is a tray or a template where the bone is placed, in or on, to allow for complete solidification and better healing. The material used is polyetherurethane impregnated poly(ethylene terephthalate) (Dacron) which is known as osteomesh. In few instances, the osteomesh was removed with complete solidification of the bone, resembling the skull in the case of a skull defect. For over 10 years, since the success in experimental animals (1), this composite graft has been used in clinical situations with varying degrees of success (2).

INTRODUCTION

Defects due to facial skeletal abnormalities, congenital or acquired, are customarily treated by repositioning of the facial bones, by autogenous bone grafting, or by the use of inorganic implants. The materials for implanting may also include a choice of method of reconstruction depends upon the site and size of the defects as well as functional stresses that may be placed upon the implant.

Autogenous bone graft provides one of the best methods for reconstruction. However, for large defects it is difficult to shape solid bone

---

(c) To whom correspondence should be sent at 4211 Carrollwood Village Drive, Tampa, Florida 33624.

grafts in order to achieve a smooth contour. Fore massive defects, as in gunshot wounds to the neurocranium, no donor site is suitable to provide an adequate solid piece of bone to restore the defect in one piece. For mandibular defects, where intermaxillary fixation is impossible and where the anterior aspect of the mandible, which is highly contoured, is missing, the device in this paper provides sufficient rigidity and an excellent contour so that particles of corticocancellous bone can be used to achieve an esthetic and functional result in restoring the continuity of the mandible. Particulate corticocancellous bone grafts, rich in osteogenic potential, as an inductive matrix, are especially suitable for the restoration of large defects, except that the bone graft material lacks internal strength. For stress bearing bones, external fixation or implantable rigid splints must be employed to immobilize the reconstructed area during the period of bone healing (Figures 1 and 2).

The use of the composite graft allows the reconstructive surgeon to use autologous bone graft and yet contour it to fix the defect over in an area of stress and mobility which the soft graft will be impossible to use without the use of the supportive template. The osteomesh had pores that are large enough to allow both fibrous tissue ingrowth and new blood vessels to penetrate the graft for the good and perfect "take" that no other composite can offer (Figures 3 and 4).

The purpose of this report is to present a decade's experience in the use of osteomesh in the reconstruction of over 100 discontinuity defects. In the craniofacial regions the bioimplant used alone, as in orbital defects, or with bone grafts as a carrier also used with material or with a synthetic material such as hydroxyapatite.

CLINICAL MATERIALS

Initial experience was using a composite made of poly(ethylene terephthalate)(Dacron) mesh stiffened with polyetherurethane elastomer. Each material, employed individually, has been successfully used in other implantation applications (Figure 2). Observations on 52 implants placed in various anatomical sites in 45 laboratory animals over time periods ranging from 1 month to 27 months, have now been published (1-5).

Later on, in man, the implant device used in conjunction with particulate cancellous or corticocancellous bone grafts, has been used for reconstruction of discontinuity defects of the mandible in over 100 patients, in 29 medical centers. Sixty patients with neurocranial defects have been reconstructed in nine medical centers during the past decade. Other uses include chin augmentation or recontouring, restoration of continuity of the orbital floor, and to establish symmetry in the malar region.

THE MANDIBLE

To reconstruct mandible discontinuity defects there must be adequate soft tissue and the oral mucosa must be intact. The bone ends in the graft site are exposed from an extraoral approach and the bone margins are freshened to the point of creating a bleeding surface. Suture line tension and intraoral contamination should be avoided (Figure 5). The osteomesh mandibular reconstruction tray is trimmed with scissors so that there is adequate overlap, ideally an inch, of the ,mandibular stumps and the tray is securely fastened with two or more 24-gauge stainless steel sutures impaling the tray and both buccal and lingual cortices of the mandible. The tray must be packed fully with bone graft material. In the

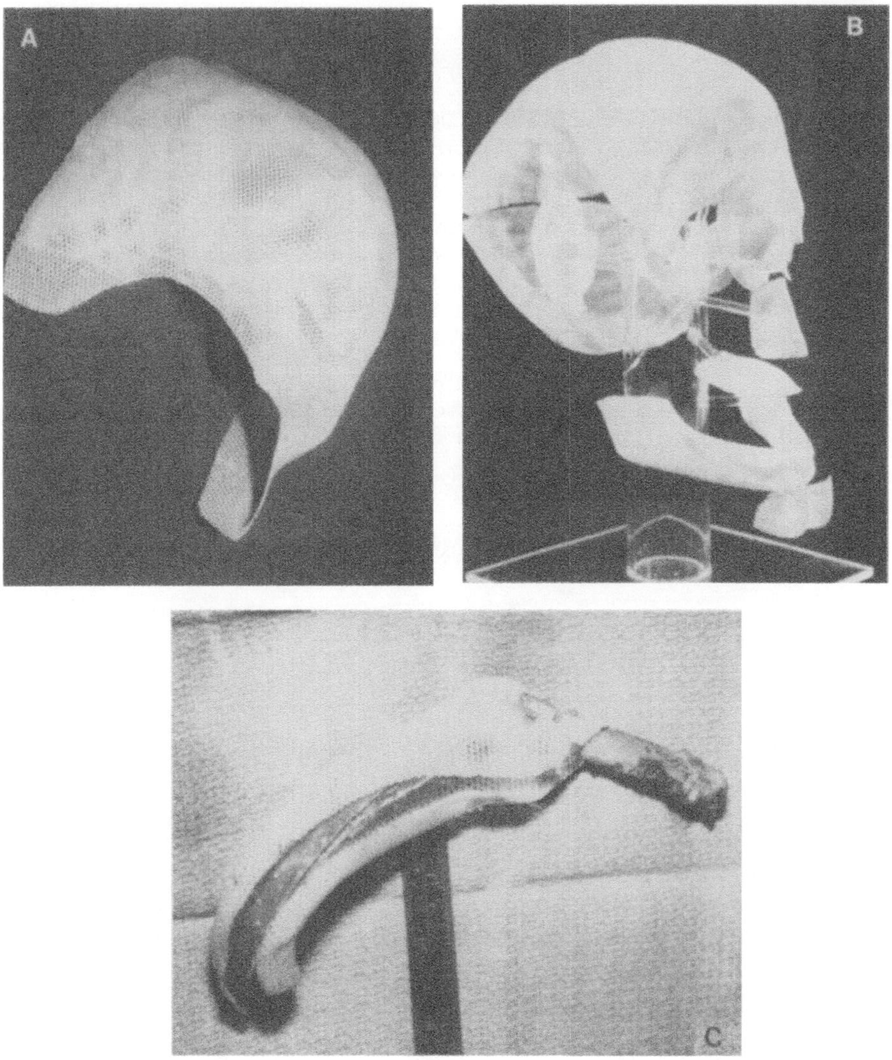

Figure 1. The different components used in facial reconstruction of the osteomesh. The facial, superorbital prosthesis should have all margins covered by hair-bearing skin producing a "fairing" or smoothing effect on the forehead. The mandibular implant requires a temporomandibular joint which is reconstructed from cortical cartilage on a rib and fixed to the prosthesis with a titanium 5 mm screw.

presence of teeth in one or both fragments of the mandible, intermaxillary fixation should be achieved prior to making the extraoral incision. This may be accomplished by fabricating an acrylic splint and determining the ideal occlusion preoperatively. If the patient is edentulous, no attempt at intermaxillary fixation is necessary. When the tray is completely packed with cancellous bone, new bone formation occurs to the outline imposed by the tray and thus a smooth esthetic restoration of mandibular continuity is possible. This method of reconstruction has been

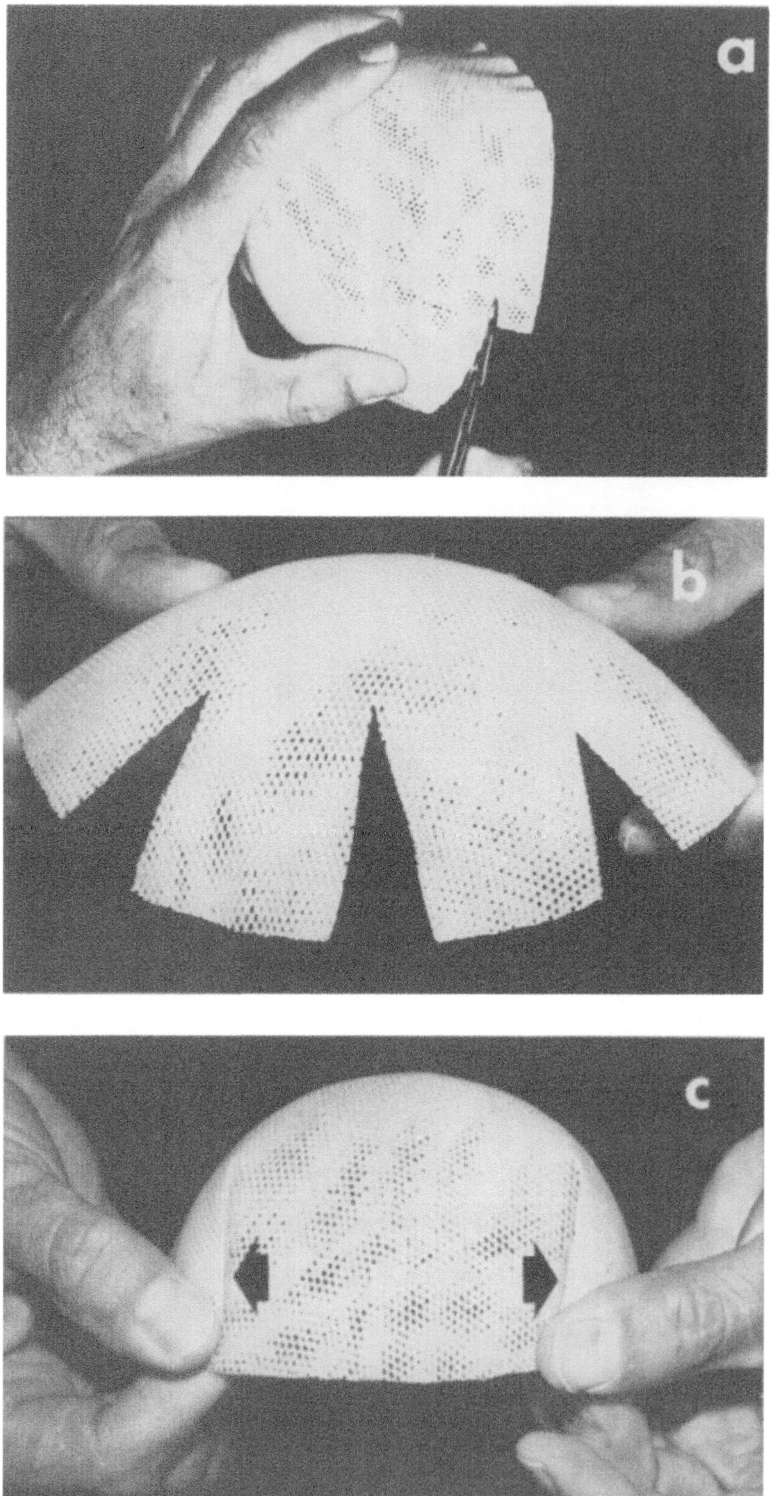

Figure 2. The tailorability of the prosthesis is well demonstrated. It can be increased or decreased in size.

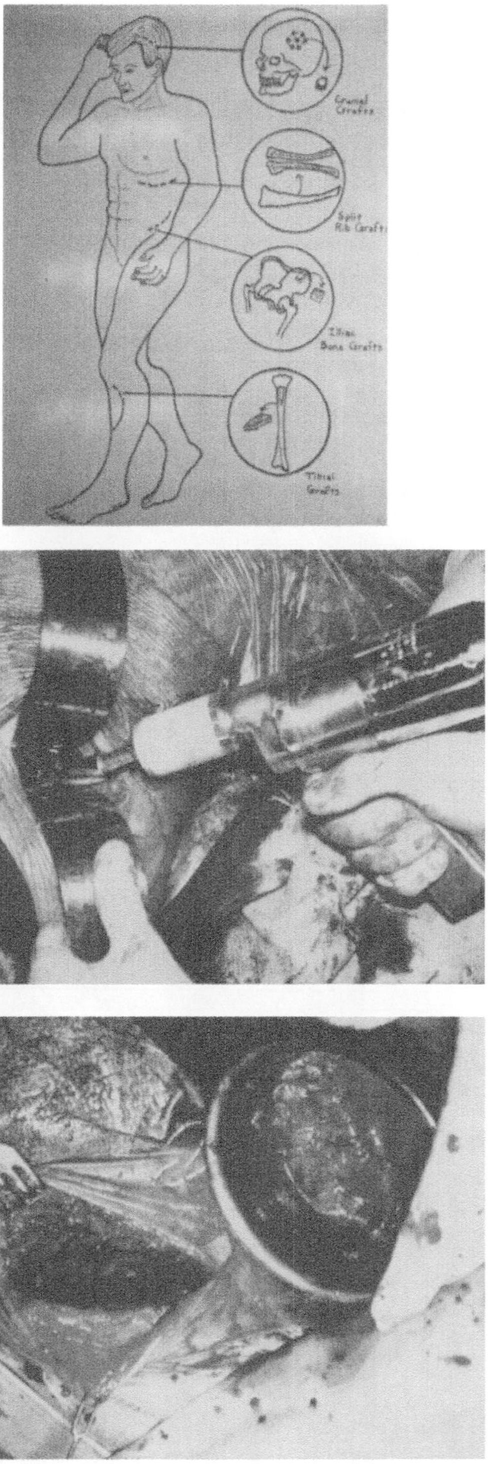

Figure 3. Different donor sites for autogenous bone graft. A craniotome is used to grind the bone which, when mixed with blood has the consistency of a paste that is usually used with the prosthesis as a composite.

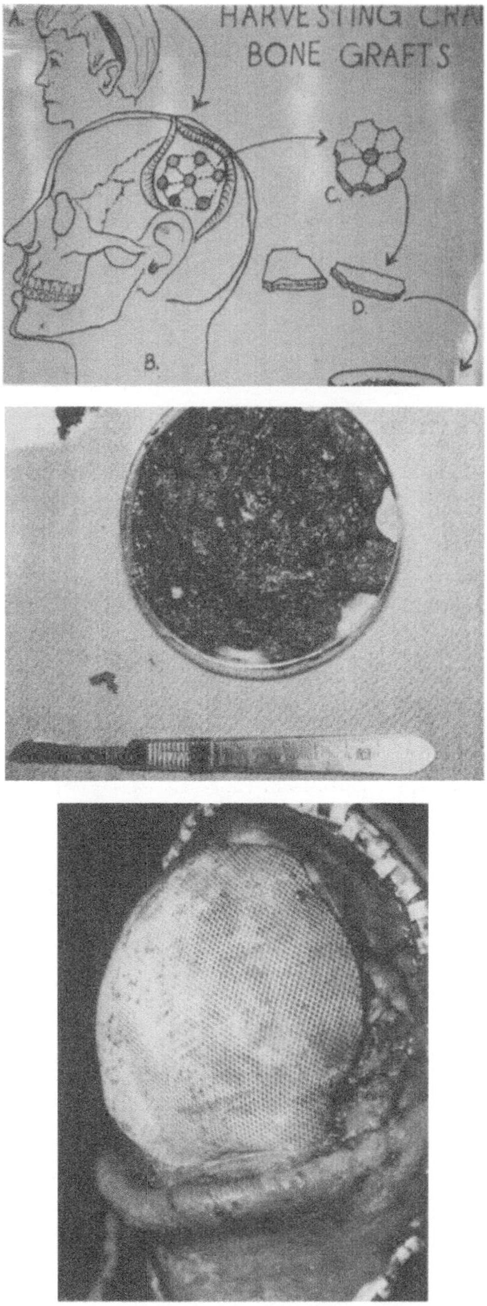

Figure 4. Cranial bone is harvested from the cranium via a hair bearing incision and the bone paste is placed under the prosthesis. The prosthesis act as a template and help during the healing and osteogenesis stage.

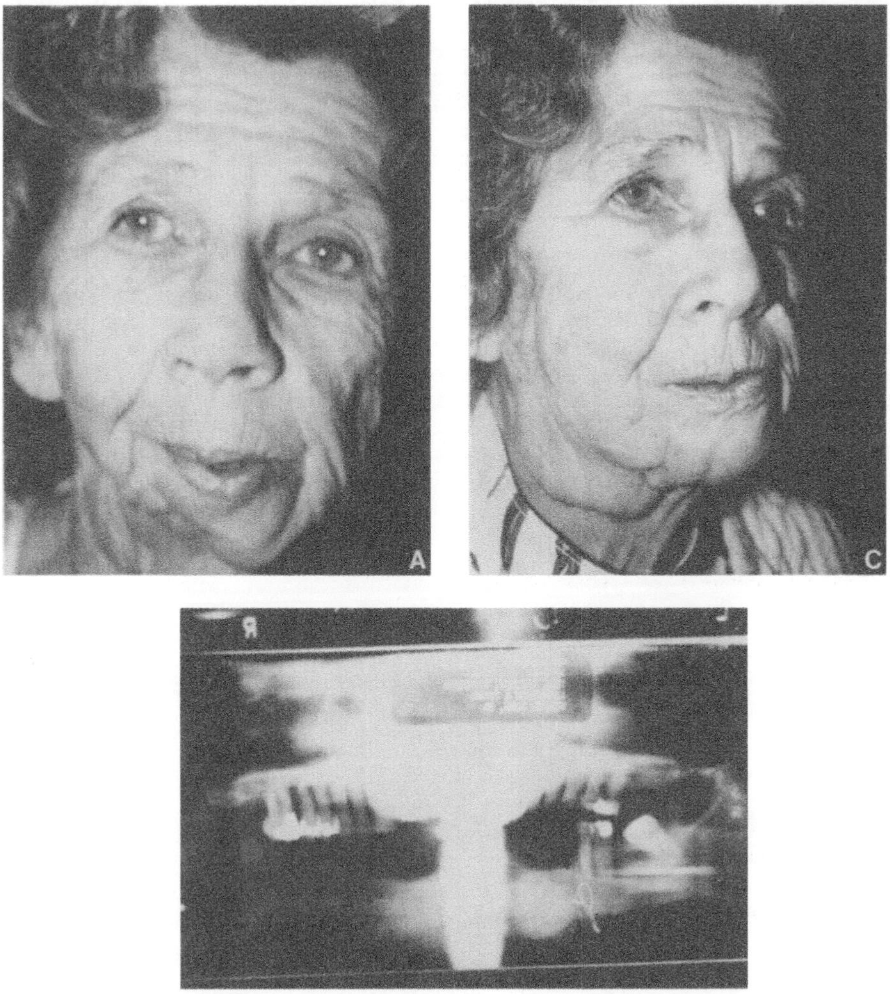

Figure 5. Patient with mandible defect. The defect reconstructed with a prosthesis, bone grafts had normal contour and function.

particularly useful in restoring the anterior portion of the mandible, an area generally challenging because of the contours, and the continous movements from the pull of the musculature of the floor of the mouth and the suprahyoid group. In secondary or delayed reconstruction, additional pull on the proximal fragments of the mandible, the ascending rami, tend to displace the graft and nonunion or malunion may result in the absence of adequate fixation of the implant. Sectioning the coronoid process, thus detaching the temporal muscle, may help reduce the pull on the proximal fragment of the mandible.

CHIN AND MALAR AUGMENTATION

Malformations in static, non-stress areas, marked by asymmetry may be reconstructed by using a prefabricated Osteo-Mesh implant. The implant must be adequately trimmed and filled with bone or a combination of bone graft material and hydroxyapatite. Fixation usually is achieved with intraosseous wires. Augmentation or recontouring of the chin is an excellent example of this use of the material. Restoration of symmetry in the malar area is another example. It is important to expose the malar areas bilaterally so that proper trimming and alignment of the implant can be achieved, establishing on the deficient side the same bulk and contour as the normal side.

NEUROCRANIUM

Loss of bone in the neurocranial or craniofacial areas is very disfiguring and difficult to reconstruct because of the need for a smooth contour (Figure 6). Osteomesh implants for restoration of a number of large defects of the skull, including the forehead, involve a bitemporal coronal incision made above the hairline or through a preexisting scar which provides an excellent surgical approach (Figure 7). The forehead flap thus created exposes the defect which is prepared for grafting by

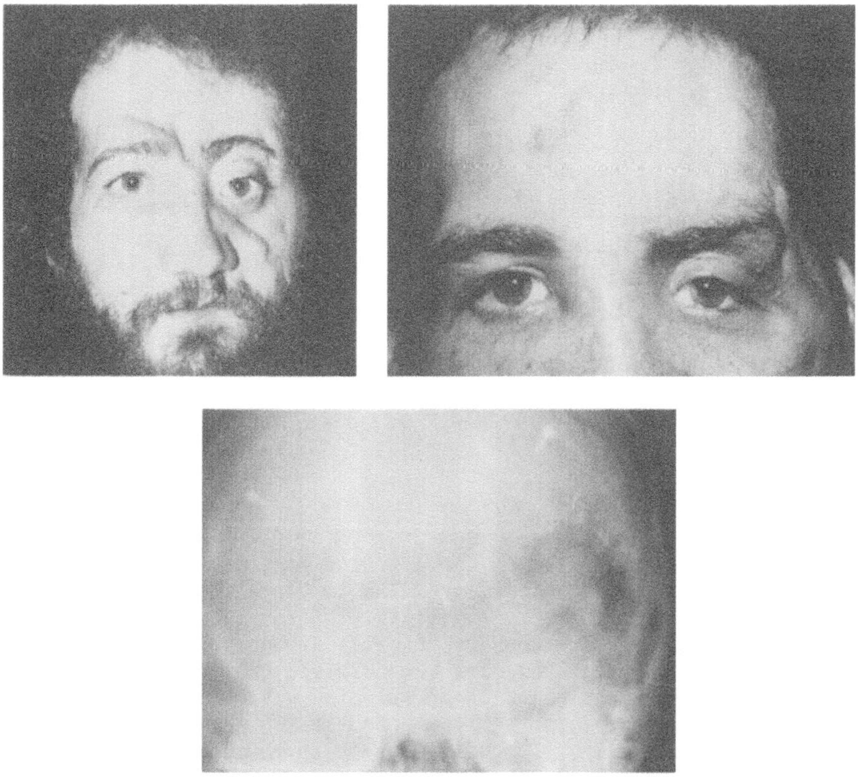

Figure 6. A cranio-orbital defect corrected with bone graft, note the complete smoothness of the forehead after reconstruction and the ground glass appearance of the bone paste after healing. The screws are used to fix the prosthesis in the skull.

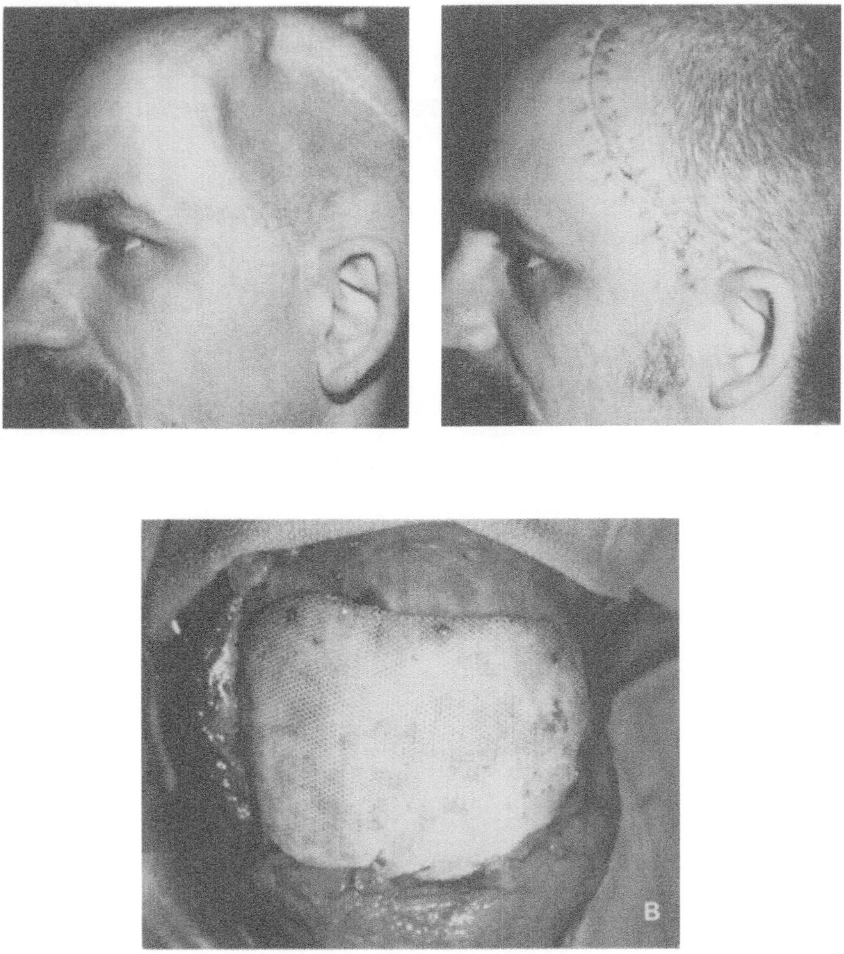

Figure 7. Large post traumatic defect of the skull treated with the composite graft producing complete restoration of the discontinuity defects in the patient.

freshening the bone edges with an air powered burr. Large defects are restored using two or three split ribs, contoured and mortised into the defect. Additional bone is taken from the ilium as a particulate bone or from the adjacent skull as bone powder; this is mixed with blood to form a paste-like material that is placed between the ribs to provide a complete restoration (Figure 8). The bone is placed directly on the dura. Osteomesh covers the bone graft material and overlaps the skull by a centimeter or more. It is secured in place directly on the dura. It is secured in place by countersinking, with a drill, a hole to receive 5 mm Vitallium or titanium screws. The area of overlap is positioned subperiosteally on the skull. In the course of a year, or more, a solid cranial table forms underlying the implant restoring bony continuity to the skull (Figure 9).

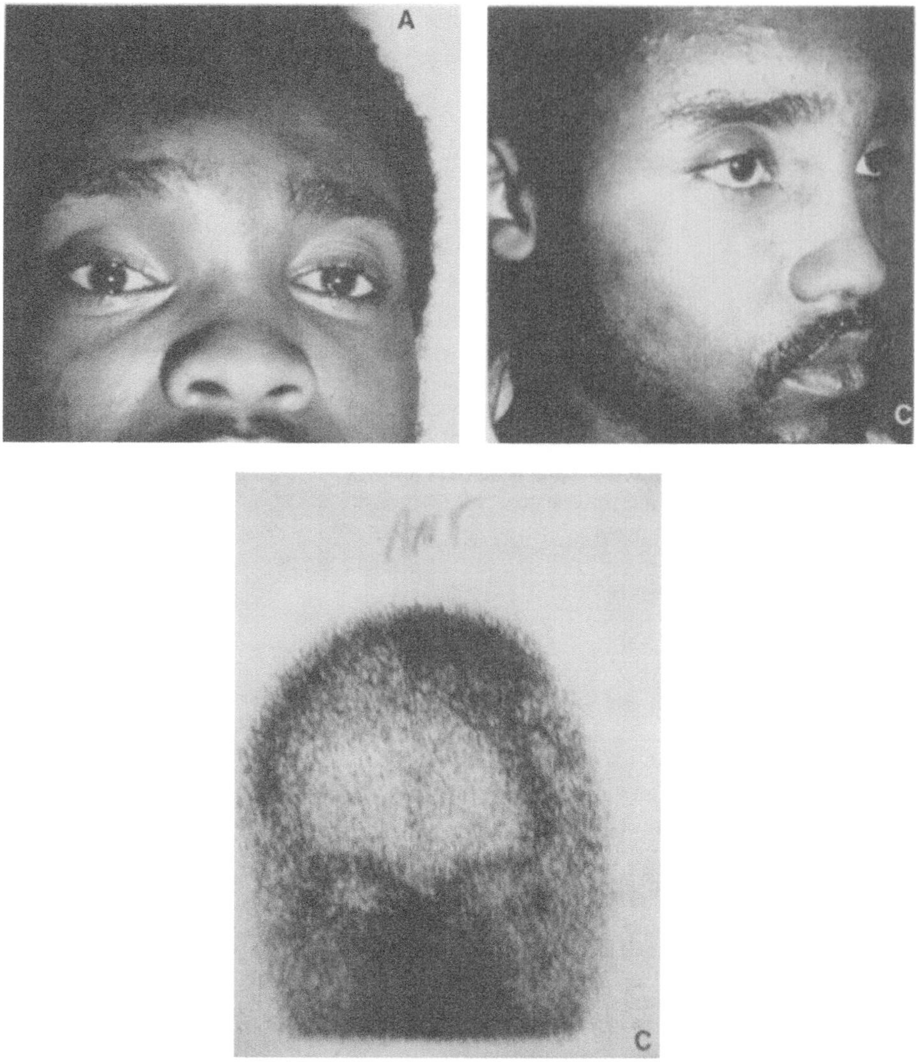

Figure 8. Large tumor of the cranio-orbital region corrected by excision and immediate reconstruction with the composite graft. The technesium scan (Tc99) glows in the reconstructed area as a demonstration of osteoneogenesis.

PROSPECTIVES

Two new possibilities for applications has been worked on in the last few years. The first being the possibility of using the principle of composite grafting in long bone discontinuity defects (6). The possibility of this being successful in the experimental design awaits the results of clinical trials. Also, the use of the bone paste has been successful as a fixture for the cementless prosthesis in long bones. The bone paste will allow ingrowth to the prosthesis and reduce the open interface between the prosthesis and the bone to which it comes into

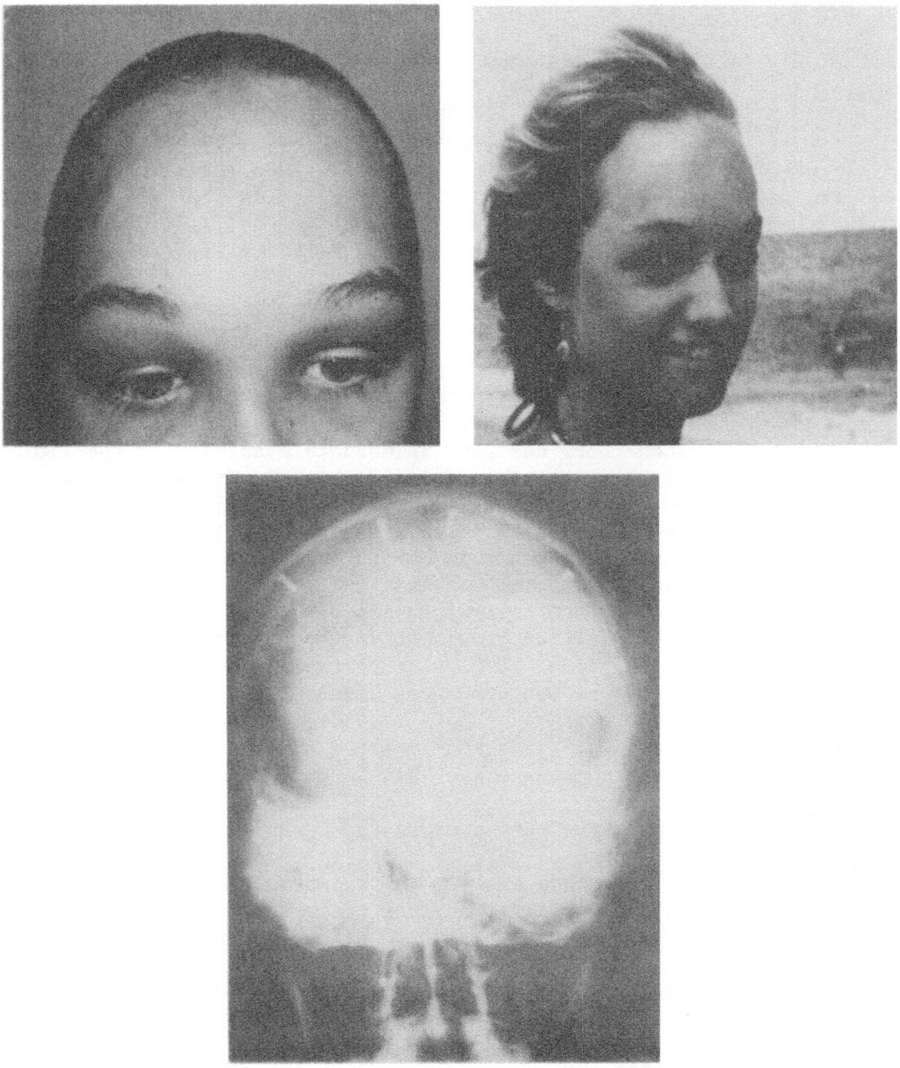

Figure 9. Patient with large defect in the forehead immediately after reconstruction and about 6 years later. Note the complete restoration of the forehead tc normal contour. The radiograph demonstrate the ground glass appearance in the area of good healing and osteogenesis.

contact. Perhaps in the future we will see more utilization of composite bone grafts in all forms of reconstruction.

The use of the osteomesh prosthesis in combination of cartilage grafts to produce a template for ear reconstruction had not been very successful. The cartilage grafts undergo transformation to bone, and enchondral ossification sets in producing complete solidification of the newly formed ear. Such will not make it usable in the clinical circumstance desired for total or near total ear reconstruction.

CONTRAINDICATIONS FOR THE USE OF OSTEOMESH

The use of osteomesh for reconstruction of any bony defect is contraindicated in the presence of infection, including resolved osteomyelitis. It should not be used if there is insufficient tissue vascularity, as for example, in heavily irradiated tissue. The use of radionuclide scanning may be helpful in evaluating soft tissue perfusion. The prosthesis should not be placed in contact of any open air sinus in the head and neck regions, since that usually leads to infections and loss of the graft, and also contamination of the oral cavity and the salivary contents.

COMPLICATIONS

Complications are most often related to mucosal dehiscence and/or infection. Dehiscence is likely to occur when the implant has been placed under tension on the suture line, where there is inadequate soft tissue for coverage, or when there has been inadequate stability of the implant. If dehiscence or formation occurs after the bone graft material has been adequately invaded by connective tissue and by capillaries, the implant can be selectively trimmed back to an area where soft tissue covers the margins of the implant with possible survival of the graft. Appropriate bacterial culture and antibiotic sensitivity tests, combined with vigorous antibiotic therapy and meticulous concern for oral hygiene, have often resulted in the ultimate success of the bone graft. The success rate in static applications, e.g., neurocranial reconstruction, chin augmentation and malar augmentation has been better than in mandibular reconstruction. The reasons for this include the fact that stability is much easier to achieve in static applications and soft tissue coverage is more readily achieved in these areas than in mandible.

In conclusion, over the past decade, the technique described here for the reconstruction of contour defects of facial bones has had wide clinical applications. The method employs an implantable device prefabricated from polyetherurethane and poly(ethylene terephthalate) cloth mesh with the addition of autogenous bone grafts. Although experience thus far suggests that the implant can be left in indefinately, it is possible, after the bone graft area has matured adequately, to remove the implant without compromising the esthetic result or the function of the reconstructed area. The prospects for the future will be a better understanding of the allografts and bone implants with the osteomesh to minimize donor site morbidity.

REFERENCES

1. D. L. Leake and M. Rappoport, "Mandibular reconstruction: Bone induction in an alloplastic tray", Surgery, 34, 585 (1972).
2. D. L. Leake, M. B. Habal, H. Schwartz, S. Michieli and A. Pizzoferrato, "Urethane elastomer coated cloth mesh: A biocompatible composite material for reconstructive surgery", in: "Mechanical Properties of Biomaterials", G. W. Hastings and D. F. Williams, eds., John Wiley and Sons Ltd., 1980, pp 419-430.
3. D. L. Leake, "A mesh for reconstructive maxillofacial and plastic surgery", in: "Biocompatibility in Clinical Practice", Vol 2, David F. Williams, ed., CRC Press, Boca Raton, Florida, 1982, pp 183-194.
4. D. L. Leake and M. B. Habal, "Osteo-mesh: 10 year experience in reconstructive surgery of the mandible and other bone defects", Translations of the International Society of Plastic Surgeons, 8, 355-356, (1983)
5. M. B. Habal, D. L. Leake, J. E. Maniscalco and J. Kim, "Repair of

major cranio-orbital defects with an elastomer-coated mesh and autogenous bone paste", Plastic and Reconstructive Surgery, 61, 394-404 (1978).
6. M. B. Habal, R. U. Velasco, P. G. Spiegel and D. L. Leake, "Autogenous corticocancellous bone paste for long bone discontinuity defects: An Experiemntal approach", Annals of Plastic Surgery, 15, 138-142 (1985).
7. D. Leake and M. B. Leake, "Mandibular reconstruction and craniofacial fairing", British Journal of Oral Surgery 16, 198-206 (1978-79).
8. H. C. Schwartz, M. Wollin, D. L. Leake and A. R. Kagan, "Interface radiation dosimetry in mandibular reconstruction", AMA Arch. Otolaryngol., 105, p 293, (1979).

POLYPHOSPHAZINE FLUOROELASTOMER (PNF) AS A PERMANENT SOFT LINER FOR REMOVABLE DENTURES

Lawrence Gettleman, Janet M. Vargo, Paul H. Gebert, Charles L. Farris, Ralph J. LeBoeuf, Jr., and H. Ralph Rawls

Gulf South Research Institute
P.O. Box 26518
New Orleans, LA 70186-6518

A semi-organic elastomer, polyphosphazine fluoroelastomer, has been compounded to make it bond to hard dental acrylic resins for use as a liner for removable dentures. The material retains its elastic properties indefinitely, resists swelling, can be machined and polished, and is opaque to X rays. It is cured in gypsum flasks against old or new denture bases using ordinary dental laboratory techniques at no more than boiling water conditions. The PNF material is undergoing a controlled clinical trial at the present time.

INTRODUCTION

Dentistry has been remarkably successful in the past two decades, reducing the incidence of dental disease and providing more durable treatment mechanisms, materials, and devices. There is still an unmet need for dental care in the United States, however, often involving the lower socio-economic population, those fearful of dental services, and the elderly. The unmet dental need is even greater outside of the developed countries. Although more teeth are being preserved today, the aging of the population essentially negates the gains of prevention and conservation, and there are more people alive today who wear removable partial or complete (full) dentures than ever before (see Figure 1). Estimates of completely edentulous Americans number more than 23 million.

Patients who are partially or totally edentulous (without teeth) are usually forced to wear removable dentures. The baseplates of these dentures are made from pink-colored poly(methyl methacrylate) which often causes difficulty for patients by pinching the soft tissues of the mouth between the bone and the denture. This is especially true for older patients who lack the tolerance or tissue strength, or proper bone support in the aveolar ridges of the maxillae or mandible. Denture wearing is often painful for these patients because sores will develop from the constant motion of the denture. Increasingly common surgical procedures to rebuild the alveolar ridges with artificial bone results in greatly stretched tissues which are needed to cover the augmented bone; such tissues are easily irritated by a denture.

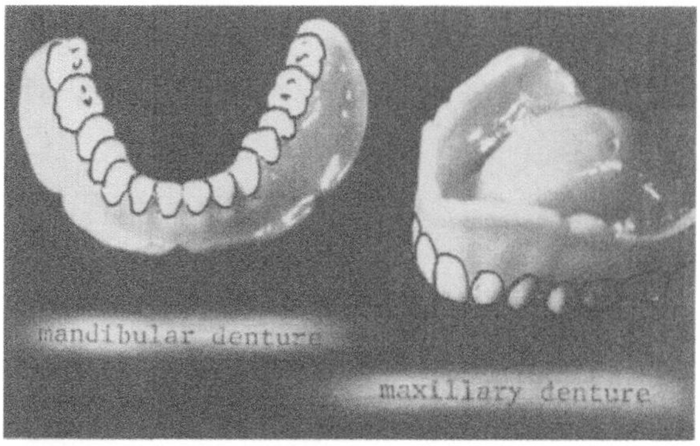

Figure 1. Removable complete (full) dentures for maxillary and mandibular edentulous arches. Teeth made of PMMA or porcelain; the denture base is usually made of poly(methyl methacrylate).

Due to individual variations in denture wearers, many patients can tolerate very poor prostheses regardless of the competence of the dentist constructing them, changes in the oral tissues since delivery of the dentures, fractures or wear of the structure, or use of large amounts of over-the-counter denture adhesives. Conversely, many patients cannot tolerate even well-constructed dentures for a variety of physical or psychological reasons. In many cases temporary soft liners placed between the hard denture base and the mucosal tissues of the mouth have several advantages:

(a) cushioning of the soft tissues between the bone and the hard denture base;

(b) the liner may creep a small amount, providing better adaptation of the denture to the tissues. This is especially useful in cases of extreme alveolar ridge resorption, more often in the mandible than the maxilla (see Figures 2 and 3). A soft liner is especially useful after surgical intervention due to traumatic fracture of the jaw or alveolar ridge augmentation with autogenous or manmade materials;

(c) latitude for imperfections or inaccuracies in the dimensions of the denture; and

(d) fewer adjustments required of the denture after delivery, reducing the number of appointments for the dentist and the patient.

Materials in use today as soft liners include temporary and permanent varieties. The temporary liners are self-curing (RTV) in the mouth and are used to help restore the oral mucosa to a healthy condition through frequent replacement and/or as a functional impression material in the construction of a new or relined denture. The permanent materials are heat-cured in a dental laboratory and are intended for longer-term service, usually months or years. They include:

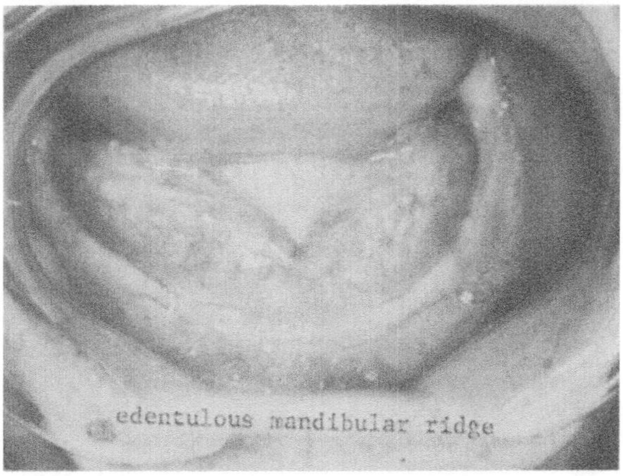

Figure 2. Intraoral photograph of edentulous mandibular ridge, showing extreme of resorbed alveolar (tooth-supporting) bone. Tongue is at top of figure.

(a) plasticized acrylics, usually made from ethyl methacrylate powder and methyl or ethyl methacrylate liquid, plasticized with alcohols or phthalates. They are usually heat-curing. These materials bond well to the acrylic denture base but harden, shrink and crack over a period of weeks or months. Some material is usually wasted during use of plasticized acrylics;

(b) silicone rubbers, either self- or heat-curing, are considered to be permanent. They do not bond as well to acrylic denture bases and often require the use of chemical bonding agents. They will not harden but will absorb oral and dietary fluids which then encourage the growth of fungus within their substance and on the surface; and

(c) polyurethane elastomers have found limited use because of the relatively extreme conditions in the laboratory needed for bonding to the acrylic base. This brings the technology beyond the methods commonly used in commercial dental laboratories. They will not harden, they bond well to the denture base, but absorb stains from foods, beverages, or tobacco products and change color to a dark brown in a matter of months.

There is clearly a need for an improved soft liner that will withstand oral conditions for years before replacement is indicated.

EXPERIMENTAL METHODS

Work at GSRI, starting in 1979, involved examination of novel industrial elastomers in an effort to discover if a better material could be found for use as a long-term denture liner. The following materials were chosen for initial evaluation:

(a) Kraton 1107 isoprene-stryene-isoprene block copolymer (Shell Chemical);

(b) chlorinated polyethylene (CPE 3615), a 36% chlorinated polyethylene made by Dow Chemical U.S.A.: and

(c) polyphosphazine fluoroelastomer, a semi-organic elastomer originally made by Firestone Tire & Rubber Company as PNF-200™, but now synthesized by Ethyl Corporation as Eypel™-F.

Kraton was rejected due to depolymerization on exposure to ultraviolet light. The CPE was found to be too hard to be a soft liner, and it was difficult to form the material in varying thicknesses.

The polyphosphazine was chosen for further development. This elastomer is a semi-organic rubber with a phosphorus-nitrogen backbone and fluorocarbon side groups (see Figure 4).

Curing is by crosslinking of the side groups, using peroxides and multifunctional acrylates. This also results in a good bond to the acrylic baseplate which is further promoted by wetting with methyl methacrylate monomer. Radiopacity is achieved with $BaSO_3$ so that fragments of the denture, if fractured in an accident or other trauma, may be identified within the tissues or if swallowed or inhaled (see Figures 5 and 6).

Fabrication of the denture takes place in the usual dental fashion by first packing the methyl methacrylate-poly(methyl methacrylate) dough in a gypsum mold. The dough is packed against a spacer made from a putty silicone rubber and precured at 74°C. for 45 minutes. This hardens the acrylic resin sufficiently to resist deformation when the soft liner dough is inserted subsequently. The spacer is then removed and the PNF dough is packed into place. Very little material is wasted because the spacer and then an equivalent amount of PNF is weighed in order to use just enough material to fill the mold. A few trial packs may be necessary. Just prior to packing the liner, a small amount of methyl methacrylate is applied to the acrylic dough to promote adhesion.

Representative properties of the cured PNF liner material are:

Durometer A Hardness:	45
7-day Water Sorption:	1.5% w/w
Peel Strength to Acrylic:	4700 N/m (26.9 lbf/in)
Tensile Strength:	1.9 MPa (276 lbf/in^2)
Elongation:	240%
Stress at 200% Strain:	1.8 MPa

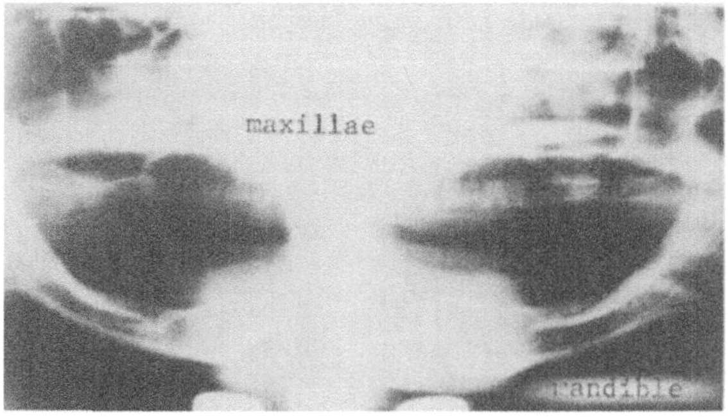

Figure 3. Panoramic radiograph of patient in Figure 2, showing extreme loss of mandibular and maxillary bone. Construction of a removable denture on this foundation is challenging to the dentist and the patient.

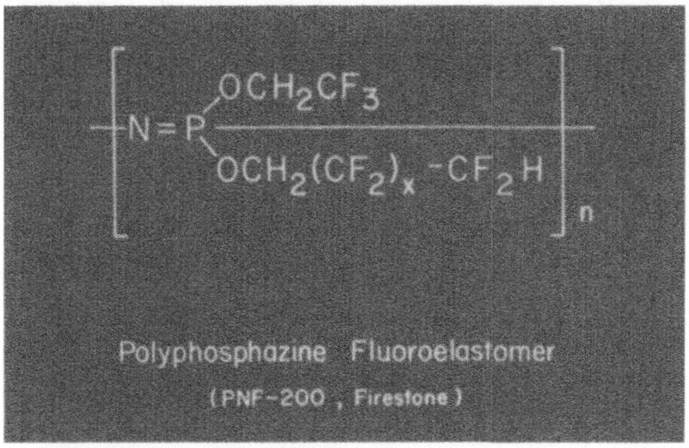

Figure 4. Poly(fluoroalkoxyphosphazine), now made by Ethyl Corporation as Eypel-F™.

Acute toxicity tests have indicated that the material has little potential to stimulate biologic reactions and is less toxic than other commercial products. Tests by the University of Tennessee's Materials Science Toxicology Laboratories yielded a Cumulative Toxicity Index of 56 and 69 out of 1500 for GSRI-PNF-soft and GSRI-PNF-firm, respectively. For comparison, Molloplast-B, a commercial silicone soft liner, measured 200 on this scale, and Super-soft, a plasticized acrylic, measured 108.

Microbiological growth tests reveal no acceleration of _Candida albicans_ or _Staphylococcus aureus_ growth around or under the PNF material when incubated on nutrient agar.

DISCUSSION

At the present time, detailed clinical trials are underway at Charity Hospital of New Orleans and at Northwestern University Dental School in Chicago. An A-B comparison is being made between GSRI-PNF and Molloplast B, a commercial silicone rubber soft liner, in both maxillary and mandibular sites in the same patient. It is recognized that if a reliable and longlasting soft denture liner existed, then most dentists and prosthodontists would tend to use them routinely rather than just for problem cases. The ultimate success of these concepts would be the development of a permanent soft liner that could be applied by the dentist directly in the mouth under new or existing dentures. At present, the best materials still require processing and curing in a dental laboratory, which means that the patients must be without their dentures for some days, a disadvantage for most patients. It is for these reasons that the soft liners available today find only limited usefulness.

ACKNOWLEDGEMENT

Supported by research grant No. R01-DE04814 from the National Institute of Dental Research, National Institute of Health, Bethesda, Maryland.

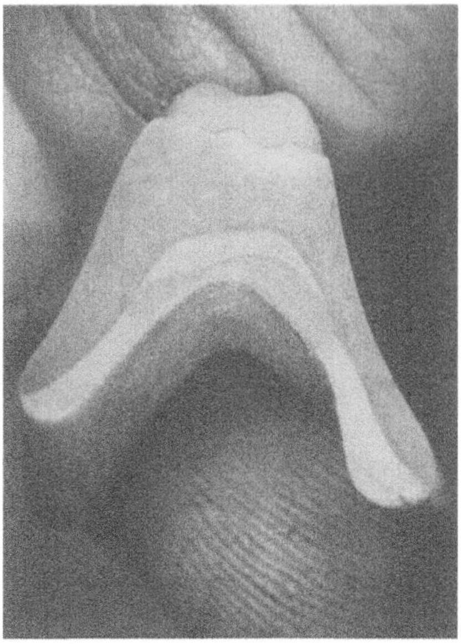

Figure 5. Cross-section of GSRI-PNF-lined denture. Denture tooth is to the top; denture liner is toward the bottom.

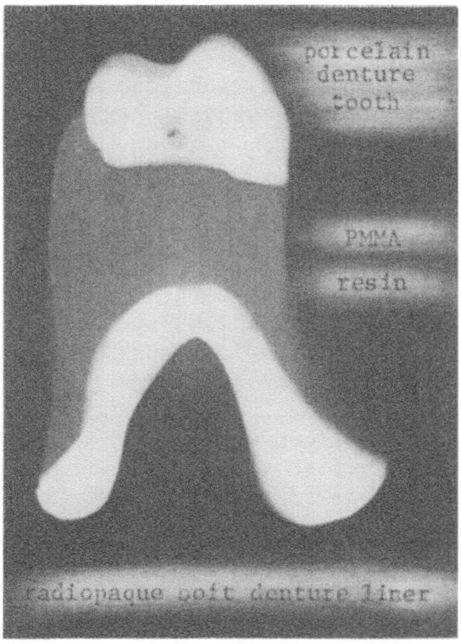

Figure 6. Radiograph of same denture cross-section as in Figure 5. A porcelain denture tooth was used in this instance. Note that the denture soft liner is as opaque to x rays as the tooth, and is useful when searching for fragments of the denture after a traumatic accident.

REFERENCES

1. P. May and L. R Guerra, U.S. Patent 4,251,215, "Phosphonitrilic Fluoroelastomer Lined Denture", Feb. 17, 1981.
2. L. Gettleman, C. L. Farris, H. R. Rawls and R. J. LeBouef, Jr., U.S. Patents 4,432,730 and 4,543,379, "Soft and Firm Denture Liner for Composite Dentures and Method for Fabricating", Feb. 21, 1984 and Sept. 24, 1985.
3. L. Gettleman, D. L. Fischer and C. Farris, "Self-sanitizing Soft Denture Liners: Paradoxical Results", J. Biomed. Mater. Res., 17, 731-734 (1983).
4. P. D. May, C. L. Farris, L. Gettleman, F. F. Holland and I. Cabasso, "New Elastomers for Soft Denture Liners", J. Dent. Res., 60:A,437, No. 507 (1981).
5. L. Gettleman, C. L. Farris, R. J. LeBoeuf, Jr., H. R. Rawls and E. O. Dillingham, "Improvement of Experimental Elastomers for Permanent Soft Denture Liners: Physical and Biologic Properties", Biomat., Med. Dev., Artif. Organs, 9, 345-347, 1981.
6. L. Gettleman, C. L. Farris and S. B. Ruiz, "Properties of Experimental Elastomers for Permanent Soft Denture Liners", J. Dent. Res., 61, 262, No. 753 (1982).
7. C. L. Farris, L. Gettleman and H. R. Rawls, "Improvement of Bonding and Strength of a Polyphosphazine Elastomer for Denture Liners", J. Dent. Res., 61, 285, No. 957 (1982).
8. L. Gettleman, R. J. LeBoeuf, Jr., and H. R. Rawls, "Soft and Firm PNF Denture Liners", J. Dent. Res., 62, 250, No. 725 (1983).
9. L. Gettleman, R. J. LeBoeuf, Jr., and H. R. Rawls, "Polyphosphazine Fluoroelastomer as a Permanent Soft Liner for Removable Dentures", J. Dent. Res., 62, 682, No. 290 (1983).
10. L. Gettleman, R. J. LeBoeuf, Jr., and H. R. Rawls, "Polyphosphazine Fluoroelastomers as a Permanent Soft Liner for Removable Dentures - Improvements in Radiopacity and Soft/Firm Properties", Society for Biomaterials Trans., No. 71, (1983).
11. P. H. Gebert, M. R. Capezza, H. R. Rawls, J. K. Smith and L. Gettleman, "The Effect of Methacrylate on the Physical Properties of Poly(Fluoroalkoxy)Phosphazine Elastomers", Second World Congress on Biomaterials, Trans., No. 612 (1984).

NEW FLUOROPOLYMERS FOR POTENTIAL MEDICAL APPLICATIONS

J. R. Griffith and J. G. O'Rear

Chemistry Division
Naval Research Laboratory, Code 6120
Washington, D.C. 20375-5000

A series of fluoropolymers has been synthesized in recent years which hold promise for use in medical or dental prostheses or as components of medical devices. In the present work, new fluororesins are synthesized which extend the range of functional content of such materials. The first fluoroepoxy containing a branched perfluoroether in a side chain and the first trifunctional fluoroepoxy resin are reported. These advances are expected to have an effect upon the oxygen transport properties and mechanical rigidities, respectively, when used in conjunction with previously reported fluororesins.

INTRODUCTION

In previous publications, we have described the synthesis and polymerization of some fluorine-containing monomers of the epoxy and acrylic types (1-3). These materials have a number of properties which promise effective use in dental and medical prostheses, such as, very low fluid absorption, color stability, easy moldability, and in some cases, permanent lubricity. In this paper, we will describe the preparation of two new fluororesins which extend the range of functional variation possible with this class of materials.

EXPERIMENTAL

Materials

Dimethyl sulfoxide (DMSO) from Fisher Scientific Company (Certified A.C.S.) was used as received. The branched-chain perfluoroether iodide $(CF_3)_2CFO(CF_2)_4I$, (II), was supplied by the Allied Corporation; b.p. 40-43°/25 torr. (4); redistilled b.p. 64°/50 torr.; n_D^{20} = 1.3180.

Synthesis of 1,3-Bis-(2-hydroxyhexafluoro-2-propyl)-5-iodobenzene (I)

This material was acquired by direct iodination of the precursor diol, as previously reported (5).

Synthesis of 1,3-Bis-(2-hydrohexafluoro-2-propyl)5-(4-heptafluoroisopropoxy-1,1,2,2,3,3,4,4-octafluoro-1-butyl)-benzene(III):

This synthesis is outlined in scheme I. The reaction was carried out in a 500ml, three-necked round bottom flask fitted with a mechanical stirrer, thermometer, water-cooled condenser and dry argon bleed. The flask was charged with (I) (97.2g, 0.181 mole), (II) (102.4g, 0.200 mole), activated copper (57.6g) and DMSO (100ml). The mixture was stirred and heated at 122-126° for 16 hours, after which the excess copper and copper salts were removed by vacuum filtration. The filtrate was poured in 2.5 l of water. A heavy oil which separated was extracted with 400ml of ethyl ether. The ether solution was contacted with sodium sulfite crystals (4g), washed with saturated sodium chloride solution, dried ($MgSO_4$) and filtered. The filtrate was flash evaporated to remove ether. The residue distilled to yield analytical III as a viscous clear liquid; 117.3g (99.7% purity by GLC, 81.6% yield; b.p. 95°/1.0 torr. n_D^{20} = 1.3664; IR (neat) 3620(OH), 3125(CH), 1615, 1467, 1263(CF), 1205(CF), 1150(CF), 991, 982, 730, 708 cm-1;

Analysis:

Calculated for $C_{19}H_5F_{27}O_3$: C,28.73; H,0.63; F,64.59.

Found:

C, 28.72; H,0.65, F,64.45.

Synthesis of diglycidyl ether of III (IV)

This synthesis is outlined in scheme 2. A mixture of the fluorodiol (III) (60.0g, 0.0756 mole), epichlorohydrin (41.9g, 0.453 mole) and trimethylammonium bromide (0.82g) was stirred in a closed system at 60° for 144 hours. The mixture was flash evaporated, finally at 90°C/3 torr. to remove epichlorohydrin and the viscous liquid residue (73g) was dissolved in 90g 3:1 toluene; ethyl alcohol. Resultant mixture was dehydrohalogenated by successive contacts with 3.00 N sodium hydroxide solution. The first contact was accomplished by stirring the solvent mixture at 50°C for 30 minutes while the sodium hydroxide solution (20.0ml) was added dropwise. The heavy oil layer from this operation was replenished with ethyl alcohol (10ml) and another similar caustic contact repeated. The fifth contact showed no further caustic consumption. Work-up of the final oil layer left 63.6g of clear, viscous liquid analyzing > 99% IV. Short path distillation of the residue furnished analytical IV as a clear viscous liquid: 58.1g, 84.9% yield, 99.8% purity by GLC; b.p. 135-140°.0.5 torr. n_D^{20} = 1.3745; IR (neat) 3100(CH) 3080(CH), 3018(CH), 2952(CH), 2910(CH), 2860(CH), 1465, 1258(CF), 1225(CF), 1205(CF), 1153(CF), 1130(CF), 992, 738cm^{-1};

Analysis:

Calculated for $C_{25}H_{13}F_{27}O_5$: C, 33.13; H, 1.45,; F, 55.60;

Found:

C, 33.10; H, 1.39; F, 55.89.

Synthesis of 1,3,5-Tris-(2-hydroxyhexafluoro-2-propyl)-benzene (VI)

This synthesis is outlined in scheme 3. Demethylation of (V), (7), was accomplished in a 300ml flask equipped with a thermometer, magnetic stirrer, air condenser and external heating bath (diethylene glycol). The flask was charged with aluminum chloride (24.6 g, 0.088 mole), chlorobenzene (220ml) and V (53.0g, 0.00877 mole). The reaction mixture was stirred for 1 hour at 70-75°C. The resultant black mixture was cooled and then hydrolyzed by pouring it slowly into water (500ml). Suction filtration separated the hydrolyzate into a crystalline residue and a filtrate comprised of an aqueous layer and a heavy chlorobenzene layer. The crystalline residue was extracted with 10% sodium hydroxide solution (400g), and the chlorobenzene layer was extracted with 10% sodium hydroxide solution (200g). The combined caustic extracts were extracted with ether (2x250ml) to remove chlorobenzene. The resulting caustic mixture was acidified with excess 12 N hydrochloric acid (135ml). The acidified mixture was extracted with ether (2x200ml), the extract washed with saturated sodium chloride solution, the extract dried ($MgSO_4$) and flash evaporated at 65°/5.0mm Hg to give 39.04g crude triol analyzing 87% VI by GLC (67.2% yield). The crude triol was distilled to obtain 27.8g of a fraction analyzing 98.5% triol VI; b.p. 101-102°C/12 torr.; mp 95-101°. Three recrystalizations from heptane (300ml) at -20° led to 22.0g of VI (99.8% purity by GLC); mp 103-105°; 43.5% yield. The triol VI was observed to be hygroscopic and to lose its water of hydration under desiccation at 0.5 torr. The triol VI corresponds to that reported by Jones (6) (resublimed mp 105-106°), but prepared by the direct reaction of 1,3-bis-(2-hydroxyhexafluoro-2-propyl)-benzene with hexafluoroacetone and boron trifluoride at 200-250°C and 20 kiloatmospheres.

IR (KBr) peaks found: 3620(OH), 3599(OH), 3518(OH), 3132(CH), 1468, 1272(CF), 1227(CH), 1169(CF), 1132(CF), 1118(CF), 1013, 982, 891, 730, 709cm^{-1}.

Synthesis of triglycidyl ether of VI (VII)

This synthesis is outlined in scheme 4. A mixture of the triol V (16.00g, 0.0278 mole), epichlorohydrin (38.54g, 0.417 mole), and tetramethylammonium bromide (0.62g) was stirred and heated in a closed system at 60°C for 100 hours. Flash evaporation of the reaction mixture, finally at 90°C/5 torr., left 23.4g of crude chlorohydrin ethers of VI. This mixture was dissolved in 54g 3:1 toluene: ether alcohol and dehydrohalogenated by five successive 1 hour contacts at 50-55°C with 3.00 N sodium hydroxide solution, replenishing ethyl alcohol lost to the aqueous layer each time. The clear oil layer from the fifth wash was separated, dissolved in ether, and the extract washed, dried and concentrated by flash evaporation (80°C/5 torr.) to obtain VII as a clear viscous liquid that crystallized: 18.6g (90.0% yield); 99% purity by GLC; mp 47-50°. Recrystillizations from heptane led to the analytical VII; mp 48-50°C; IR (KBr) 3090(CH), 3020(CH), 2957(CH), 2916(CH), 2860(CH), 1058, 1018, 740, 711cm^{-1};

Analysis:

Calculated for $C_{24}H_{18}F_{18}O_6$: C, 38.72; H, 2.44; F, 45.94;

Found:

C, 38.65; H, 2.20; F, 46.00.

DISCUSSION

These two new fluoroepoxy resins can be cured in the manner previously disclosed for similar materials to yield strong, stable plastics. The

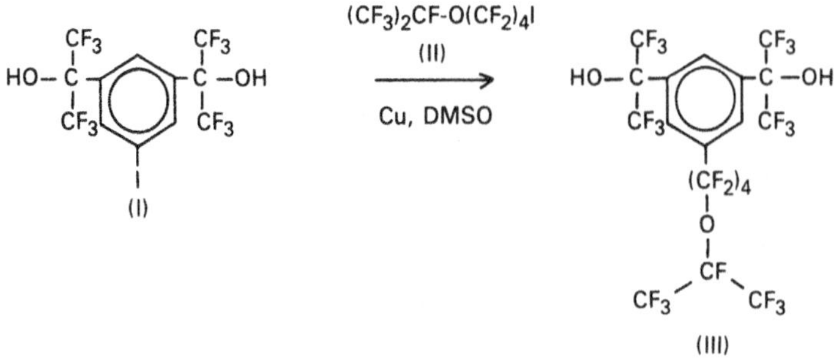

Scheme 1. Reaction scheme for diol with branched perfluoroether side chain.

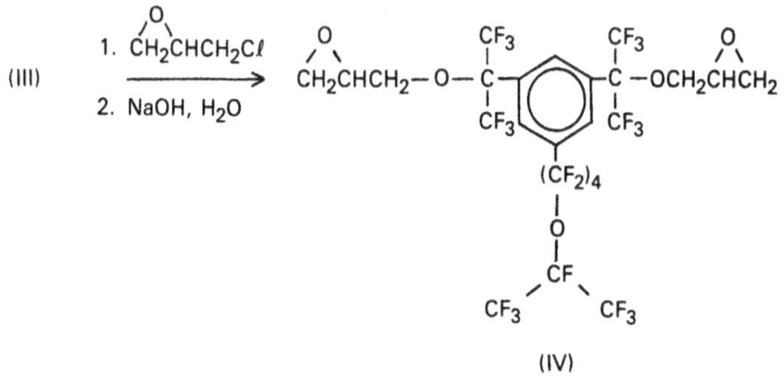

Scheme 2. Reaction scheme for epoxy resin with branched perfluoroether side chain.

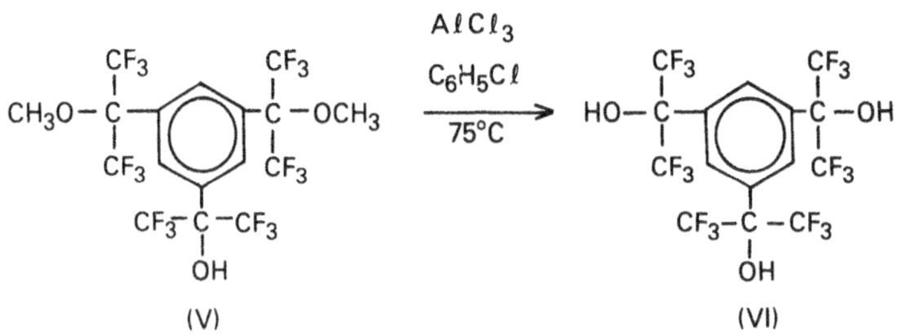

Scheme 3. Reaction scheme for a triol.

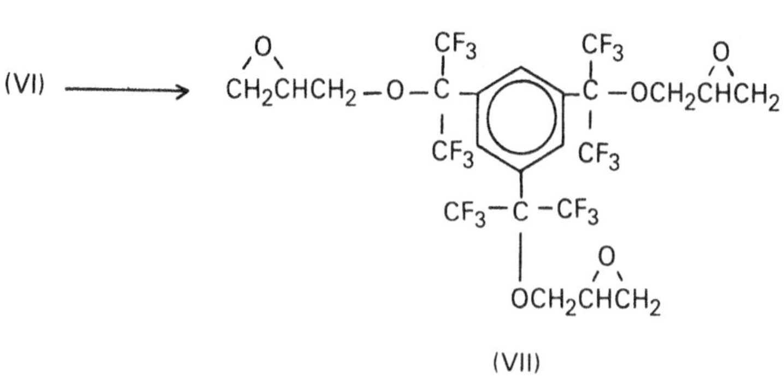

Scheme 4. Reaction scheme for a trifunctional fluoroepoxy resin.

trifunctional resin will indoubtedly yield products of higher glass transition temperatures and higher mechanical moduli than the difunctional resins synthesized earlier. The resin with branched ether structure may form effective oxygen-transport membranes which suggests possible applications in blood oxygenators or, eventually, artificial lungs.

REFERENCES

1. J. G. O'Rear and J. R. Griffith, Org. Coatings and Plastics Chem., 33 (1), 657 (1973).
2. J. R. Griffith and J. G. O'Rear, in "Biomedical and Dental Applications of Polymers", C. G. Gebelein and F. F. Koblitz, eds., Plemun, New York, 1981, p. 373.
3. J. R. Griffith, Polymeric Materials Science and Engineering, 50, 304 (1984).
4. L. G. Anello, R. F. Sweeney and M. H. Litt, French Patent No. 1,574,801 (18 July 1969); Chem. Abst. 72, 100023 (1970).
5. J. G. O'Rear and J. R. Griffith, U.S. Patent No. 3,879,430 (22 April 1975).
6. E. S. Jones, U.S. Patent No. 3,304,334 (14 Feb, 1967).
7. J. R. Griffith, Polymeric Materials Science and Engineering, 50, 422 (1984).

THE EVOLUTION OF POLYURETHANE ORTHOPAEDIC POLYMERIC CASTING SYSTEMS

Z. Oser

Research Division
Johnson & Johnson Products, Inc.
501 George St., New Brunswick, N.J. 08903

Within the past several years polyurethane polymeric casting systems have been increasingly used by orthopaedic medical personnel as an alternate material to the traditional plaster of Paris in immobilizing various parts of the body. This is occurring because of advances being made in the utilization of polyurethane technology for this purpose. Over the past 25 years a number of different polymeric systems have been evaluated for orthopaedic cast use with only limited success due to one or more deficiencies present making the use of them less desirable than the standard plaster of Paris casting material. This paper traces the evolution of orthopaedic casting materials starting from plaster of Paris. The chemistry of the curing reaction of the several polymeric materials is discussed together with a definition of the chemical requirements for a functional casting material. Special emphasis is given to the polyurethane systems which have proven to be quite adequate in their function and are predicted to replace a significant portion of plaster of Paris casting materials in the future.

INTRODUCTION

The history of plaster of Paris can be traced back to at least the time of the ancient Egyptians, (4000 BC) who used it in their palaces and pyramids. It was also used in ancient Greece (500 BC) by plasterers and stuccoers and has been found in archeological digs corresponding to ancient Judea where it had been used to construct tables and benches at the beginning of the Christian era. The first use of plaster of Paris for orthopaedic use appears to have occurred in the 9th century where the Arabs used plaster of Paris admixed with the white of egg to immobilize fractured limbs. Napoleon's surgeons first introduced the use of plaster of Paris for fracture immobilization into Europe having learned of it during their Egyptian campaigns in the early 1800's. Their technique involved using a wooden box as a mold around the limb. It was not until about 1850 however, that a Dutch army surgeon named Mathysen rubbed plaster of Paris into open mesh cloth strips and used these to wrap around a fractured limb becoming the first person to use this material in the form in which it is used today. In fact, polymeric casting systems, which are now becoming widely used and the subject of this presentation, still are presented to the medical professional in roll form and are applied by wrapping around the body member to be immobilized.

Chemistry of the Plaster of Paris Setting Reaction

Calcium sulfate dihydrate, in various mineral forms, is generically called gypsum ore. This is mined, ground and calcined to drive off water and yield calcium sulfate hemihydrate, i.e., plaster of Paris. When used in bandage form around a body member, after water immersion, the calcium sulfate hemihydate is converted to dihydrate, as shown below:

$$CaSO_4 \cdot 2H_2O \xrightarrow{\text{heat}} CaSO_4 \cdot 1/2H_2O + 1.5\ H_2O$$

$$(CaSO_4)_2 \cdot H_2O + 3H_2O \longrightarrow 2CaSO_4 \cdot 2H_2O$$

Advantages/Disadvantages of Plaster of Paris for Casts

Gypsum is readily available world-wide, and is relatively low cost to the end user and patient. The wrappable bandage is extremely easy to apply and is very conformable to the body member being cast. However, the resultant cast is very heavy due to the high density of plaster of Paris and suffers from poor impact strength since the material is brittle. In addition, plaster of Paris casts have poor water resistance which often leads to early cast breakdown and the need for recasting of the fracture. Nevertheless this material has been in continuous use for fracture immobilization for approximately 1000 years.

POLYMERIC CAST SYSTEMS

During the 15 year period of 1965-1980, various polymer chemists, mostly in industrial research laboratories, turned their attention to applying the technology of polymerization reactions toward attempting to develop a functional polymeric cast. The driving force for this work came from the general recognition of the disadvantages of plaster of Paris, cited above. Thus, the advantages that a polymeric cast should offer became defined as those of light weight, water resistant and impact resistant. The technical requirements to be met included: room temperature fast reaction(s), monomer(s) safe to use on a person, application techniques no more complex than those used with plaster of Paris, reaction(s) that could be initiated at the time of use by water, light, oxygen or a simple technique for introducing a second component, and finally the polymer, formed with a low exotherm, should be tough and preferably crosslinked.

Non-Urethane Based Polymeric Cast Systems

Light Activated Polymerization Systems

One of the first systems commercialized was a U.V. light activated system (1,2) formulated from vinyl toluene, an unsaturated polyester and benzoin methylether, as sensitizer. The viscous liquid was coated onto a knit fiberglass narrow tape and supplied in roll form to the end user. After wrapping the body member to be immobilized, a UV light source (3650-3925 A^{o}) was used for polymerization initiation. While this system provided a finished cast that, in principle met the requirements cited above, in practice, wide spread displacement of plaster of Paris did not occur, probably because of the requirement for special UV lamps which were relatively expensive. Also, the volatile, noxious monomer in this system required a very well ventilated room for the casting procedure. Similar systems requiring visible light initiation (3) were also developed somewhat later than the UV light systems; however, none of these were commercialized, probably because of the need to use a high intensity lamp and the presence of volatile monomers.

Polyionomer Polymerization Systems

A novel polymeric system based on a formulation of a polycarboxylic acid and a multi(metal) ion leachable powdered glass (4,5,6) was developed and has been commercialized. The chemistry of the setting reaction involves the leaching of multi-valent ions from glass when water is added. These form metal salts between carboxyl groups, which act as crosslinks. The reaction schematic is shown below.

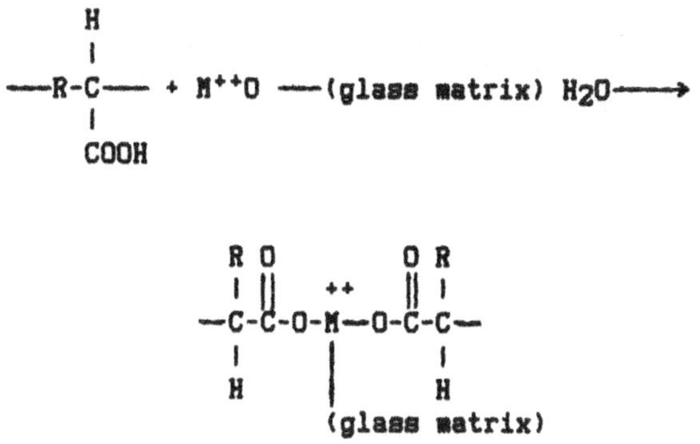

Wide spread displacement of plaster of Paris by this system has not occurred, probably because of cost factors.

Poly(Diacetone Acrylamide) System

A novel vinyl monomer resin system was developed based on diacetone acrylamide (7) and a novel redox polymerization catalyst (copper acetylacetonate and sodium sulfite)(8). Immersion in water caused dissolution of the catalyst which then initiated polymerization of the monomer. The structure of the monomer is shown below:

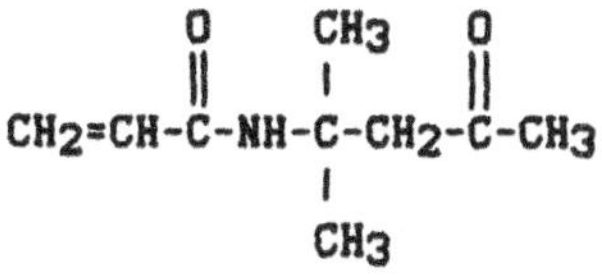

This system was not commercialized because the commercialization of the polyurethane systems discussed below showed them to be far superior in performance.

Polycaprolactone

A thermoplastic polymer, polycaprolactone, impregnated on a knit polyester fabric has been commercialized as a casting system (9). This is based on the principle of heating the roll in hot water to soften the material (160-180°F). At the higher temperature it can be wrapped about a body member since it is quite malleable. After cooling to below body temperature it becomes rigid and will immobilize a fracture. The technique of handling this system is quite different from that of plaster of Paris since it involves handling a hot roll of material plus the need to strip away an interliner film used to keep the hot material from sticking to itself while in roll form.

Urethane Based Polymeric Cast Systems

Two different polyurethane systems for polymeric casts have evolved both of which now have been commercialized. They differ in that one utilizes an internal tertiary amine catalyst contained in the polyol component, an aminopolyol (10), while the other utilizes an external tertiary amine catalyst added to the quasiprepolymer formulation during synthesis (11,12). These systems are prepared by reacting excess di- or poly-aromatic isocyanate with a polyol which is usually prepared from either ethylene or propylene oxide.

When these are dipped into water the isocyanate-water reaction occurs leading to high polymer formation containing both urethane and urea linkages. Crosslinking occurs where triol or higher polyols have been used. Schematic representations of the reactions are shown below (aminopolyol example).

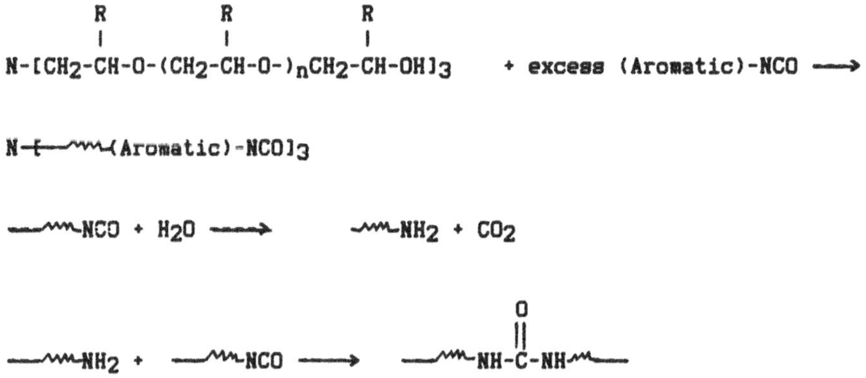

Since the preferred setting time for an orthopaedic cast is well under 10 minutes much work has been done in optimizing catalysts for the isocyante-water reaction. Tertiary amines appear to function better than metal salts, which are too active. Foam catalysts work well, but foam formation must be suppressed for best results. Very active foam catalysts at low concentrations give good setting times, however, they also catalyze allophanate formation leading to short shelf-life for the product. Stabilizers appear to help to enhance shelf-life. Several catalysts that have been utilized in these formulations include:

B-dimethylaminoethylether (11)

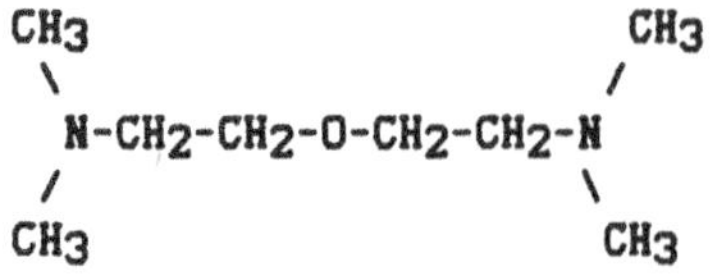

B-dimorpholinodiethylether (12)

```
   ___                              ___
  /   \                            /   \
 O     N - CH2-CH2-O-CH2-CH2 - N     O
  \___/                            \___/
```

CONCLUSIONS

The polyurethane systems described above have been utilized in preparing polymeric casting systems which are gaining wide-spread acceptance in the medical community. Changes in formulations, as well as in fabric substrate, continue to be made as manufacturers fine tune their products in response to user experiences. The great versatility of polyurethane chemistry continues to permit development chemists working in industrial laboratories to vary the performance characteristics of these casting products, almost at will.

After approximately 1000 years of use plaster of Paris is being displaced to a significant extent by polyurethane.

REFERENCES

1. J. A. Cozi, and D. C. Garwood, US Patent 3,881,473 (1975).
2. L. E. Beightol, US Patent 3,421,501 (1969).
3. C. J. Buck, US Patent 4,512,340 (1985).
4. W. D. Duncan, A. C. Barclay, R. Dunning and R. J. Parry, US Patent 4,043,327 (2977).
5. W. D. Potter, US Patent 4,243,567 (1981).
6. C. H. Beede and R. N. Zirnite, US Patent 4,324,591 (1982).
7. F. Boardman, US Patent 3,630,194 (1971).
8. A. Evans, J. M. Lesniak and F. Boardman, US Patent 4,344,423 (1982).
9. B. Phillips, D. F. Pollart and J. V. Koleski, US Patent 3,692,023 (1972).
10. F. A. Straube and G. Lenhnert, US Patent 4,376,438 (1983).
11. K. S. Baron and D. W. Wood, US Patent 4,427,002 (1984).
12. H. K. Yoon, US Patent 4,433,680 (1984).

TOTALLY RESORBABLE HIGH-STRENGTH COMPOSITE MATERIAL

Benjamin S. Kelley, Richard L. Dunn, and Robert A. Casper

Southern Research Institute
2000 Ninth Ave., South
Birmingham, AL 35255-5305

Metal plating is the traditional technique used to repair fractured bones invasively. Because these plates can be up to 15 times as stiff as normal bone, they act as a mechanical shunt, however, and direct forces around the bone that are normally transmitted through the bone at the fracture site. This condition usually results in stress protection which may lead to osteoporosis and atrophy. Also, a second surgical procedure is required to remove metallic or other nonresorbable plates. At this time, the healed bone may not be strong enough to support normal loads, making refacture a frequent occurrence. We have developed a new, totally resorbable composite material with mechanical properties similar to that of healthy bone. Because the material mimics the behavior of bone, it allows primary osseous union to be achieved. Postsurgically, the material begins to soften and resorb, allowing the healing bone to be exercised so that normal bone remodeling may occur. The material is eventually totally resorbed after healing is complete.

INTRODUCTION

Good clinical success in the internal fixation and stabilization of bone fractures has been achieved for many years through the use of metallic plates. This method or other invasive methods are used by the orthopedic surgeon when he is unable to achieve acceptable fracture reduction through the use of bandages, braces, or casts. Today, metallic fixation plates are generally made from stainless steel, titanium alloys, or cobalt/chromium alloys, which frequently have an elastic modulus greater than 10 times that of bone. Advocates of this system rely on the stiffness of these plates to hold the fracture in close apposition, leading to little callus formation during healing.

There are several potential disadvantages inherent in metallic plating systems, however. The rigidity of metal plates is an advantage for achieving primary osseous union during the early stages of healing, but this phenomenon is an extremely slow method of restoring the strength of the broken bone to its original level when compared to callus formation (1). In addition, because the generation of external callus is inhibited by the absence of mechanical stimulation at the fracture site, radiological assessment of the extent of healing by primary osseous union is difficult. The second problem associated with rigid fixation is related to the bone's actively remodeling itself to most efficiently perform the load bearing required of it. If a rigid metallic plate is not removed once the fracture has healed, it continues to carry a majority of the load over its span. Forces which would normally be directed through the bone at the fracture site are transmitted around the fracture instead. As a consequence, osteoporosis and atrophy occur (2,3). Although this change is attributed by some to vascular compromise to the bone, most investigators believe it to be associated with the altered stress distribution. Whereas increased strain within the bone acts by stimulating bone formation, decreased strain acts through bone resorption. The result is stress protection osteoporosis, characterized by an increase in bone formation at the site of stress transfer at the ends of the plate, a shift in the neutral axis of the bone-plate construction, and a net bone loss in the stress-shielded regions (4,5). If removal of the plate becomes necessary at a later date, the remaining bone may not be sufficiently strong to support loads, and refracture is likely unless the bone is protected by reduced load bearing.

The major difficulty in determining the optimum mechanical properties of a fracture fixation appliance involves the conflicting requirements existing during the two phases of healing (6). Initially, the plate must have sufficient stiffness to allow bony union with little bending or angulation. But some degree of mechanical deformation at the fracture site is required to promote the formation of external callus, which is the most rapid way of achieving bony union and restoring the strength of the fractured bone to its original level (1). Also, a less precise anatomical reconstruction is required with the generation of callus (7,8).

In an effort to alleviate problems with stress shielding, extensive research efforts have centered on the use of less rigid materials for internal plate fixation of fractures. The objective of this research has been to allow for more implant deformation at a given load and thus raise the load level in the bone. More flexible metals (9), polymeric back metal plates (10,11), fiber-reinforced composites (7,12,13), and thermoplastics (14,15) have been investigated as alternative materials.

Several investigators (6,14,16,17) have compared plates of various stiffness to define the parameters of the optimum mechanical properties for bone plates. Although these investigations clearly indicate that more flexible plates promote callus formation and induce less stress protection effects than metallic plates, variations in the type of mechanical testing and animal model used make it difficult to correlate the results and define the precise material properties of bone desired for each stage of fracture healing. Nonetheless, the continuous presence of a bone plate, regardless of the composition, results in loss of bone mass and delayed remodeling (8).

While the major thrust of much research is toward the development of bone plates that reduce the degree of stress protection, the ideal fixation appliance should possess enough initial rigidity to allow primary union and callus formation followed by a gradual reduction in stiffness commensurate with the healing bone's ability to serve in a load-bearing capacity. To achieve this goal without the need for additional surgical intervention, the appliance must necessarily biodegrade into byproducts which are neither toxic nor harmful to the host. It is desirable for these products to be either eliminated through normal physiological pathways or actually incorporated into the healing bone itself. An added advantage would be the use of materials which possess an osteogenic potential. While there are several biodegradable polymers available with strengths commensurate to that of virgin bone, they are much more deformable and cannot by themselves provide enough rigidity and at the same time satisfy the desired size restraints (18). Reinforcing these polymers with fibers of the same or different biodegradable polymer increases their strength but not to the point where they are viable for demanding fracture fixation applications (19,20).

Several investigators have successfully used nonresorbable reinforcing fibers to increase the strength and rigidity of biodegradable polymers for internal fracture fixation or stabilization (21,22,23). Carbon fibers are the most commonly used reinforcing medium. The polymer matrix resorbs while the reinforcing fibers disintegrate in the body and therefore gradually transfer the load from the appliance to the healing bone. These composites, however, are not truly biodegradable because the reinforcing component stays behind after the absorbable matrix is gone. While carbon is known to be biocompatible, it is not resorable, and the spread of carbon-fiber fragments into the surrounding tissues is clearly undesirable.

We have developed a totally resorbable, biocompatible fiber-reinforced composite material as an alternative to rigid metallic plates for use in invasive fracture fixation (patent pending). The matrix material can be almost any biodegradable polymer although we have primarily been using poly(DL-lactide) (DL-PLA). The reinforcing fiber is a biodegradable form of calcium metaphosphate (CMP). The composite has enough initial strength and rigidity to allow for primary fracture union but does not produce the stress-protection atrophy associated with the use of rigid metal plates. As healing progresses, the composite plate is slowly absorbed and eventually disappears entirely.

The concept of using CMP reinforcing fibers for strength and rigidity allows the material properties and, to a limited degree, the degradation rate of the appliance to be adjusted to meet the particular application. Similarly, the judicious selection of the biodegradable polymer matrix influences the attendant rate of decline of mechanical strength. Because most biodegradable polymers are thermoplastic, the potential exists for the device to be injection molded easily and then thermally contoured to conform to the bone.

This material also shows great promise for use in other orthopedic applications as well, where its high strength is only a temporary requirement until healing has progressed to the point where natural function has been restored. Included among potential end products are fixation screws, intermedullary rods, bone replacement or augmentation materials, surgical meshes, and bone ingrowth/mechanical interlocking coatings. The objective of the present investigation was to determine the <u>in vitro</u> behavior of DL-PLA/CMP composite plates and to assess the biocompatibility of the CMP fibers.

EXPERIMENTAL AND RESULTS

Preparation of Absorbable High-Strength CMP Fibers

CMP fibers were prepared by drawing or extruding from a melt using a proprietary process. Fibers up to 250 μm in diameter were easily obtained by this method.

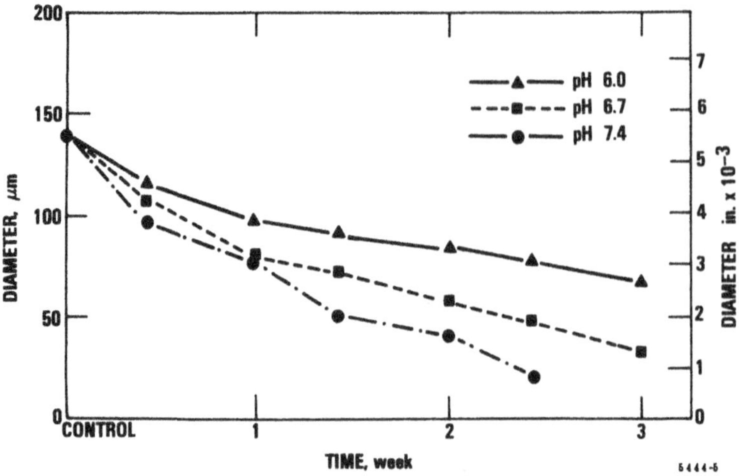

Figure 1. CMP fiber diameter vs incubation time in saline at 37°C.

CMP-Fiber Evaluation

The tensile strength and modulus of CMP fibers were determined using an Instron Model TMS. The span length was approximately 5 cm, and the crosshead speed was 5 cm/min. Tensile tests were performed on freshly drawn fibers and fibers which had been incubated in phosphate-buffered saline at 37°C. Saline baths with pH values of 6.0, 6.7, and 7.4 were used. Fibers were accurately weighed and measured before and after incubation to determine their resorption rate. The reduction in fiber diameter varied nearly linearly with incubation and is shown in Figure 1. The reduction in the cross-sectional area of the fibers is shown in Figure 2. The fiber resorption rate was found to increase with increasing pH of the saline bath.

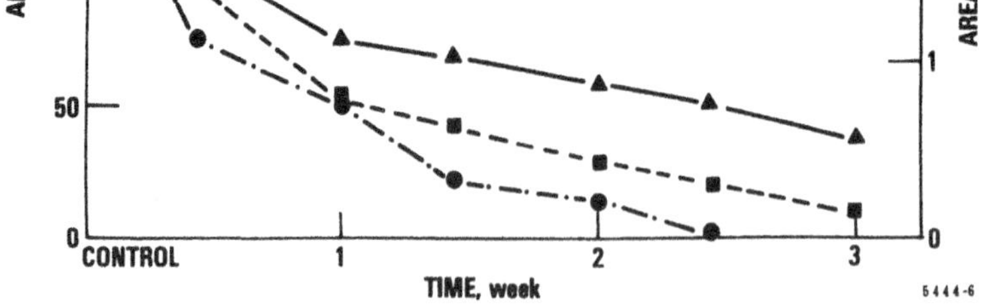

Figure 2. CMP fiber cross-section area vs incubation time in saline at 37°C.

The tensile strength of CMP fibers varied greatly with fiber diameter, with the fibers of smaller diameter having greater tensile strengths. This same trend held true even after incubation in saline. Because of the nonlinear relation between CMP fiber diameter (or cross-sectional area) and corresponding tensile strength, a simple averaging of data would be misleading when presenting these results. We chose, therefore, to match each set of tensile test data with a best-fit curve. The corresponding tensile strength for the average fiber diameter under a given condition was then read directly from these curves. The fiber diameter losses shown in Figure 2 were subtracted from the initial overall CMP fiber diameter of 140 μm and selected as the point on the best fit curves from which the tensile strengths were read. The relatively high initial fiber diameter of 140 μm was chosen so that the time duration of the study could be extended to weeks instead of days and so that all test fibers were initially of similar size.

We have previously found that CMP fibers with diameters well below 100 μm have tensile strengths exceeding 7×10^3 kg/cm^2 (100×10^3 psi), while fibers with diameters in the 200 μm range fail below 1.4×10^3 kg/cm^2 (20×10^3 psi). Control CMP fibers with a diameter of 186 μm had a tensile strength of about 1.1×10^3 kg/cm^2 (16×10^3 psi). Incubation in saline resulted in a gradual decline in the force necessary to break the CMP fibers, as illustrated in Figure 3. This trend, however, did not hold true when the tensile stress required to cause failure was calculated. As the incubation time increased, and thus the CMP fiber diameter decreased, the tensile strength of the fibers continually increased, as shown in Figure 4. After three weeks in saline, fiber tensile strength increased to over three times that of control fibers.

This phenomenon is probably a result of the erosion of micro flaws on or near the fiber surface caused by differential cooling during the fiber forming process. Figure 1 indicates that CMP fiber degradation is primarily caused by surface erosion, as the curves show a linear relationship with time. The erosion process is also evidenced by the scanning electron micrographs of Figure 5 comparing the roughened surface of a CMP fiber after three weeks in saline to the smooth surface of a fiber exposed only to air. As the erosion process occurs, an increasing number

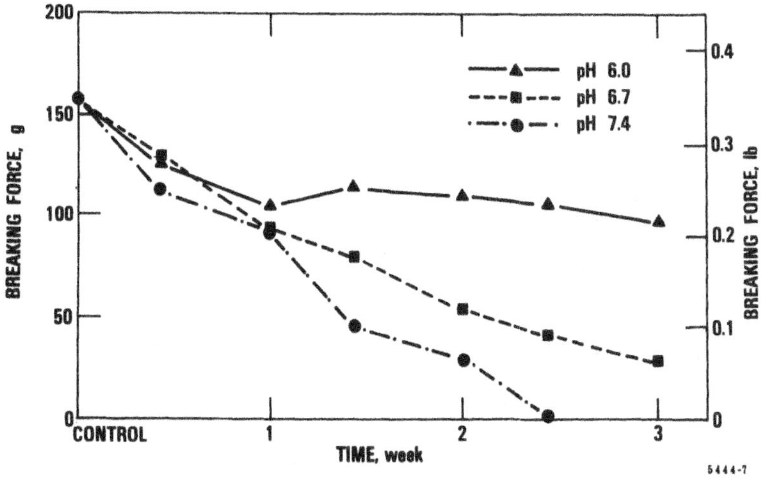

Figure 3. CMP fiber ultimate breaking force vs incubation time in saline at 37°C.

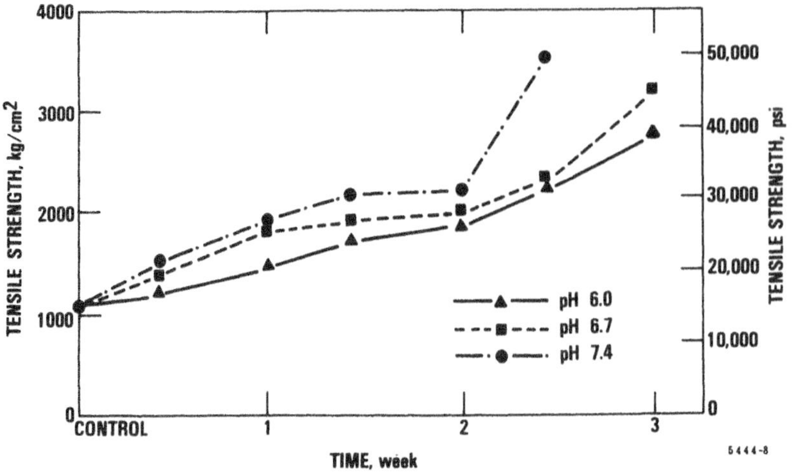

Figure 4. CMP fiber ultimate tensile strength vs incubation time in saline at 37°C.

of these micro flaws are eliminated, leaving the higher quality core of the fiber with relatively fewer flaws.

Preparation of DL-PLA

The apparatus used for the polymerization of DL-PLA consisted of a three-neck, round-bottom flask equipped with a mechanical stirrer and a gas inlet tube. DL-lactide monomer of at least 99.5% purity was charged into the flask and polymerized with the use of a suitable catalyst. After polymerization, the flask was broken away, and the polymer was dissolved in dichloromethane, filtered through a sintered glass filter, and precipitated by the addition of methanol. The purified DL-PLA was then dried in vacuo at 30 to 40°C, ground, and stored in a desiccator until needed.

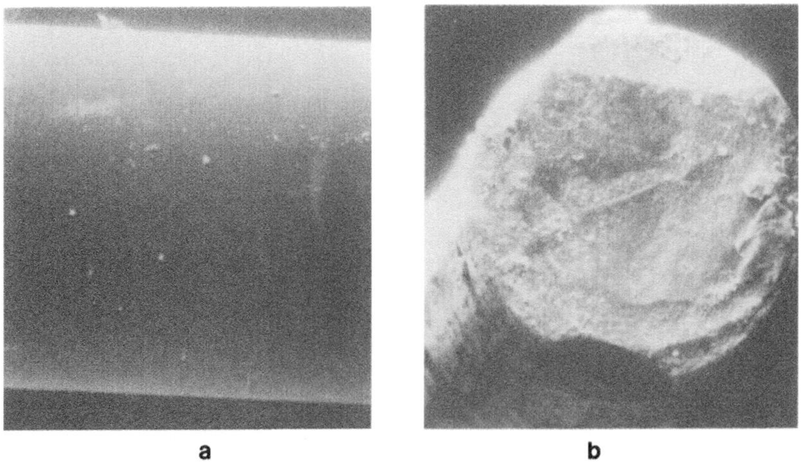

a b

Figure 5. Scanning electron micrograph of CMP fiber after (a) three weeks in air, (b) three weeks in saline at a pH of 6.7.

Preparation of Polymeric Films

Films of DL-PLA were prepared for use in fabricating composites. A dilute solution of DL-PLA in dichloromethane was prepared and injected into Mylar-lined, 4-inch spin cups rotating at 3600 rpm. Solvent was allowed to evaporate overnight at room temperature. To circumvent potential problems associated with residual solvent remaining in the polymer, films were also prepared by pressing the DL-PLA between Teflon-lined plates of a heated hydraulic press. Both techniques produced high quality, uniform DL-PLA films. Once prepared, the films were cut to the desired dimensions for use in composite fabrications.

Fabrication of Composites

Composite fixation plates were prepared by a lamination technique. The composite was laid up by alternating layers of DL-PLA film separated by layers of 7.6-cm-long CMP glass fibers. The fibers were brushed with a dilute solution of DL-PLA in dichloromethane prior to layup to improve binding of the fibers to the polymer films. The laminate was then placed in a mold and pressed. Vacuum was applied to remove residual solvent. Composite plates were also laminated without the aid of a solvent by following the procedures described above and pressing at a temperature above the glass transition temperature, but below the melting point. After the laminate plates were formed, they were removed from the mold and trimmed. Examination of the cross section of the composite plates revealed uniformly distributed reinforcing fibers with no evidence of individual laminates or delamination.

Composite Evaluation

The CMP/DL-PLA composites were tested in three-point bending according to ASTM D 790 both before and after incubation in physiologic saline. Testing was done on an Instron Model TMS with a crosshead speed of 1.3 cm/min. The attendant decline in flexural modulus and strength of composite specimens incubated in physiologic saline (pH 7.4) at 37°C for up to seven weeks is shown in Figures 6 and 7. After five weeks in saline, the flexural modulus of the 0.3 x 1.3 cm composite samples dropped from an initial value of about 160 x 10^3 kg/cm^2 (2.3 x 10^6 psi) to a final value of 26 x 10^3 kg/cm^2 (0.4 x 10^6 psi). Similarly, the flexural strength of composites dropped from 2.0 x 10^3 kg/cm^2 (28 x 10^3 psi) to 0.2 x 10^3

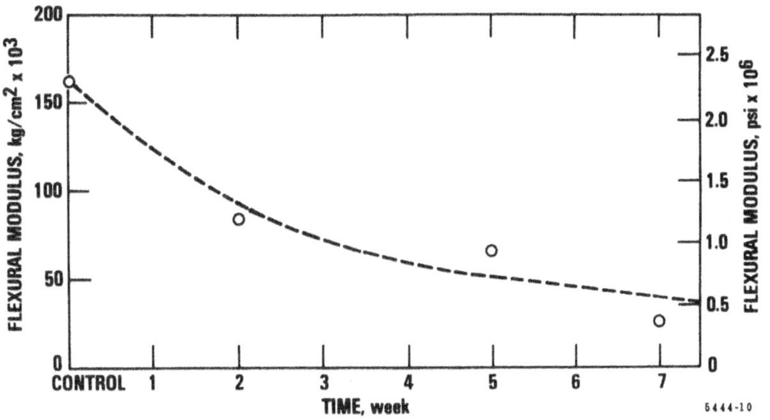

Figure 6. Flexural modulus of CMP reinforced DL-PLA plates vs incubation time in saline at 37°C.

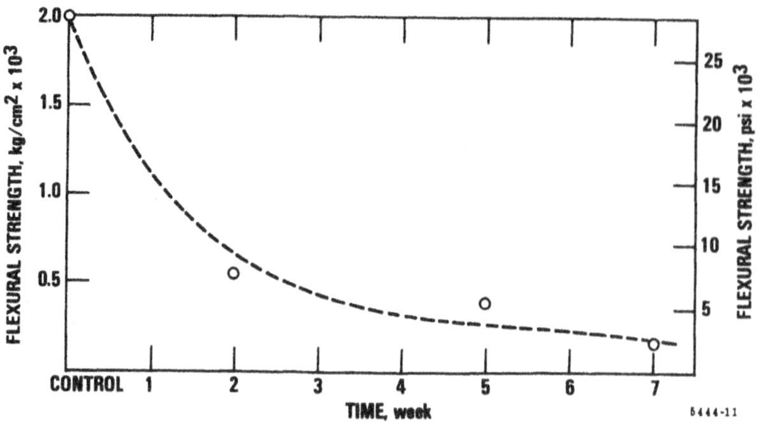

Figure 7. Flexural strength of CMP reinforced DL-PLA plates vs incubation time in saline at 37°C.

kg/cm^2 (2.8 x 10^3 psi) after seven weeks in saline. The decline in mechanical strength and stiffness proceeds at a somewhat slower rate in vivo. The results of these in vitro studies will be used as benchmark data for comparison with in vivo studies and to aid in evaluation of future composite modifications.

Biocompatibility of CMP Fibers

Growth inhibition of Vero cell cultures and cytotoxicity as determined by Cr^{51} release were used to evaluate the biocompatibility of CMP fibers (24). All assays were performed and all cells cultured in DMEM supplemented with 10% fetal bovine serum and 100 mM L-glutamine and antibiotics. Quantitative cytotoxicity assays resulted in no significant difference between CMP exposed cultures and control cultures. Cells concultured with CMP fibers were observed by phase contrast microscopy and found to exhibit no evidence of growth inhibition over a period of 99 days. Residual CMP fibers evaluated by scanning electron microscopy appeared entirely biocompatibile inasmuch as healthy Vero cells could be seen attached to the degraded fiber surfaces.

DISCUSSION AND CONCLUSIONS

There is an obvious need for internal fracture fixation appliances capable of providing initial rigid stabilization but which gradually lose their mechanical strength commensurate with the healing bone's ability to serve in a load-bearing capacity. Until now, only permanently rigid plates, partially biodegradable plates, and compliant biodegradable plates have been investigated for use in fracture fixation and stabilization. The high-strength, totally resobable fixation plate described here has an initial strength somewhat similar to that of virgin bone and relinquishes its mechanical strength in a time frame similar to healing bone's ability to resist external forces.

In addition, the fiber reinforcing component of the composite, which is chemically and crystallographically similar to the mineral constituents of bone, allows for the tailoring of mechanical properties to suit the particular application. Also, preliminary animal implant and tissue culture studies indicate that CMP exhibits excellent biocompatibility. No toxic reactions have been observed in these studies. DL-PLA, the matrix of the composite, has long been known to be biocompatibile, with its degradation products being eliminated via the Krebs cycle. Calcium phosphate materials with compositions and breakdown products similar to those of CMP have been shown (25,26) to be biocompatible as well as osteogenic and are actually utilized by healing bone. Hollinger (27) has also suggested that DL-PLA exhibits some degree of osteogenic potential. Because of the biocompatibility and bioresorbability, the composite material, with its high strength, has many potential orthopedic and medical applications.

There are many questions, however, concerning the kinetics of resorbable composite degradation which must be considered before an optimum design for any application can be obtained. Although smaller diameter CMP fibers result in higher tensile strength, one must contend with the conflicting requirements of high composite strength versus both the surface area of the fibers required for no slip-composite bonding and the retention of sufficient composite mechanical integrity until the defect being repaired has healed. While the smaller diameter CMP fibers exhibit a higher tensile strength per unit of cross-sectional area, they also present a higher surface area of which the polymer matrix must bond so that minimal matrix fiber slippage occurs. A compromise must therefore be met between the strength and size of reinforcing fiber so that optimal composite strength is attained.

Additionally, consideration must be given to the requirements the composite appliance must meet at points in time after implantation. If the consistuents of the composite are optimized to provide maximum bending and torsion resistance at the time of implantation, it may not be capable of supplying adequate support for the 8 to 12 weeks generally required for most fractures to heal. If extremely small fibers are used for reinforcement so that initial strength is optimized, they may totally degrade before the composite device has served its purpose, thus compromising its utility. Conversely, larger reinforcing fibers may not provide for maximum initial strength but may result in an appliance which surrenders its mechanical strength at a relatively slower rate so as to provide support for a longer period after implantation.

A third consideration which further complicates the degradation mechanics of this novel concept is the influence of diameter and packing density of the reinforcing fibers on the degradation kinetics of the polymer matrix. The initial phase of polymer breakdown occurs as a result of hydrolytic degradation as bonds between polymer chains are disrupted and then the polymer chains are cleaved. Before actual resorption of the polymer begins as a result of phagocytosis, the polymer begins to soften and lose its mechanical integrity. After steady state conditions are reached, polymer breakdown occurs at a nearly uniform rate throughout the sample. A transient response occurs, however, as body fluids diffuse through the boundaries of the polymer matrix at a rate that is influenced by the spacing and frequency of reinforcing fibers. Likewise, the degradation of the reinforcing fibers, which is a surface erosion process, is influenced by the gradient of calcium and phosphate ions throughout the polymer matrix and the size of the local sink.

All of these factors play an important role in the overall behavior of resorbable composites, and each must be accounted for to achieve an appliance with the optimum characteristics for a particular application. Although composite mechanics analysis techniques are necessary tools for the optimization of properties, the unique circumstances associated with resorbable composite require that modifications be made to conventional analysis schemes. Each application and set of circumstances must be evaluated individually to ensure that the best possible design results.

ACKNOWLEDGEMENTS

The aid of Mr. James P. English in preparing the DL-PLA, Ms. Amelia G. Potter in preparing the composite samples, and Mr. David N. Ellis for mechanical testing was greatly appreciated. This work was supported in part by the U.S. Army Institute of Dental Research, Washington, DC.

REFERENCES

1. B. McKibbin, J. Bone Jt. Surg., 60B, 150-162 (1978).
2. L. Stromberg and N. Dalen, Acta Chir. Scand., 142, 115-122 (1976).
3. H. K. Uhthoff and F. L. Dubuc, Clin. Orthop. Rel. Res., 81, 165-170 (1971).
4. P. Slatis, P. Paavolainen and E. Karaharju, Canad. J. Surg., 23, 247-250 (1980).
5. A. Szivek, G. C. Weatherly, R. M. Pillar and H. U. Cameron, J. Biomed. Mater. Res., 15, 853-865 (1981).
6. G. W. Bradley, G. B. McKenna, H. K. Dunn, A. U. Daniels and W. O. Stratton, J. Bone Jt. Surg., 61A, 866-872 (1979).
7. K. Taylor, C. Johnson-Nurse, B. McKibbin, J. Bradley and G. W. Hastings, J. Bone Jt. Surg., 64B, 105-111 (1982).
8. H. K. Uhthoff and M. Finnegan, J. Bone Jt. Surg., 65B, 66-71 (1983).
9. D. M. Nunamaker and S. Perren, in: "Current Concepts of Internal Fixation of Fractures", H. Uhthoff, Ed., Springer Verlag, Heidelberg, 1980, pp. 389-397.
10. G. W. Bagdy, D. L. Korvick, J. E. Newbrey, G. D. Pettit, J. D. Lincon and R. D. Sande, Transactions of the 31st Annual Meeting of the Orthopedic Research Society, Las Vegas, NV, 1985, p. 164.
11. L. L. Daniels, Cone, G. H., Kenner and R. S. Schultz, Transactions of the 31st Annual Meeting of the Orthopedic Research Society, Las Vegas, NV, 1985, p. 164.
12. N. Gillett, S. A. Brown, J. H. Dumbleton and R. P. Pool, Biomaterials, 6, 113-121 (1985).
13. S. L. Y. Woo, W. H. Akeson, R. D. Coutts, L. Rutherford, D. Doty, G. F. Jemmott and D. Amiel, J. Bone Jt. Surg., 62A, 68-78 (1980).
14. S. A. Brown and J. Vandergrift, Biomat. Med. Dev. Art. Org., 9, 27-35 (1981).
15. A. J. Tonino and P. J. Klopper, in: "Current Concepts of Internal Fixation of Fractures", H. Uhthoff, Ed., Springer Verlag, Heidelberg, 1980, pp. 342-347.
16. D. R. Carter, R. Vasu and W. H. Harris, Acta Orthop. Scand., 52, 241-248 (1981).
17. K. Taylor and J. Bradley, J. Bone Jt. Surg., 65B, 312-315 (1983).
18. D. C. Tunc, M. W. Rohovsky, W. B. Lehman, A. Strongwater and F. Kummer, Transactions of the 31st Annual Meeting of the Orthopedic Research Society, Las Vegas, NV, 1985, p. 165.
19. R. L. Dunn, D. H. Lewis and D. R. Cowsar, Annual Technical Report, Year One, SoRI-EAS-79-505. Prepared by Southern Research Institute, Birmingham, AL, under Contract DAMD17-78-C-8059 for the U.S. Army Institute of Dental Research, U.S. Army Medical Research and Development Command, Fort Detrick, MD, 1979, p. 23.
20. M. Vert, H. Garreau, M. Audion, F. Chabot and P. Christel, Transactions of the Eleventh Annual Meeting of the Society for Biomaterials in Conjuction with the Seventeen International Biomaterials Symposium, San Diego, CA, 1985, p 218.
21. K. R. St. John, in: "Biocompatible Polymers, Metals, and

Composites", M. Szycher, Ed., Technomic Publishing Co., Inc., Lancaster, PA, 1983, pp. 861-871.
22. R. A. Casper, R. L. Dunn and D. H. Lewis, Annual Technical Report, Years Two and Three, SoRI-EAS-81-799. Prepared by Southern Research Institute, Birmingham, AL, under Contract DAMD17-78-C-8059 for the U.S. Army Institute of Dental Research, U.S. Army Medical Research and Development Command, Fort Detrick, MD, 1981, p. 59.
23. J. R. Parsons, H. Alexander and A. B. Weiss, in: "Biocompatible Polymers, Metals, and Composites", M. Szycher Ed., Technomic Publishing Co., Inc., Lancaster, PA, 1983, pp. 873-905.
24. G. C. Battistone, J. W. Vincent, R. S. Borowski, M. A. Jackson and W. A. Gray, Proceedings of the 63rd General Session, International Association for Dental Research/Annual Session, American Association for Dental Research, Las Vegas, NV, 1985, p. 980.
25. J. M. Brady, and D. E. Cutright, J. Biomed. Mater. Res., 10, 977-979 (1976).
26. H. U. Cammeron, I. MacNab and R. M. Pilliar, J. Biomed. Mater. Res., 11, 179-186 (1977).
27. J. O. Hollinger, J. Biomed. Mater. Res., 17, 71-82 (1983).

BODY ABSORBABLE OSTEOSYNTHESIS DEVICES

D. C. Tunc[a], M. W. Rohovsky[a], B. Jadhav[a],
W. B. Lehman[b], A. Strongwater[b] and F. Kummer[b]

(a) Johnson & Johnson Products, Inc., Orthopaedic Research
U.S. Route #1 South, North Brunswick, NJ 08902

(b) Hospital for Joint Diseases, Orthopaedic Institute
301 E. 17th Street, New York, NY 10003

The need for the development of the totally body absorbable internal bone fixation devices has been documented in the literature during the recent years. Such a system would eliminate the second surgical procedure of removing these devices from the body after the bone is healed and, secondly, these devices would not have to be as stiff as the currently used metallic devices and therefore would not cause osteoporosis. A process has been developed to synthesize high molecular weight, low monomer content polylactide. This polymer has been characterized by chemical and mechanical tests, after simulated *in vitro* tests and *in vivo* implants. Biocompatibility of the polymer has been shown in rabbit and canine models. Body absorption of the polylactide bone plates and screws as well as their efficacy have been shown in the repair of the canine radius and tibia models.

INTRODUCTION

Although traditionally used metallic internal fixation devices such as plates, screws and rods have achieved relatively high degrees of success in the management of long bone fractures, they are not completely satisfactory for two reasons. First, these nonabsorbable devices require a second surgical procedure for their removal after the fracture is healed and before bone resorption, loosening, corrosion or infection becomes a problem. Second, the use of internal fixation devices of high stiffness such as those made from stainless steel has been identified as a cause of stress protection induced ostaopenia beneath the plate. This weakening of the bone underneath the plate may, and in certain percentage of cases, does lead to refracture of the bone after plate is removed.

Ideally, one would like to have a device whose strength and stiffness is compatible to bone and whose strength will gradually decrease as the fracture gradually heals thus eliminating the bone resorption. Sometime after the complete healing of the bone, this device should be absorbed by the body, thus eliminating the second surgical procedure.

As a result of our on going research, a high strength, body absorbable polymer was developed and totally body absorbable bone fixation devices were designed and fabricated from this polymer. The safety, biocompatibility and body absorption of these devices were established. Finally these devices were successfully tested in repairing the fractures of radii and of tibiae in the dogs. Presently our research is continuing towards further improvements and the eventual use of these products in humans.

EXPERIMENTAL DETAILS AND RESULTS

The experimental part of this work can be presented in six sections. These are:

I. Development of a process for synthesizing high molecular weight polylactide.

II. *In vitro* testing of the HMW Polylactide.

III. *In vivo* testing of the HMW Polylactide.

IV. Toxicity and biocompatibility testing of the polylactide.

V. Design and fabrication of osteosynthesis devices from the HMW polylactide.

VI. Efficacy testing of the osteosynthesis devices.

I. Development of a process for Synthesizing High Molecular Weight Polylactide

Commerically used absorbable polymers such as polylactide and polyglycolide or the copolymer of these are of relatively low molecular weight. Therefore, they are suitable for use in the manufacturing of absorbable sutures which serve their function and resorb quickly all within a few weeks. Absorbable polymers for orthopaedic use on the other hand are required to function for as long as 12-20 weeks. Therefore these polymers are required to retain their strength for a longer period of time and to be absorbed by the body at a much slower rate compared to the polymers used in sutures.

It has been established by our earlier studies (1) that in order to maintain sufficient strength long enough in the body, the -hydroxycarboxylic acid type, absorbable polymers should meet the following requirements:

1. High molecular weight.
2. Low unreacted monomer content.
3. High crystallinity.
4. Very high enantiomeric purity.

Since our previously reported studies (1) showed that none of the commercially available absorbable polymers, such as polylactides and polyglycolides as well as their copolymers, met the above requirements a new synthesis procedure had to be developed to produce polylactides and or other poly-α-hydroxy acids.

In order to obtain high enantiomeric purity in the polylactide one has to use highly purified lactide in the synthesis. We have chosen L(-) lactide in the synthesis as the starting material. The lactide should have a minimum optical rotation of -288° as 1% w/v lactide in benzene. Furthermore its water content should be less than 0.05%.

In order to synthesize polylactide which has high inherent viscosity and low unreacted monomer, two independent variables of polymerization were chosen to control the two dependent variables of the polylactide. The independent variables were: the reaction temperature, and the monomer to catalyst ratio. The dependent variables were: the inherent viscosity of polylactide, and the percent unreacted monomer in polymer.

A series of PLA batches were synthesized at temperatures varying from 105°C to 170°C at monomer to catalyst ratios of 1000 to 300,000. The polylactides that were synthesized under conditions of various combinations of these two variables were analyzed for the inherent viscosity and for the percent unreacted monomer in the polymer. The results were expressed as two equations, one for the inherent viscosity versus the temperature and the monomer to catalyst ratio and the other for the percent unreacted monomer versus the temperature and the monomer to catalyst ratio. The results were plotted as three dimensional graphs which are shown in Figures 1 and 2. From these graphs one can choose the optimum polymerization conditions for any polylactide.

II. In Vitro Testing of the HMW Polylactide

In vitro testing of the HMW-PLAs were done by incubating standard ASTM test specimens of PLA in Buffer-7 at 37 ± 0.5°C and removing six samples from the test environment at each predetermined time periods of 0, 2, 4, 8, 12, 24 weeks and immediately testing them with instron using standard ASTM test procedures. The results are shown in Table I.

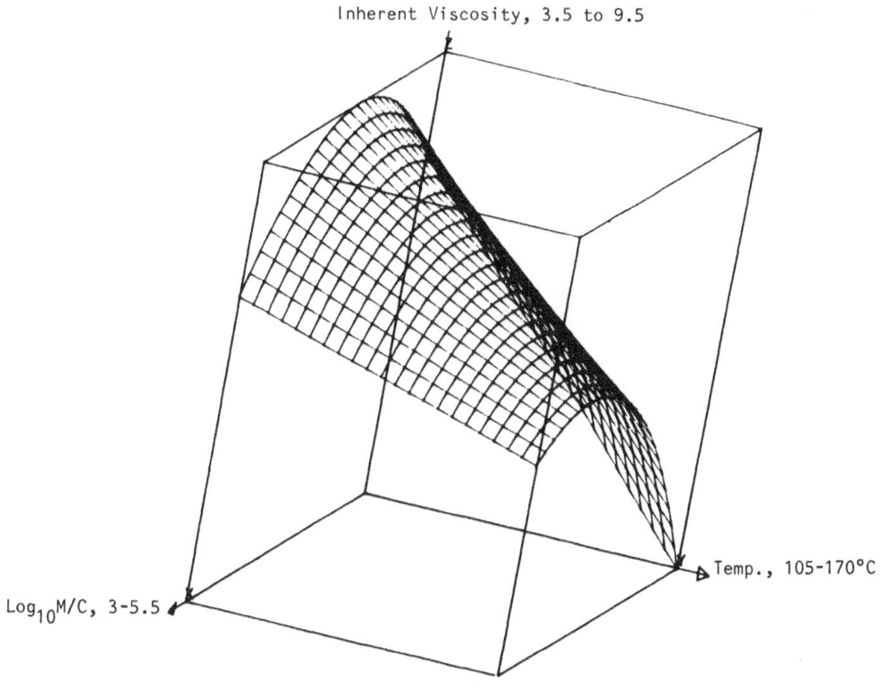

Figure 1. Inherent viscosity as a function of reaction temperature and the monomer to catalyst ratio.

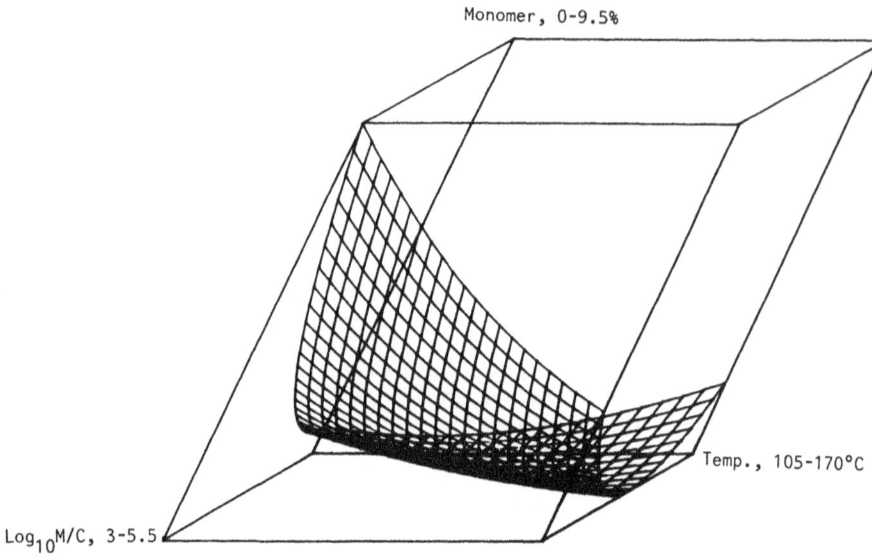

Figure 2. Percent unreacted monomer in the polymer as a function of the reaction temperature and the monomer to catalyst ratio.

It had also been determined that about 90% of the initial UTS and of the initial yield strength could be retained at the end of 24 weeks of incubation if the polymer was further purified. This data is shown in Table II.

III. In Vivo Testing of the HMW Polylactide

Test specimens measuring 20 x 3 x 1 mm were prepared from PLA. Four samples were implanted in the dorsal subcutaneous tissues of rats. At predetermined surgical time periods, 4 rats were sacrificed at each of 2, 4, 8, 12 and 24 weeks. Half of the retrieved samples were used for the tensile testing and the other half were used for flexual testing. These results are presented in Table III.

Table I. *In Vitro* Test Results of PLA After Incubation in Buffer 7 at 37°.

	Number of Weeks				
	0	4	8	12	24
Bending Modulus, MPa	5098	4076	4174	3899	3032
U.T.S., MPa	57.1	41.6	26.6	9.3	----
Yield Strength, MPa	48.2	39.0	23.0	7.8	----
Inherent Viscosity	7.1	3.4	2.2	1.5	0.9

Table II. In Vitro Test Results of Purified PLA After Incubation in Buffer 7 at 37°.

	Number of Weeks				
	0	4	8	12	24
Utimate Tensile Strength, MPa	72	71	71	69	68
Yield Strength, MPa	--	52	--	--	55
Elongation at Elastic Limit, %	4.6	--	6.1	--	4.4
Bending Modulus, MPa	4400	4800	---	--	4900
Impact Strength Joules/Mater	105	73	49	--	43
Shear Strength, MPa	54	58	56	55	57

Table III. In Vivo Test Results of PLA After Subcutaneous Implantations in Rats.

	Number of Weeks				
	0	4	8	12	24
Bending Modulus, MPa	5098	4516	4156	3570	2757
U.T.S., MPa	57.1	29.9	21.9	17.8	----
Yield Strength, MPa	48.2	26.4	18.4	----	----
Inherent Viscosity	7.1	3.0	1.9	1.4	0.9

IV. Toxicity and Biocompatibility Testing of Polylactide

Cytotoxicity testing of dry powdered PLA was done using L929 mouse fibroblast cells in culture. No cytotoxic effects were observed. Saline and sesame oil extracts of the PLA were prepared and tested by systemic injection and intracutaneous tests (ref. USPXX, p. 952) in the mouse and in the rabbit. Negative results were yielded in all instances. It was therefore concluded that no toxic or irritating ingredient leached in a measurable quantity.

Biocompatibility testing was done by implanting 3.95 mm diameter 30 mm long intramedullary rods bilaterally in the femurs of 36 rabbits and implanting 3.95 mm diameter cortical plugs bilaterally into the femurs of additional 36 rabbits. In both cases, high density polyethylene (HDPE) was used as the control in each rabbit. In addition a 3 x 1 x 20 mm PLA sample was implanted into the dorsal muscle of each rabbit receiving a cortical plug. At each end of the following periods, 4 rabbits with IM rods and 4 rabbits with cortical and intramuscular implants were sacrificed. The sacrifice periods were 2, 4, 8, 12 weeks, 6 months, 1 year, 1.5 years, 2 years and 2.5 years post surgery. At each period, 2 of the explanted IM rods and the subcutaneous samples were used for determining the percent PLA absorbed and the inherent viscosity of the PLA at that time. The other samples were used for histological evaluation.

Figure 3 shows the change in inherent viscosity as a function of time of implantation as intramedullary rods and as subcutaneous implants. It is apparent from this graph that the rate of degradation of PLA is the same in both cases. The absorption of PLA as a function of time of implantation as intramedullary rods and as subcutaneous implants is shown in Figure 4. This graph suggests a slightly slower rate of absorption in the case of intramedullary rods, compared to the subcutaneous implants.

Figures 5 and 6 are photomicrographs of histological slides of IM rods of PLA and HDPE at 2 years post surgery. The biocompatibility study demonstrated the following similarities between the PLA and HDPE.

1. Endosteal bone formation.
2. Fibrous capsule.
3. Fibrous capsule with osteogenesis.
4. Bone replacing fibrous capsule.
5. Normal bone marrow appearance.

The difference between the two materials was in gradual degradation and phagocytosis of PLA while the HDPE remained intact.

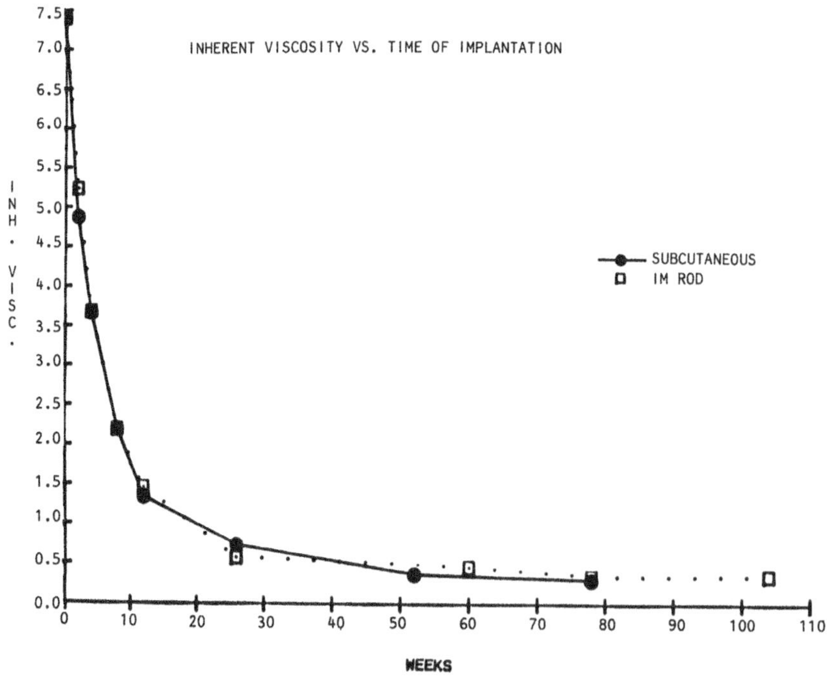

Figure 3. Inherent viscosity change in PLA intramedullary rods and in PLA subcutaneous implants as a function of duration of implantation in rabbits.

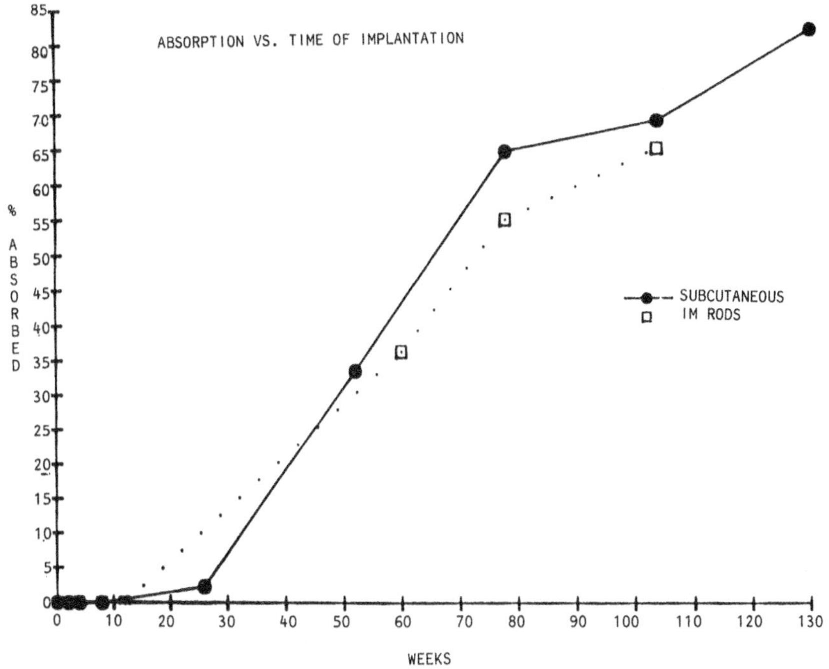

Figure 4. Absorption of PLA as a function of the duration of implantation as intramedullary rods and as subcutaneous implants.

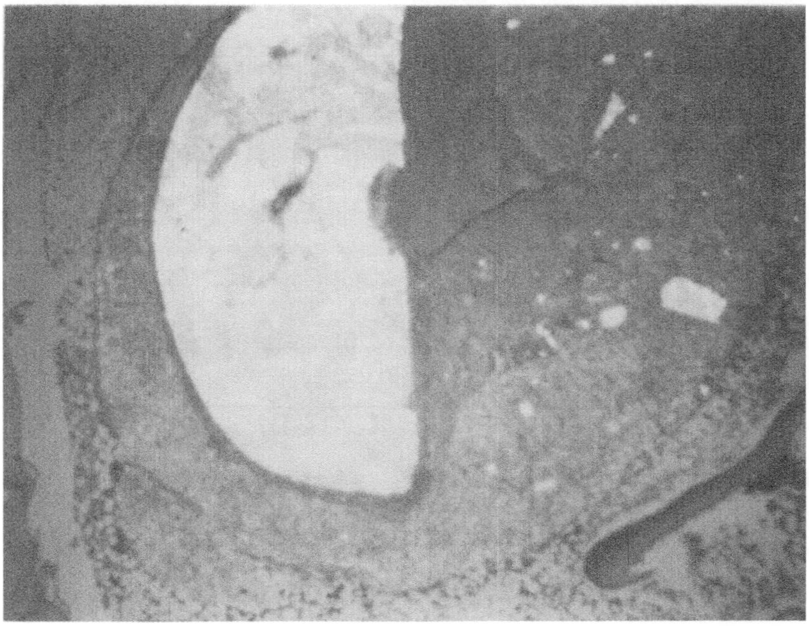

Figure 5. PLA IM rod after 2 years of implantation. 16x Magnification.

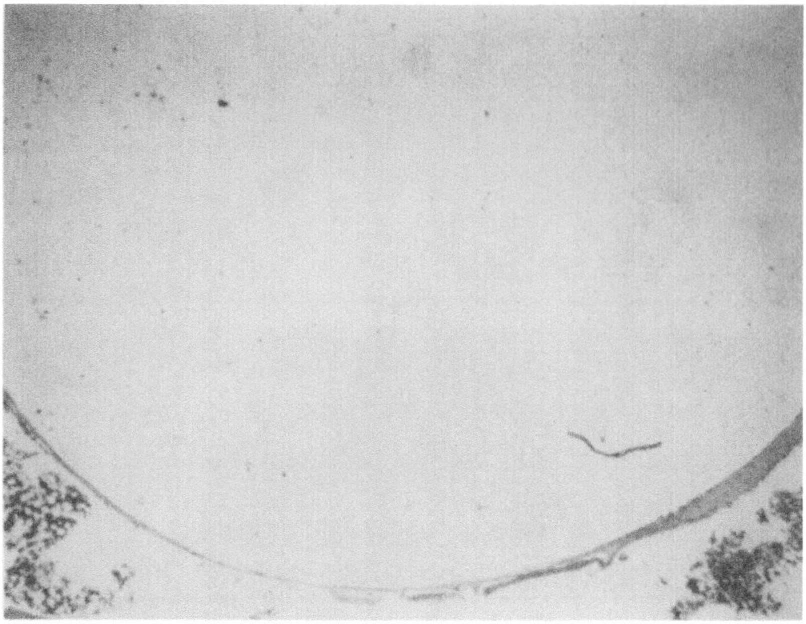

Figure 6. HDPE IM rod after 2 years of implantation. 16x Magnification.

V. Design and Fabrication of Osteosynthesis Devices from the HMW PLA

A new bone plate design was developed specifically for the PLA. The plate used in the tibial osteotomy model was a 6 hole, 10 mm thick plate. The radial osteotomy plate was a 4 hole, 6.4 mm thick plate. The PLA bone screws were 5 mm diameter, 25 mm long, 0.8 mm pitch screws. The stainless steel screws were 4.5 mm cortical screws. Mechanical tests results of the 6.4 mm thick PLA bone plate and its comparison with the two stainless steel plates are shown in Table IV.

Table IV. Mechanical Test Results of Stainless Steel and Polylactide Bone Plates.

	Four Point Bending Test		
Plate identification	Rigidity, EI NM^2	Yield Moment NM	Max. Moment NM
PLA,DCT-2, 6 holes	0.91	2.7	3.6
SS,Semitubular, 6 holes	0.28	0.91	1.4*
SS, DCP, 6 holes	3.33	6.3	14.3*

* Did not fail.

VI. Efficacy Testing of the PLA Osteosynthesis Devices

Two different animal models were used to evaluate the PLA plates with steel screws and PLA plates with PLA screws: (A) Canine tibial osteotomy model; (B) Canine radial osteotomy model:

Model A. Canine Tibial Osteotomy Model

The canine tibial osteotomy model was used to evaluate the PLA plates with stainless steel screws. In this study four beagles of 1.5-2.5 year age were used. Each animal received a mid-shaft tibial osteotomy and mid-shaft fibular fracture. Tibiae of 3 beagles were reduced using PLA plates with 4.5 mm stainless steel cortical screws and the tibia of one dog was reduced using a 6-hole stainless steel DCP plate with 3.5 mm stainless steel screws. In both cases, the operated leg was put in a cast for about a week. Healing of the fracture was followed radiographically up to 2 year post surgery period. At this time the dogs were sacrificed. The operated tibiae and the contralateral tibiae were excised. The implants were removed and the tibiae were tested in a four point bending test to determine the strength of the operated tibiae and to compare them with the unoperated (contralateral) tibia of the same dog. The results are shown in Table V. This pilot study data suggests that the tibiae plated with the polylactide (PLA) plates possessed about 100% of the strength of the unoperated tibiae at 2 years post surgery whereas the tibia plated with the stainless steel plate had only 23% the strength of the unoperated leg.

To assess the effect of the screw holes in the bone on the strength of the bone, freshly excised tibiae from other beagles were drilled and tapped just as it was done in surgery. These drilled and tapped tibia from each dog was compared with undrilled contralateral tibia of the same dog in a four point bending test. The results are shown in Table VI.

Model B. Canine Radial Osteotomy Model

The canine radial osteotomy model was used to evaluate the efficacy of the PLA plates with PLA screws. In this study 28 female beagles of 1.5-2.5 year age groups were used in 4 groups as shown in Table VII. Each animal received a radial osteotomy (left leg). Groups I and IV received 4 hole PLA plates and 5 mm PLA screws; Group II received 4 hole stainless

Table V. Mechanical Test Results of the Canine Tibia Two Years Post Surgery.

Implant Plate/Screw	Dog No.	Four Point Bending Test Failure Load, Newtons: Control	Operated	% of Control
PLA/SS	663	1737	1670	96
PLA/SS	669	1572	1853	118
PLA/SS	671	1572	1443*	92*
SS/SS	661	2477	561	23

* Broke through screw hole.

Table VI. Mechanical Test Results of the Drilled and Tapped Canine Tibiae.

Tapped Hole Size mm	Dog No.	Four Point Bending Test Failure Load, Newtons Control	Drilled	% of Control
3.5	816	1354	829	61
4.5	815	2209	873	40
4.5	726	1715	726	42

Table VII. Design of the Canine Radial Osteotomy Model Study.

Group	Number of Dogs	Trauma System	Sacrifice
I	10	PLA	12 wks(5); 52 wks(5)
II	9	Stainless Steel	12 wks(5); 52 wks(2) 104 wks(2)
III	5	Not Plated	12 wks(5)
IV	4	PLA	104 wks(2); 156 wks(2)

Table VIII. Radius Model at 12 Weeks

Implant Plate/Screw	Dog No.	Four Point Bending Test Failure Load, Newtons Control	Operated	% of Control
SS/SS	735	802	196	24
SS/SS	731	1091	167	15
SS/SS	732	869	174	20
PLA/PLA	777	1225	187	15
PLA/PLA	767	967	223	23
PLA/PLA	769	864	129	15
None	726	802	67*	8
None	823	739	129	15
None	726	802	67*	8
None	823	739	165*	22

* Pulled apart rather than broke.

Table IX. Mechanical Test Results of the Canine Radii at 52 Weeks Post Surgery.

Implant Plate/Screw	Dog No.	Four Point Bending Test Failure Load, Newtons: Control	Operated	% of Control
PLA/PLA	717	891	316*	35
PLA/PLA	719	1354	815*	60

* Broke through screw hole.

steel Plate and 3.5 mm stainless steel screws. Group III was not plated. All groups were kept in cast for 12 weeks. Healing of the fractures were followed radiographically. At the time of sacrifice, radii were excised from both the operated and the contralateral legs. The plates and screws were removed and the radii were tested for fracture load in 4 point bending. The results from the dogs that reached the 12 week post surgery test period are shown in Table VIII, and the results from the dogs that reached the 52 week post surgery period are shown in Table IX.

In order to assess the effect of the screw holes on the strength of the radii, fresh left radii of 6 beagles were drilled and tapped as it was done in surgery and the strength of each radius was compared with the strength of undrilled right radius of the same dog. The results are shown in Table X. The results in Tables VIII, IX, and X indicate that substantial healing of osteotomies were obtained at 12 and 52 weeks. There were no significant differences in the strengths of the PLA and SS plated radii at 12 weeks. Radiographally they were all healed. Longer term part of the study is still in progress. Figures 7 and 8 show the two radiographs of a tibia, one at post surgery, one at 382 days post surgery respectively.

Table X. Mechanical Test Results of the Drilled and Tapped Canine Radii.

Tapped Hole Size mm	Dog No.	Four Point Bending Test Failure Load, Newtons: Control	Drilled	% of Control
5.0	819	882	267	30
5.0	820	891	263	30
5.0	815	1105	437	40
5.0	818	1354	330	24
3.5	817	1541	797	52
3.5	816	971	517	53

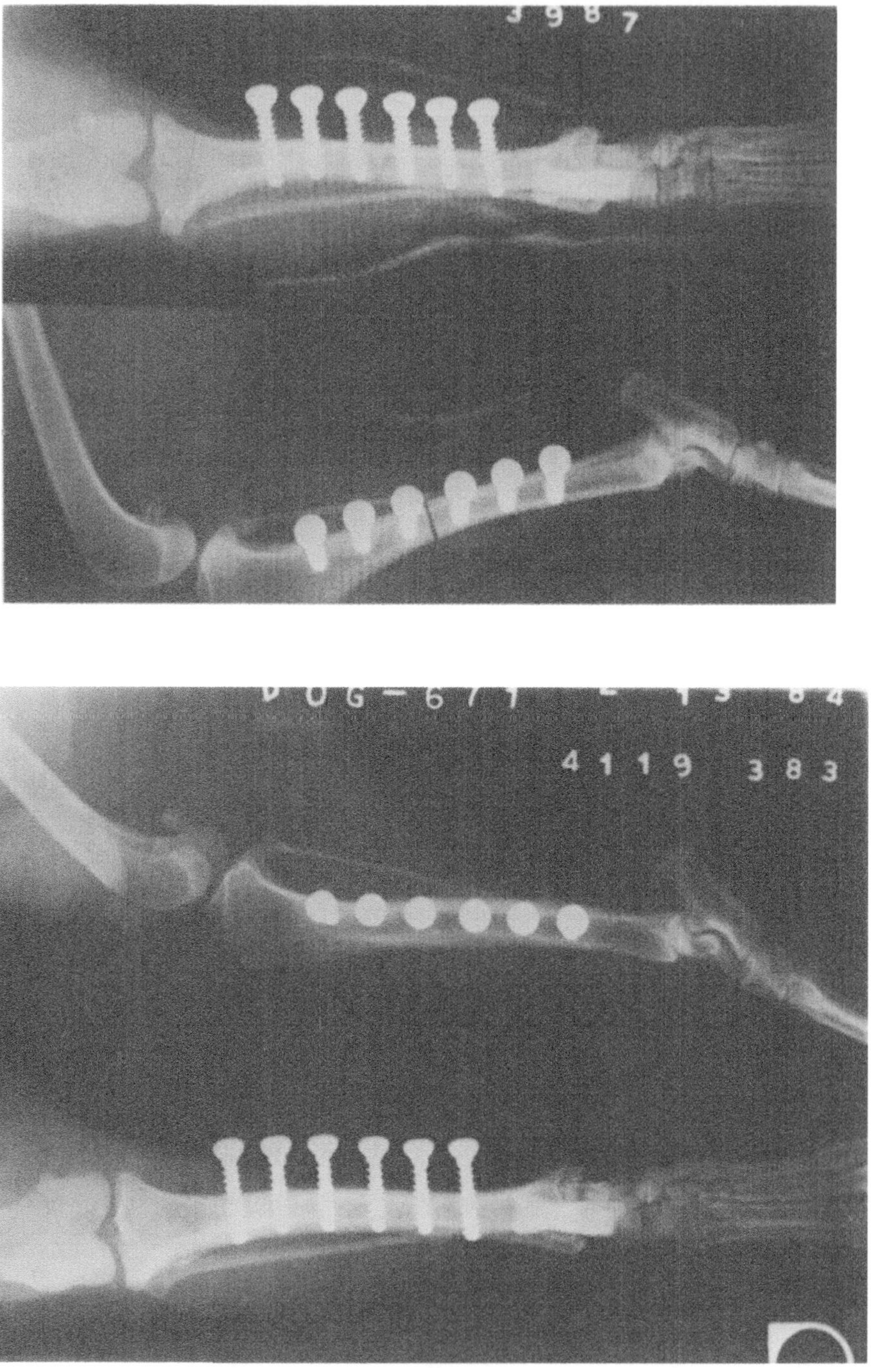

Figure 8. Radiograph of the same tibia as in Figure 7, 382 days post surgery.

CONCLUSIONS

High strength PLA has been synthesized. Biocompatibility of PLA IM rods and cortical plugs has been shown. Body absorption of PLA has been shown. Preliminary results indicate efficacy of PLA plates and screws in canine radius and tibia models.

REFERENCES

1. D. C. Tunc, Transactions 9th Annual Meeting of the Society for Biomaterials, 47, April 27-May 1, 1983.

CROSSLINKABLE POLYESTERS FOR BIOMEDICAL COMPOSITES (1)

Samuel J. Huang, Peter G. Edelman and J. A. Cameron

Institute of Materials Science
University of Connecticut
Storrs, Connecticut 06268

A low molecular weight poly D,L-lactide containing carbon carbon double bonds was synthesized from D,L-lactide, 2-butene-1,4-diol and fumaric acid. This functionized polymer can be thermally crosslinked with lauroyl peroxide initiator. It is potentially a useful thermosetting resin for biodegradable composites for implant applications.

INTRODUCTION

There are considerable research activities in the area of polymeric materials in medicine (2-7). Of special interest to us are polymers suitable for the construction of artificial organs. Many of the existing polymers with good chemical and physical properties for nonmedical applications often display undesirable side effects once implanted in the human body. The specifications which apply to polymers used as substitutes for internal organs are very severe. The polymers must have the following properties: appropriate chemical and physical structures to provide desirable useful life time in the body; ability of being sterilized and absence of contaminant. Moreover, the polymers must not induce tumor formation, autileukotactic response, thrombus formation, and inflammatory encapsulation or cell modification in the surrounding tissue. All these requirements greatly reduce the number of polymers which are potential canidates for biomedical applications.

We have investigated the possibility of using composite materials for implants. Structural fiber composites with very high specific properties have been successfully utilized in many engineering applications. Since the fibers used are much stronger than ordinary polymeric materials they normally impart strength to whatever matrix they are put in. When the fibers are all aligned in one direction, maximum strength is achieved in the composite material along the direction of the fiber. It is possible to tailor-make specific properties in different directions by placing or weaving the fibers into specific orientation, and thereby satisfy the design requirements of structures subject to a variety of multiaxial stresses. The metal plates currently used in osteoplasty for fracture fixation present an elastic modulus much higher than the bones to which they are connected. This difference of rigidity between plates and bones prevents the healing by proliferation of primary callus and results in lower strength of the repaired bone due to excess stress protection (8). It is therefore desirable to design a composite bone plate that degrades slowly to stimulate regenerative tissue growth, and eventually be completely absorbed by the body, and thus eliminate the necessary second surgery to remove the plate after the bone is healed.

Our approach to the composite bone plate is a continuous fiber reinforced crosslinked matrix system. Preferential hydrolysis of amorphus regions of a semi-crystalline polymer implant often results in catastrophic strength loss (9). Instead of having a two-phase semi-crystalline system as the resin it will be much better to have a gradually degrading one-phase crosslinked system as resin. Our results on the degradation of crosslinked poly(ε-caprolactone) showed that a porous morphology resulted as the uncrosslinked segments were preferentially degraded before the crosslinked segments. The shapes and volumes of samples remained unchanged up to 70% weight loss due to degradation. This should be very beneficial as it has been shown that a porous morphology stimulates regenerative tissue growth (10-12). It should also be easier to obtain composites of better physical properties than those derived from thermoplastics. There are inherent difficulties in interfacial wetting between linear thermoplastics and reinforcing fibers which limits optimum stress transfer between fibers and matrix (2).

We chose to incorporate crosslinkable carbon-carbon double bonds into the structures of poly(α-hydroxy acids). The biodegradation of polyglycolate and polylactate and their copolymers in vitro and in vivo are well documented. We report here our recent results.

EXPERIMENTAL

Materials

D,L-lactide (m.p. 127°C), kindly donated by the American Cyanamid Company was sublimed at 100°C under vacuum just before use. Cis-2-butene-1,4-diol from General Aniline and Film Corporation was distilled under reduced pressure. Furmaric acid (trans-2-butenedioic acid) was 99+% from Matheson Coleman and Bell and used as received. Lauroyl peroxide from Pfaltz and Bauer was recrystallized from chloroform by the addition of methanol. Stannous octoate [tin (II)-2-ethyl hexanoate] was used as received from Polysciences. Diethyl ether and chloroform were reagent grade from Baker and used as received. THF was refluxed over sodium and then distilled.

Instrumentation

Infrared spectra were recorded on a Nicolet 60SX FTIR at a resolution of 4 cm^{-1}. An IBM WP-200 SY FT-NMR was used for ^{1}H and ^{13}C-NMR. Differential Scanning Calorimetry (DSC) was performed on an Omnitherm QC-25 thermal analysis unit. Dilute solution viscosity was done on filtered solutions at 25.0 $\pm$ 0.05°C in a constant temperature bath.

Procedure

Poly(D,L-lactic acid)diol with 2-butene-1,4-diol

2-Butene-1,4-diol (5.00g, 5.67 x 10^{-2} mole) was mixed with D,L-lactide (16.36g, 1.13 x 10^{-2} mole) and stannous octoate (1.13 ml of 0.10 M stannous octoate in ethyl ether to make 0.1% catalyst based on D,L-lactide). The reactants, in a single neck 50 ml round bottom flask with 2 magnetic stirring bars, were evacuated slowly at room temperature to 0.02

mm Hg. Vacuum was maintained for 12 hours after which the pressure was released with an argon purge. The flask was submerged in a silicone oil bath at 135°-140°C for 3 hours with a very slow argon purge flow. The product was cooled and the IR and NMR were taken at this point. Based on ^{1}H-NMR peak integration, the molecular weight was around 450 g/mol corresponding to 2.5 dilactide units per butene diol.

Poly(D,L-lactic acid) diol + fumaric acid

To the oligomeric D,L-lactide diol (15.82g, 3.6×10^{-2} mole) was added fumaric acid (4.6g, 4.0×10^{-2} mole) which totals 10% excess fumaric acid. The flask was equipped with an argon purge and submerged into the oil bath at 135°-140°C. After 40 hours the melt was observed to appear slightly yellow. After 112 hours the temperature was slowly raised over a period of 1 hour to 160°C. Simultaneously a vacuum of 0.02 mm Hg was slowly attained. These conditions were maintained for 12 hours yielding a light brown clear glassy resin.

Crosslinking of poly(D,L-lactide)diol/fumarate

The above brown resin was dissolved in THF to form a 7.76% solution. To the solution (5.0g) was added lauroyl peroxide (0.0194g, 5% w/w) which was stirred until dissolution. The solution was poured into a 5 cm petrie dish and placed under an argon purge on a hot plate mainainted at 55°-60°C for 10 hours. The temperature was finally raised to 100°C for 2 additional hours. The resulting product was a crosslinked film insoluble in THF.

RESULTS AND DISCUSSION

2-Butene-1,4-diol was used as a difunctional initiator with stannous octoate as catalyst for the ring opening polymerization of D,L-lactide to give the oligomeric diol 1, Equation 1. Subsequently 1 was chain-extended by reaction with fumaric acid to give the unsaturated polyester 2, Equation 2.

The IR spectrum of the poly(D,L-lactide)diol 1 show a broad rounded absorption band at about 3500 cm^{-1} due to the OH end groups (Figure 1) The C=O band absorption is at about 1750 cm^{-1} for the 1:2-butene-diol/D,L-lactide reaction, the OH and C=O bands are shifted slightly to longer wavelengths than in the product from 1:4 butenediol/D,L-lactide. This reflects an increase in the relative extent of hydrogen bonding found in the lower molecular weight diol. After reaction with fumaric acid, the carbonyl band is broadened with some lower energy shoulders due to the conjugated fumarate carbonyl (Figure 2). There is also a weak olefin

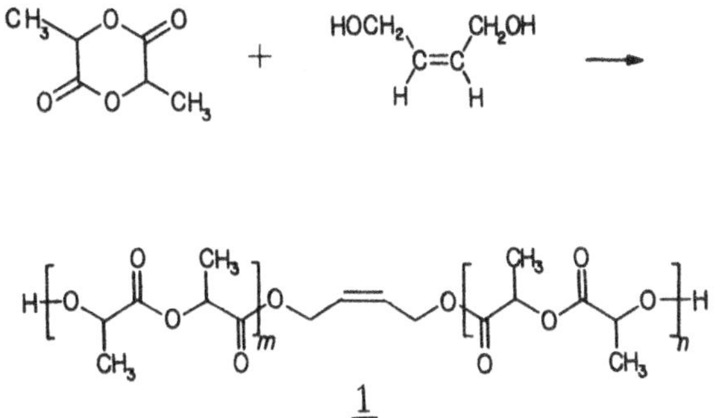

1

(Equation 1)

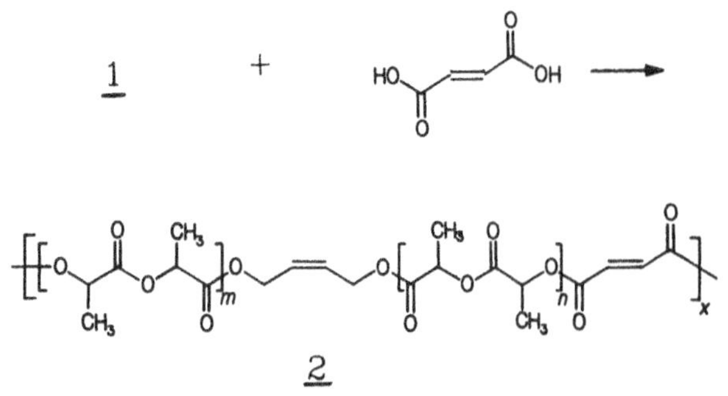

(Equation 2)

absorption at 1647 cm^{-1}. The hydroxyl band has flattened out considerably. Other changes in the spectra reflect incorporation of fumarate groups and loss of absorptions associated with free hydroxyl groups. There is a concurrent increase in absorptions associated with lactate/fumarate esters, particularly fumarate bands at 979 cm^{-1}, 1262 cm^{-1}, and 1289 cm^{-1}. ^{1}H-NMR results are very elucidative of the structure obtained (Figure 3). For the product from the reaction stoichiometry of two D,L-lactides per one butene diol, the spectra shows many combinations of multiplets for the methyl and methine protons of lactate. This is to be expected because in addition to different possible diad, triad and tetrad sequences of d- and l-isomers (13,14), there are different possibilities for either adjacent hydroxyl group, ester group or the new methylene group of the butene diol. There is a simple singlet for the hydroxyl proton. Other evidence strongly supporting the indicated structures is born out in the integration ratio of 2 to 1 for the methylene and olefin protons, respectively, of the butene diol segment of the molecule. On

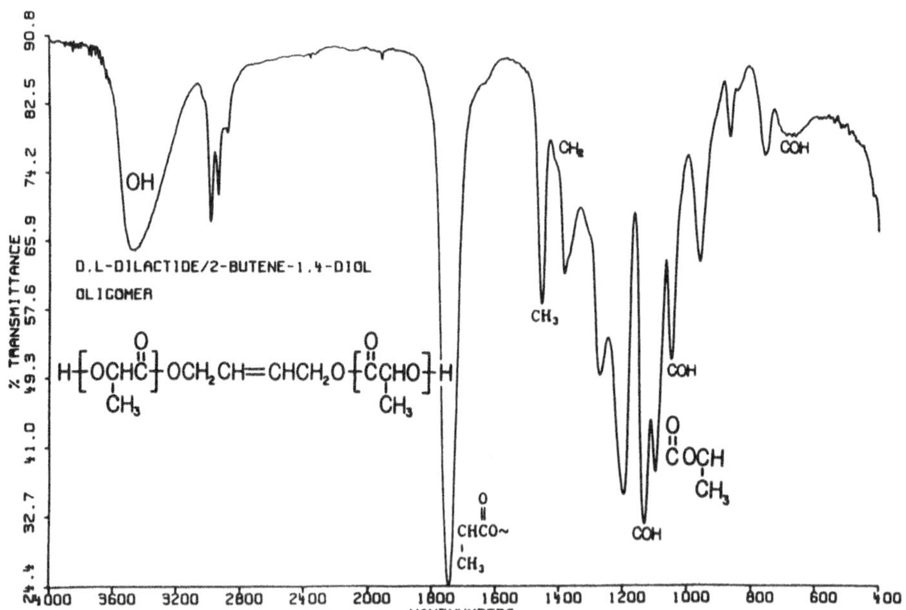

Figure 1. Infrared spectrum of butenediol/lactate oligomer.

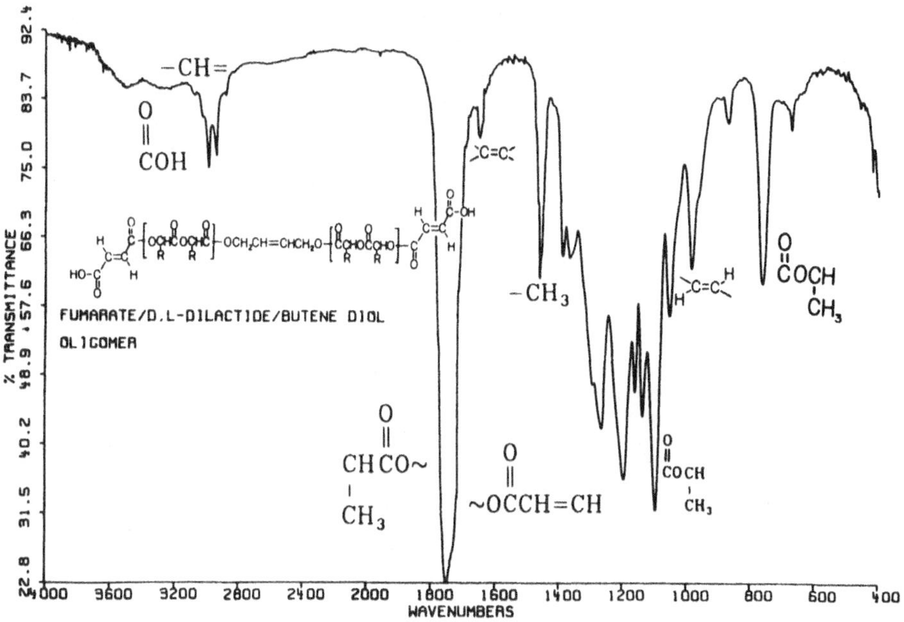

Figure 2. Infrared spectrum of butenediol/lactate/fumarate oligomer.

reaction with fumaric acid, two changes become very apparent in the ^{1}H-NMR spectrum (Figure 4). The signal for the methine proton adjacent to the hydroxyl group becomes practically imperceptable and the methyl signal simplifies considerably. Also it is evident that the product contains disubstituted and monosubstituted fumarate ester product, as evidenced by the peaks at 6.97 and 6.88 ppm respectively (15). The small fumarate peak at 6.88 ppm may be due to a fumarate adjacent to a butene diol.

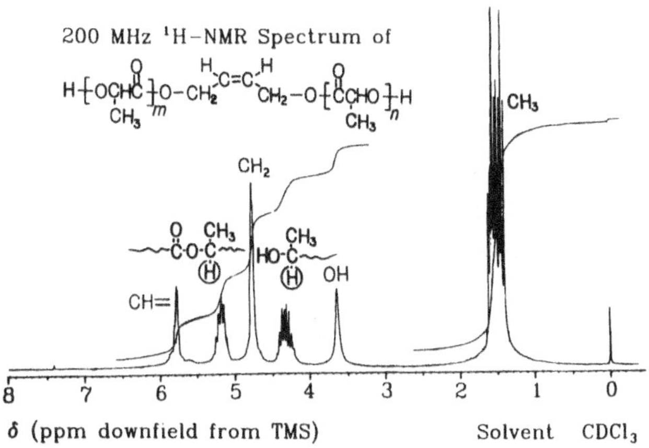

Figure 3. 200 MHz ^{1}H-NMR spectrum of butenediol/lactate oligomer.

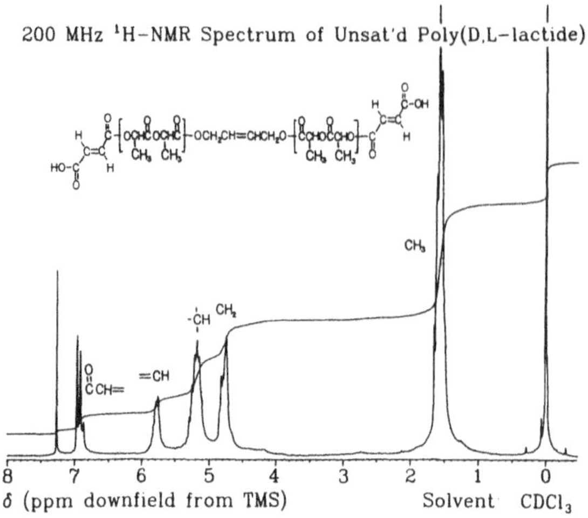

Figure 4. 200 MHz ^{1}H-NMR spectrum of butenediol/lactate/fumarate oligomer.

^{13}C-NMR results corroborate well with the ^{1}H-NMR data (Figures 5,6). Again changes are evident on going from the poly(DL-lactide)diol oligomer to the fumarate. In the carbonyl region there are two groups of peaks in each spectrum for the poly(DL-lactide)diol and the fumarate. For the diol case, the downfield set at about 175 ppm is most probably due to the ester carbonyl one carbon removed from a hydroxyl group (Figure 6). This set disappears on reaction with fumaric acid, and a new set due to the fumarate carbonyl appears at about 164 ppm (Figure 6). The peaks centered around 169-170 ppm are from the lactate carbonyl and remain relatively unchanged.

In the spectrum of the diol oligomer there is a main set of peaks at 128 ppm due to resonance of the carbon on the reacted side of the butene diol. The small peaks at 134 ppm are due to unreacted butene diol and the small peaks at 124 ppm are of uncertain origin. This region changes on going to the fumarate, yielding one peak at 128 ppm for disubstituted butene diol and three fumarate olefin peaks. The downfield fumarate olefin peak results from monosubstituted fumarate and the middle peak is from disubstituted fumarate (14). The upfield fumarate peak may be due to fumarate/butene diol ester sequences, as opposed to fumarate/lactate sequences. This is strongly suggested by comparison with spectra from products of different stoichiometry. Upfield in the aliphatic carbon region of the poly(DL-lactide)diol spectrum there are two main peak groups due to the lactate methine carbon. At 69.1 ppm are peaks from the ester methine carbon while at 66.9 ppm are peaks from the methine carbon adjacent to a hydroxyl group. The methine carbon of butene diol gives rise to a signal at 60.7 ppm due to esterification, while at 58.2 ppm there is an absorption from the unreacted butene diol methylene carbon. Similarly the methyl carbons give two main peaks, one at 20.2 ppm due to adjacent hydroxyl group and one at 16.7 ppm for the esterified lactate. This region simplifies on going to the fumarate ester (Figure 6). The aliphatic region essentially has only one peak set for the methine, methylene and methyl carbons previously mentioned. Any residual broadening can be attributed to different stereosequencing of D- and L-isomers. Dilute solution viscosities (0.5g/dl in $CHCl_3$) of 1 and 2 were found to be 0.040 and 0.24 dl/g, reflecting the increase of molecular weight through chain-extending.

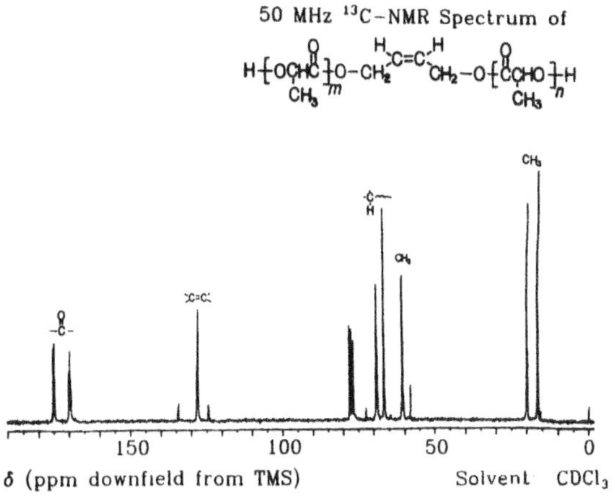

Figure 5. 50MHz ^{13}C-NMR spectrum of butenediol/lactate oligomer.

With the structures of 1 and 2 established we than proceeded to explore the crosslinking of 2. Thermal crosslinking of 2 was effected by using lauroyl peroxide as radical initiator, Equation 3. DSC study on 2 and the crosslinked insoluble polymer revealed an increase in glass transition temperature through crosslinking (Figure 7).

Studies on the biodegradation of crosslinked poly(ε-caprolactone) (16) and poly(δ -valerolactone) showed that as crosslinking increases the rate of degradation decreases. We have shown that although the rate of degradation decreases with increasing crosslinking, highly crosslinked poly(ε -caprolactone) is eventually degraded completely. The study of the biodegradation of crosslinked polyester 3 is in progress. In view of the relatively easy degradation of poly(DL-lactide) as compared with that of poly(ε -caprolactone), we expect with reasonable certainty that cross-linked 3 can be biodegraded.

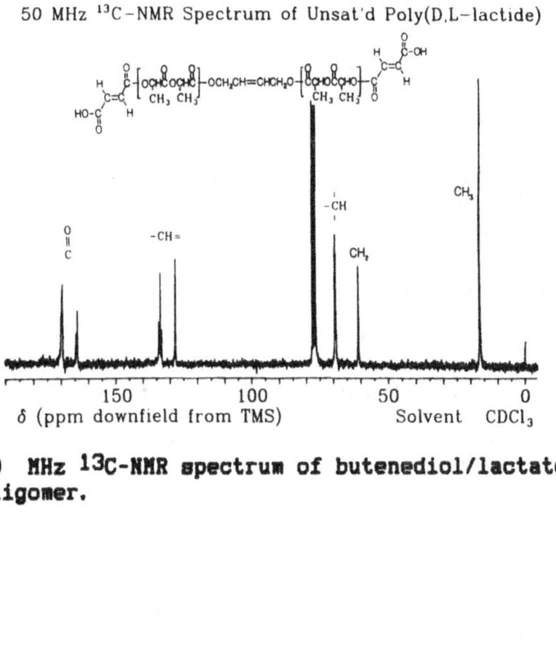

Figure 6. 50 MHz ^{13}C-NMR spectrum of butenediol/lactate/fumarate oligomer.

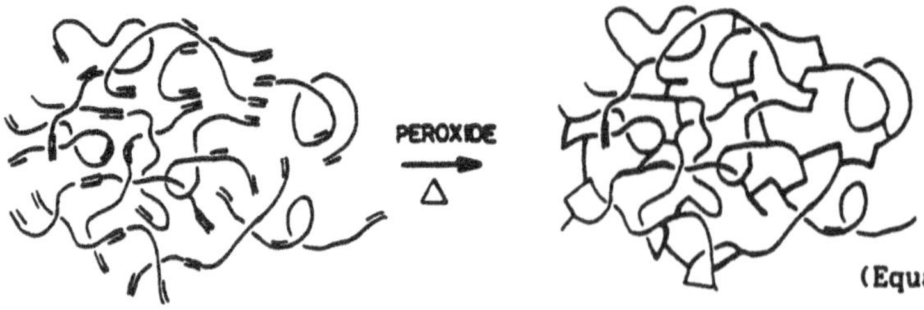

(Equation 3)

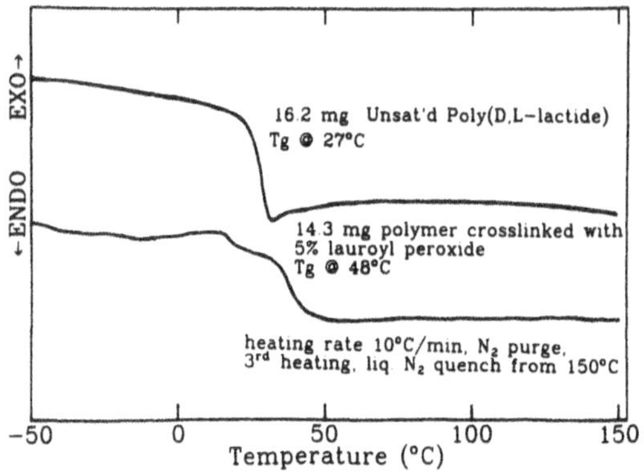

Figure 7. DSC study on butenediol/lactate/fumarate oligomer.

CONCLUSION

Low molecular weight poly(DL-lactide) containing double bonds was successfully synthesized and its crosslinking can be easily effected. The anticipated applications of this polymer as thermosetting resin for composite implants are being explored.

ACKNOWLEDGEMENT

Support of our research by U.S. Army (DAAG 29-85-K-0222) and National Science Foundation (DMR 80-13689) are gratefully acknowledged.

REFERENCES

1. Paper 34 in Biodegradable Polymers; for preceeding paper see S. J. Huang and M. S. Roby, J. Bioact. Compat. Polym., 1, 61 (1986).
2. J. R. Parsons, H. Alexander and A. E. Weiss, in: "Biocompatability Polymers, Metals and Composites", M. Szycher, ed., Technomic Publ. Co., Lancaster, PA, 1983, Chapter 38.
3. D. K. Gilding, in "Biocompatibility of Clinical Implant Materials", D. F. Williams, ed., CRC Press, Boca Raton, FLA, V. II, 1982, chapter 9.
4. R. L. Kronenthal and Z. Oser, eds., "Polymers in Medicine and Surgery", Plenum Press, 1975.
5. S. J. Huang, in: Mark, Bikales, Overberger, Menges, "Encyclipedia of Polymer Science and Engineering", Vol. 2, 2nd ed., 1985, pp. 220-244.

6. P. Jarrett, S. J. Huang, J. P. Bell, J. A. Cameron and C. Benedict, Org. Coatings and Plast. Chem., (Div. Am. Chem. Soc.), 47, 45 (1982).
7. P. Jarrett, S. J. Huang, J. P. Bell, J. A. Cameron and C. Benedict, in: "Polymers as Biomaterials", S. W. Shalaby, A. S. Hoffman, B. D. Ratner and T. A. Horbelt, Plenum Press, NY, 1984, pp. 181-192.
8. J. A. Szivek, G. C. Weatherly, R. M. Pilliar and H. V. Cameron, J. Biomed. Mater. Res., 15, 853 (1981).
9. R. J. Fredricks, A. J. Melveger and L. J. Dolegiewitz, J. Polym. Sci., Polym. Phys. ed., 22, 57 (1984)
10. J. W. Leenslag, S. Gogolewski and A. J. Pennings, J. Appl. Polym. Sci., 29, 2829 (1984).
11. S. Gogolewski, A. J. Pennings, E. Lommen, C. R. H. Widevuur and P. Nieuwehuis, Makromol. Chem. Rapid Commun., 4, 213 (1983).
12. S. Gogolewski and A. J. Pennings, Makromol. Chem. Rapid Commun., 3, 839 (1982).
13. E. Lillie and R. C. Schulz, Makromol. Chem., 176, 1901 (1975).
14. A. Schindler and D. Harper, J. Polym. Sci., Polym. Letter. ed., 14, 729 (1976).
15. M. Andreis, Z. Veksli and Z. Meic, Polym., 26, 1099 (1985).
16. A. Schindler and C. G. Pitt, Polym. Prepr., 28 (2), 111 (1982).

SYNTHESIS AND PROPERTIES OF POLYMERS FOR BIODEGRADABLE IMPLANTS

A. C. Ibay, G. C. Battistone , R. A. Miller and
H. Carr, Jr.

U. S. Army Institute of Dental Research
Walter Reed Army Medical Center
Washington, DC 20307-5300

As part of a program to develop biocompatible, biodegradable polymers, the feasibility of synthesizing such polymers from Kreb's cycle intermediates is being investigated and, where possible, formulated to produce moldable implant materials. These intermediates are all dicarboxylic acids that can be polycondensed with biocompatible diols to produce linear polyesters which are subject to non-specific hydrolysis under physiological conditions. The polymer synthesized in this investigation is poly(propylene fumarate), a polyester of fumaric acid and propyleneglycol. Various synthetic approaches to this polymer are presented along with characterization data from gel permeation chromatography, differential scanning calorimetry, IR and NMR spectroscopy. Following synthesis, poly(propylene fumarate) was investigated for its crosslinking properties with selected vinyl compounds in the presence of a peroxide catalyst. The resulting thermoset resin was then characterized for its hydrodynamic and mechanical properties.

INTRODUCTION

Research and development of biocompatible and biodegradable materials for use as temporary replacement for lost soft tissue and/or bone following trauma has been an objective of the U. S. Army Institute of Dental Research for many years. Among the first to be developed as biodegradable surgical repair materials were polymers of lactic and glycolic acids (1-3). Over the past decade, the polyesters from these organic acids have also found application as controlled release devices for biologically active agents (4-7). The biodegradability of polylactic and polyglycolic acids is attributed to their ability to undergo hydrolysis back to their respective monomeric carboxylic acids; and the presence of nonspecific esterases. These monomers, in turn, are shunted to metabolic pathways where they are enzymatically transformed into carbon dioxide and water.

While copolymers of lactic and glycolic acids have been found to be highly effective as bone replacement materials (8), their chemical natures dictate that they be used as solids which are preshaped and/or carved to fit a bone defect at the time of surgery. This can be a time consuming procedure. In the mass casualty situation that occur in combat, any simplification of surgical procedures can result in saved lives and reduced morbidity. Thus, it would be highly desirable to have a biocompatible polymer which can be prepared in a putty-like consistency, easily formed into a bone defect and rapidly hardened in situ. In this regard, selected Kreb's cycle intermediates are being considered because these dicarboxylic acids have additional functionalities which can allow crosslinking and thereby provide a mechanism for proceeding from the moldable to the solidified system described above. The concept of developing polymeric products from Kreb's cycle intermediates for use in a physiological environment was first mentioned in a patent (9). To synthesize a polyester of a dicarboxylic acid, a diol would be required as a comonomer. In this investigation, propylene glycol is the selected diol because it is inexpensive and biocompatible. The selected Kreb's cycle intermediate is fumaric acid because it possesses an unsaturated site which can crosslink with various vinyl compounds. Thus, the targeted polyester is poly(propylene fumarate) (PPF).

In preparing moldable implant appliances from crosslinkable polymers such as PPF, the prepolymer is mixed in a pre-determined stoichiometric ratio with a crosslinking agent so that a thermoset plastic having mechanical properties approximating those of bone would result. The crosslinking step would have to be preformed by the surgeon because this procedure passes through a "putty" stage before setting. This stage is critical in that it must be long enough to allow the surgeon time to mold the polymer to conform to any shape, but not so long as to cause patient discomfort (10). Thus, strict control of the number-average molecular weight (Mn) and molecular weight distribution (MWD) during synthesis of the prepolymer, and the stoichiometric ratio during crosslinking must be observed.

EXPERIMENTAL

PPF was synthesized by polycondensation of various fumaric acid derivatives with one equivalent or, in certain cases, varying excesses of propylene glycol. One of the objectives in the synthesis is to develop a means of controlling Mn and MWD since these parameters have a direct effect on the degradation rate of the polymer and on its crosslinking properties. Thus, several methods of polycondensation were investigated. Table I and Equations 1-5 summarize the methods employed PPF was first produced under contract with Dynatec Corporation using the procedure shown in Equation 1. Difficulty in reproducing this procedure led us to develop alternative methods. After the 24 hours that each reaction was allowed to proceed, the raw polymer from the bulk reactions was analyzed by gel permeation chromatography (GPC) while the polymer from the solution polymerizations was stripped of the solvents before GPC analysis. All the GPC measurements were carried out at 25°C in chloroform or methylene chloride in a Waters 6000A liquid chromatograph using mixed bed polystyrene columns in the 10 and 10^7 permeability range. ^{1}H-NMR and IR spectra were also obtained using a Varian EM 360 NMR and Perkin-Elmer 1550 FT-IR spectrometers, respectively. Differential scanning calorimetry (DSC) was also performed using a Perkin-Elmer DSC-4 scanning calorimeter to determine the glass transition (Tg) and melting (Tm) temperatures.

Following the synthesis and characterization of the polymer, crosslinking was attempted by addition of 5 weight-percent of a solution of N-vinyl-2-pyrrolidone and benzoyl peroxide to the powdered polymer and followed by kneading of the resulting "putty". In cases where the polymer was in a highly viscous state, longer time was required to attain a doughy state. The resulting thermoset was characterized by DSC and for

Table I. Summary of Polycondensation Methods.

Method	Equation	Fumarate Derivative	Solvent	Temperature °C	Time hrs	Catalyst
Transesterification	1	Diethyl Fumarate	Bulk	200	24	PTSA[d]
Acid Chloride	2	Fumaryl Chloride[c]	Bulk	0	24	none
Acid Chloride Semiester	3	Fumaryl Chloride monoethyl ester[a,c]	Bulk	0 then 200	24	none
Carbodiimide	4	Fumarate Dicarbodiimide[b,c]	THF	67	24	none
Esterification	5	Fumaric Acid	(e)	200	24	none

(a) Synthesized by treating fumaric acid-monoethyl ester with thionyl chloride.
(b) Generated *in situ* from fumaric acid and dicyclohexyl carbodiimide.
(c) Comonomer-feed technique used.
(d) Para-toluene sulfonic acid.
(e) Propylene glycol.

[1]

(Equation 1)

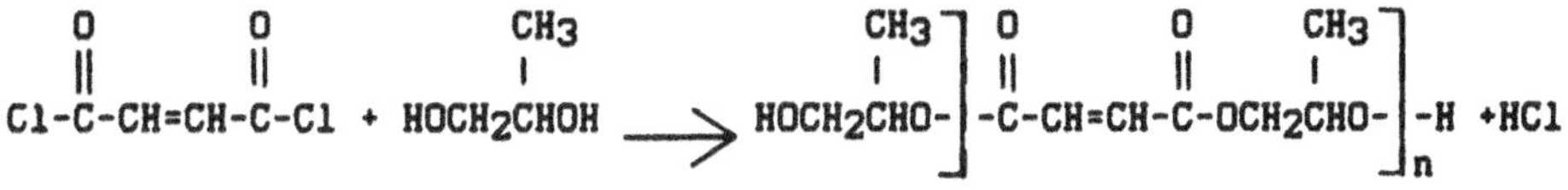

[2]

(Equation 2)

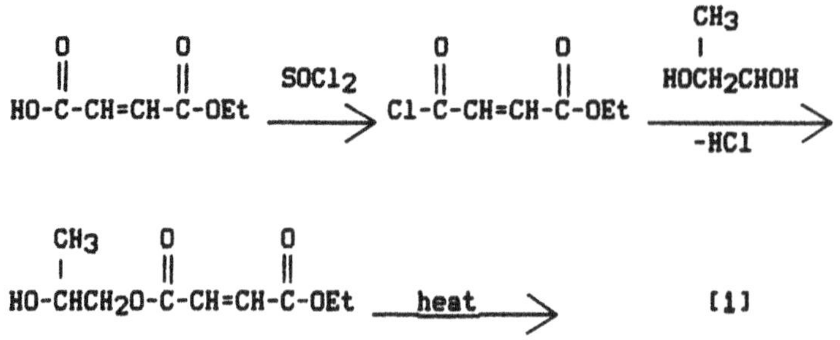

[1] (Equation 3)

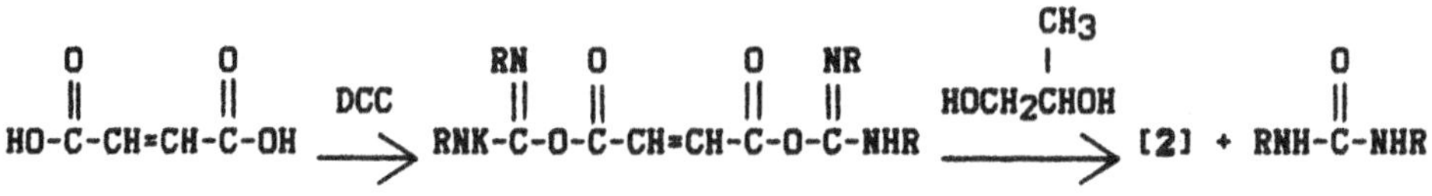

DCC = R-N=C=N-R where R = cyclohexyl (Equation 4)

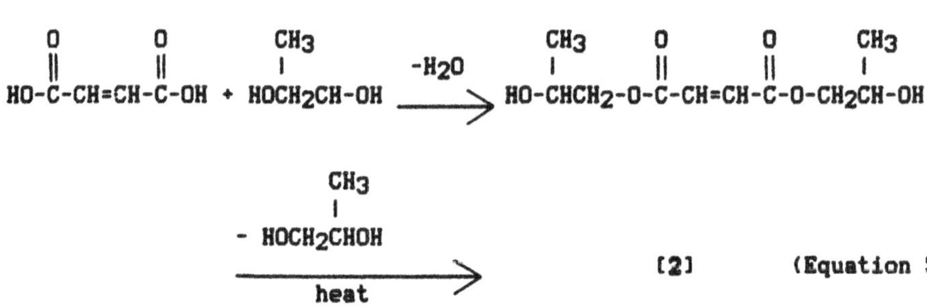

[2] (Equation 5)

its hardness with a Kentrall hardness tester. Swelling and leaching studies were performed as well.

RESULTS AND DISCUSSION

Of the five methods attempted, the polyesterification method as shown in equation 5 was the most efficient in yielding high molecular weight ($Mn = 6 \times 10^6$) PPF. However, preliminary results from crosslinking studies indicate that the high molecular weight fraction, which was isolated by leaching the raw polymer with methanol, passes through its "putty" stage so rapidly upon addition of the crosslinker/catalyst solution that the resulting resin became "crumbly" when kneaded. On the other hand, the lower molecular weight fraction ($Mn = 3,600$) required longer time with heating to induce formation of a hardened thermoset resin. The other polycondensation methods did not yield any isolable amounts of polymer having a $Mn > 1,000$, and therefore, were not pursued any further. Initial scans on DSC has shown that the uncrosslinked polymer has no distinguishable Tg and Tm between 20° to 200°C which indicates a low degree of crystallinity. This was expected since the propylene glycol used was a racemic mixture.

CONCLUSIONS

PPF appears to be a good candidate for developing biodegradable, moldable implants. Current work is now focused on formulating PPF to have a Mn and MWD that would render the polymer amenable to its intended application. The desired Mn is postulated to be between 6×10^6 and 3.6×10^3 polymer such that a narrow molecular weight distribution is still maintained at any targeted molecular weight. Each formulation will then be investigated for their crosslinking properties to determine which ones best fit the criteria for a moldable implant.

REFERENCES

1. E. J. Frazza and E. E. Schmitt, J. Biomed. Mater. Res. Symposium, 1, 43 (1971).
2. D. E. Cutright and E. E. Hunsuch, Oral Surg., 33, 28 (1972).
3. J. M. Brady, D. E. Cutright, R. A. Miller, E. E. Hunsuch and G. C. Battistone, J. Biomed. Mater. Res., 7, 155 (1973).
4. L. R. Beck, V. F. Pope, C. E. Flowers, D. R. Cowsar, T. R. Tice, D. H. Lewis, R. L. Dunn, A. B. Moore and R. M. Gilley, Bio. Prod., 28, 186-195 (1983).
5. J. Heller and R. W. Baker, in "Controlled Release of Bioactive Materials", R. W. Baker, ed., Academic Press, New York, 1980, pp 1-17.
6. C. G. Pitt, M. M. Grayzl, G. L. Kummel, J. Surles and A. Schindler, Biomaterials, 2, 215-220 (1981).
7. J. E. Potts, in "Aspects of Degradation and Stabilization", H. J. Jellinek, ed., Elsevier, Amsterdam, 1978, p. 617.
8. J. O. Hollinger, Bio. Mater. Res., 17, 71 (1983).
9. D. L. Wise, U.S. Patent 3,978,203, (to Dynatec Corp., Burlington, MA) "Sustained Release of Pharmaceuticals from Polyester Matrices", (1974).
10. E. C. Conrad, B. A. Bidlingmeyer and K. E. Alden, Jr., Medical Devices and Diagnostic Industry, 7, 124 (1985).

NEW FINGER JOINT IMPLANTABLE PROSTHESIS IN AN EX-VIVO MODEL:

BIOSTEREOMETRIC STUDIES

Mutaz B. Habal[a,b], Donald L. Leake[b] and Bruce Dunn[c]

(a) Human Resources Institute
University of South Florida
The Tampa Bay Craniofacial Center
Tampa, FL
(b) Dental Research Institute
University of California,
Los Angeles, CA
(c) Engineering Dept.
University of California
Los Angeles, CA

Finger joint arthroplasty is a method of replacing defective and diseased delicate finger joints. Multiple new finger joints have been introduced to replace function of the fingers. The ideal joint for replacement arthroplasty should have complete fixation in the long shafts of the bones in the hand and fingers, and a hinge that is durable. The design proposed here will produce these ideal conditions. Polyether-urethane-Terephthalate has those particular characteristics. Long term implantation in man has shown that bones ingrowth is feasible. The experimental work presented, suggest the cloth mesh part can be used for joint replacement as a hinge, because it withstood the mechanical changes. There was no material fracture noted, but the mechanical changes were basically of fraying, crazing, and fragmentation. Over 6 million cycles were used. The results in hand surgery is an integral part of this fast developing field.

INTRODUCTION

Replacement of the delicate finger joints is one of the challenging aspects of replacement arthroplasty. Since there are many inherent problems with small joints prosthesis, multiple attempts at replacement arthroplasty has not been very encouraging.

Artificial joint surgery has moved from the experimental stage to its present clinical status which is well sophisticated. However, the replacement of the delicate joints in the finger represents a difficult problem in spite of all the recent advances made in joint replacement

(d) To whom correspondence should be sent: Dr. Habal, 4211 Carrollwood Village Drive, Tampa, Florida 33624.

operations. Multiple new finger joints have been introduced without basically solving the inherent problems encountered, such as fatigue fracture, complete disruption, and lack of stability. The replaced finger joints in humans require a prosthetic device with a long range of durability, stiffness, and flexibility so that the normal delicate functions can be restored after implantation.

Special attention and delicate functions can be restored after implantation. Special attention is also directed toward the bony ingrowth at the shaft level to provide a form of fixation, while the hinge inbetween stays mobile to allow for the delicate movements needed.

The usual standard joint used now is the silicone joint, (Silastic Dow Corning), which by itself had many inherent problems. Other joints introduced did not meet the same success rate as the silicone joint had. The bone problem precipitates the lack of stability in the joint giving the finger a weakness that can't be altered with exercise. Rather than stop at this level of sophistication, a methodology should be prescribed to produce a challenge and improvement to the present finger joint; research should always be encouraged. The problem of the silicone eroding the shafts of the hand bones is a serious problem.

The aim of this research is to evaluate the use of a porous prosthesis with a new design for finger joint replacement in an ex-vivo experimental design, prior to implantation, to test the durability of the hinge part of the prosthesis. The bony ingrowth in the shaft is similar to that described previously (1, 2). The prosthesis encourages the bony growth particularly when in contact with a bone inducing matrix when autologous bone paste is used to fix the prosthesis in place (Figures 1 and 2).

EXPERIMENTAL DESIGN

The new joint prosthesis, made of poly(ethylene terephthalate) (Dac-

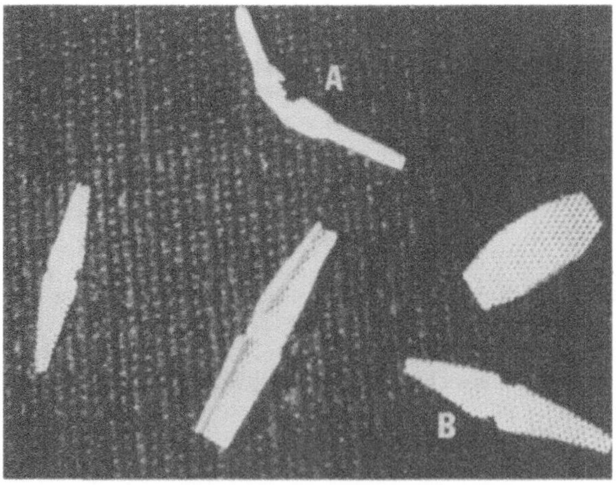

Figure 1. Different designs for the artificial joint prosthesis (actual size). A) Tubular design. B) Flat design - note the end of the prosthesis is rounded to fit into the shaft of the bone.

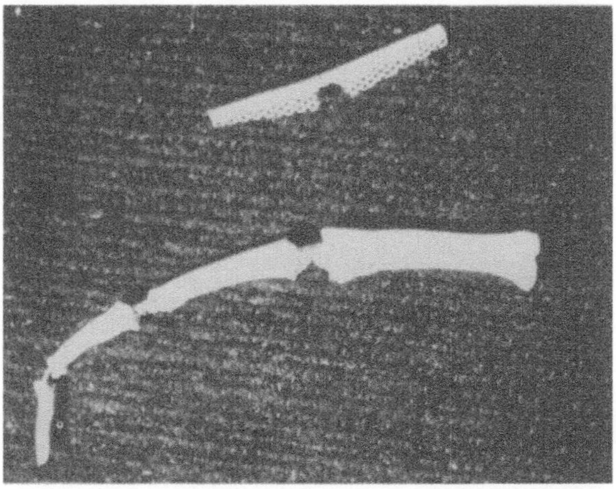

Figure 2. The finger joint prosthesis fixed in dry bones of the hand to emulate their use in hand surgery.

ron) impregnated with polyetherurethane to stiffen it up, was designed (Figure 3). This joint prosthesis has two tubular ends for bony shaft implantation and a flattened middle segment to serve as a hinge, all in one unit. Twenty-four such, joints were used in this experiment. Bone paste was used to fix the tubular ends in adjacent bones. Two designs of the joints were tested: one with a partially formed tube hinge and one with a parabolic flat strap of the neck. The ex-vivo environment, composed of physiologic solutions or in air, was set to test the function of the hinge part of the joint prosthesis prior to implantation. The two ends were fixed into a movable unit on a specially designed machine, so the flexation and extension were done on a continuous basis with a counter to count the cycles. The joints were then removed, and ultramicroscopic analysis was done on the hinge part. The prosthesis was flexed through an arc of 90° and returned to the rest position. A rigid rod on a

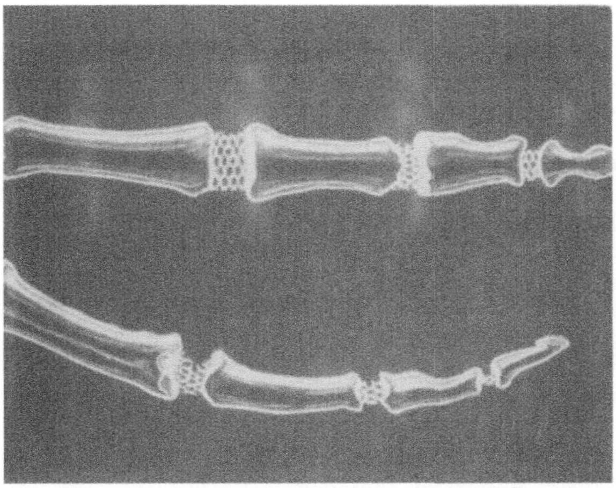

Figure 3. A diagram to show the exact application as demonstrated in Figure 2.

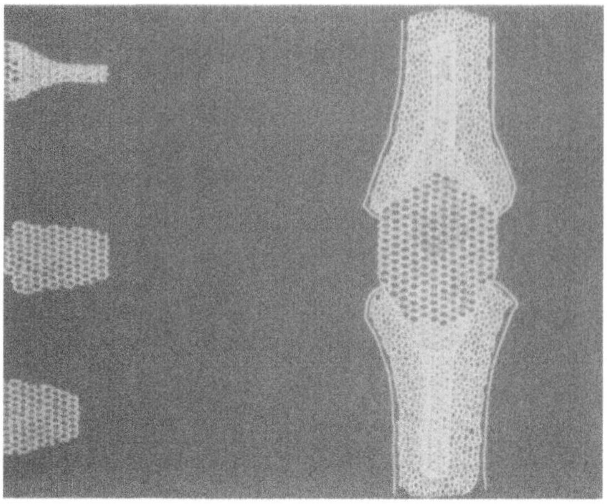

Figure 4. A diagram with close up of one joint of the hand. Note the hinge of the prosthesis producing the lateral stabilization and the end filled snuggly in the bones of the hand.

cam was attached to an AC motor. The other end of the rod was attached to a moving clamp (Figures 3 and 4).

Sixteen joint prosthesis were used to test the joint as it bent through an arc of 90°, the total translational distance through which it travels is the radius of rotation, and the diameter of the driving wheel was made equal to the radius of the arc through which the clamp travels. To test the implants in solution, eight joint prosthesis were used and the motor was placed on a platform above the tank holding the solution. The motor was attached to a timer, and by counting the number of rotations per unit time, the total number of flexes could be closely approximated. The experimenter thus subjected the prosthesis to flexing for a standard number of cycles, then examined the sample for the degree of material degradation (Figure 4).

RESULTS

Most of the damage occured in the first 100,000 to 200,000 cycles. The tubular design resisted continuous bending more than the flat design. Ringer's solution increased the fatigue-life of material. The orientation of the holes in the design of the mesh altered fatigue-life, being either horizontal or vertical (Figure 5). Although these tests determined that the highest rate of degradation occured after 100,000 to 200,000 cycles, further cycling did introduce new damage, but at a slower rate than during the first 100,000 cycles. Although complete failure had not occurred at $6x10^6$ cycles, there was enough damage to consider part of the joint inoperative. Although there was no significant microscopic differences between the orientations of the holes of the mesh, apparently the material lasted longer when the holes were oriented horizontally rather than vertically. After 300,000 cycles, the vertical orientation shows more degradation. The damage occurred in a specific area, the longitudinal area, presumably because of higher stresses in that area. When the tubular design was compared with the flat design, the tubular design outperformed the flat design. The difference was apparent almost immediately (Figure 5). In the samples examined, the flat design had a thicker poly-

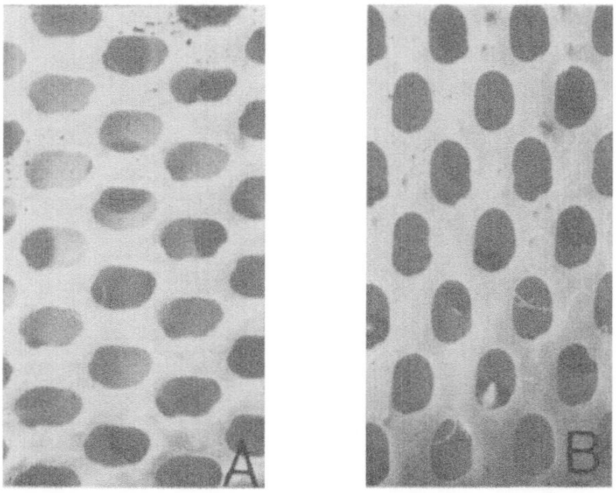

Figure 5. Close up of the prosthesis joint hinge; note the hole orientation. (A) Horizontal (B) Vertical.

urethane coating, thus it was more brittle than the tubular design which, because of its thinner coat of polyurethane, was more flexible (Figure 6). Fatigue tests were also performed in a simulated biological milieu to evaluate the lifetime of the prosthesis. The experiments were conducted in solution showed a marked increase in fatigue-life for both designs (Figure7). Some joints showed no damage after as many as 800,000 cycles. The details of the results in comparison with the milieu are expressed in Table I.

DISCUSSION

The ex-vivo fatigue test for plastic materials differs from that for metal materials in two ways: First, the effective repetitive flexing creates fraying of the material rather than a complete fracture. Fraying

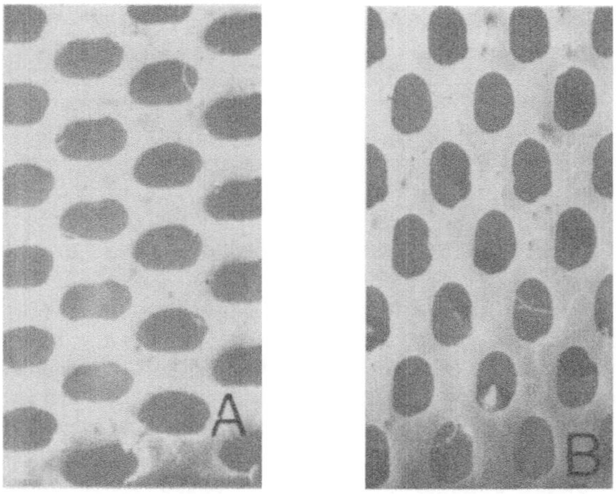

Figure 6. Early degradation, as seen in the close up, which is minimal. (A) Horizontal (B) Vertical

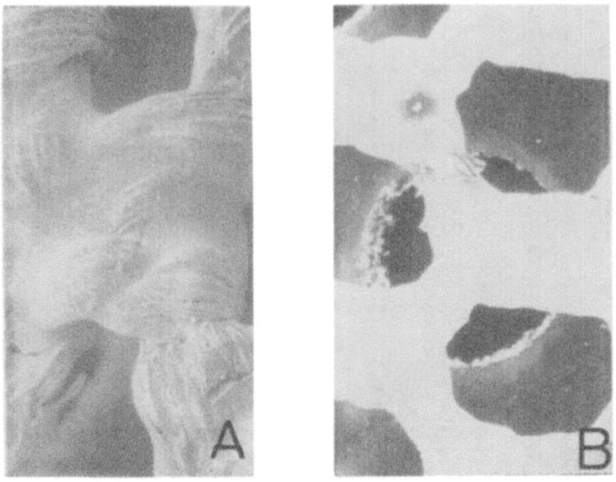

Figure 7. (A) Degradation (B) Fraying, as seen in a close up of the joint, to demonstrate the difference degrees of changes seen in the implant.

is manifested by a series of mini-fractures of the fibers of the Dacron mesh and a flaking off of the polyurethane coating. Second, the material is not elastic, so it was necessary to restore the sample to its original undeformed or straight condition, rather than a constant stress. Fatigue damage in the prosthesis material is typically initiated as a crazing-type phenomenon. The damage proceeds in the form of larger cracks in the coating of the urethane around the junctions of the fiber of the mesh. The cracks grow, then the polyurethane coating begins to flake off, exposing the Dacron fibers. This process takes place, usually, in the first 100,000 cycles, but depending on orientation of the holes, whether the material was in a solution, and on the thickness of the polyurethane coating. Some joints can cycle 800,000 times without any evidence of damage.

Basically, this experimental design proves that the material is suitable for being used as a prosthesis in its inherent physical structure which allows it to be utilized as a hinge. The bony ingrowth at the shaft of the long bone has been proven from previous experience to be feasible (3). Thus, this finger joint prosthesis proved, in the ex-vivo method, to be usable as designed. It is important to develop the methodology to try and use new biomaterials for finger joint replacement, since the real ideal joint may be far from being introduced. However, improvements on the already existing prosthesis are in progress.

In conclusion, the combination of polyetherurethane-impregnated poly(ethylene terephthalate) mesh has been used to reconstruct bony defects, when combined with bone paste. In a similar concept, when the tubular ends are placed inside the bones to allow for adequate bony ingrowth for fixation, the exposed segment acts as a hinge and allows the joint to be flexed, thus making the joint prosthesis potentially applicable to human use. This design is also strong in producing stability at the hinge. This joint prosthesis has potential as a sound replacement for delicate joints in the hand. Since it is also stabilized positionally, other possible uses are related to large joint replacement where more corticocancellous bone paste could be utilized to fix the prosthesis in long bones and allow adequate fixation with bony ingrowth

Table I. Comparison of the finger joint prosthesis in air or Ringer's solution after various numbers of flexing cycles.

	Testing Media	
Number of Cycles	Air	Ringer's Solution
1,000	none	none
100,000	cracks in coating	none
300,000	cracks in coating	cracks in coating; salt deposits in cracks
500,000	coating separates and falls off	coating separates and falls off
900,000	large area of exposed fibers	small area of exposed fibers
1,000,000	initial fraying of fibers	initial fraying of fibers
6,000,000	fracture of fibers	did not test

on fixed segments, so that good functions can solve some of the problems seen with the large joint prosthesis.

REFERENCES

1. M. B. Habal and D. L. Leake, Plast. Reconstr. Surg., 61, 394, (1978).
2. M. B. Habal and D. L. Leake, Surgical Forum, (1985).
3. M. B. Habal, Annal Plastic Surgery, 15, 138-142, (1985).

CALCIFICATION OF POLYMERIC BIOMATERIALS IN LONG-TERM CARDIOVASCULAR USES

Stephen D. Bruck

Biomedical Technology Risk Assessment Group
Stephen D. Bruck Associates, Inc.
7315 Wisconsin Avenue, Suite 727-E
Bethesda, Maryland 20814 USA

The calcification of polymeric materials in long-term medical implants such as cardiac assist devices, mechanical artificial hearts, and xenograft bioprostheses is a complex phenomenon, and apparently involves biochemical, physiological, physico-chemical, and species-related parameters. Biochemical explanation implicate the Ca^{2+} binding amino acid, gamma-carboxyglutamic acid, and other vitamin K-dependent plasma proteins and blood clotting factors, such as atherocalcin, osteocalcin, protein C, prothrombin (factor II), and factors IX and X. Physiological parameters of calcium homeostasis depend on the intricate interactions between calcitonin, parathorme, and vitamin D. These regulate the calcium and phosphorous (hydroxyapatite) in the bone and respond rapidly to various stimuli in readjusting the extracellular calcium levels from the bone. In the realm of physico-chemical parameters of polymeric biomaterials, I have proposed elsewhere that the propensity of polymers to sorption and permeation (in addition to surface phenomena of adsorption) of plasma components and drugs may be important in the calcification process of prostheses. This prediction was validated by our recent experimental findings.

INTRODUCTION

The calcification of polymeric materials in long-term medical implants such as cardiac assist devices and artificial hearts is increasingly being recognized. Calcification is a complex phenomenon governed by biomedical, physiological, physico-chemical and species-related factors. This brief critical review will try to summarize what is known about this serious problem.

DISCUSSION

1. Biochemical Origins of Calcification

Biochemical explanations of calcification implicate vitamin K in blood clotting and the special Ca^{2+} binding amino acid, gamma-carboxyglutamic acid, Vitamin K-dependent plasma proteins and blood clotting factors containing gamma-carboxyglutamic acid include atherocalcin, osteocalcin, protein C, prothrobin (factor II), factors IX and X. The bioconversion of the glutamate residue in precursor prothrombin (descarboxyprothrombin) to calcium binding -carboxyglutamate residue in prothrombin takes place under the influence of a viatamin K-requiring enzyme system (1).

Another vitamin K-dependent agent is protein C. It inhibits coagulation by inactivating factors V and VIII, and, at least in dogs, promotes fibrinolysis (2). Recently questions were raised about the fibrinolytic activity of protein C. in humans. This is based on work done with squirrel monkeys indicating that in this primate no increase of fribrinolytic activity was observed by the administration of activated human protein C or upon the activation of endogenous protein C (3). Human protein C has a molecular chain (molecular weight 41,000), and a light chain (molecular weight 21,000). The light chain contains gamma-carboxyglutamic acid residues involved in Ca^{2+}-dependent membrane association (2). In contrast to human protein C, the light chain of bovine protein C also contains one residue of erythrohydroxyaspartic acid with as yet unknown functional importance (4).

2. Physiological Factors: Calcium Homeostasis

In plasma about half of the calcium is present as free ions, the rest being largely bound to proteins, especially albumin. Phosphorus, in the forms of HPO_4^{2-} and $H_2PO_4^-$, represents about 35% of the total in plasma. An additional 12% is protein bound, 29% is in the form of $NaHPO_4^-$(5). The main regulating agents are calcitonin (produced by the "C" cells of ultimobranchial cells of the thyroid), parathormone, and vitamin D. The exchange of calcium between plasma and bone continues even in the absence of endocrine influences but serum calcium is lowered. Calcium and phosphorus occur in bone mostly as hydroxypatite $[(Ca_3PO_4)_2]_3Ca(OH)_2$. Osteoblasts, that synthesize collagen, and osteocytes, that take part in osteolysis, are the main cells responsible for new bone production. Of some 1,200 g calcium in bone, less than 1% exchange freely with extracellular calcium, normally at a rate of about 2 g/day (5). Various stimuli can cause a rapid readjustment of extracellular calcium levels from the bone. The concentrations of calcium and phosphorus also influence bone accretion and resorption. Vitamin D facilitates the absorption of calcium and the calcification of bone. Its main effect is to raise the serum calcium concentration. The kidney and the bones are the two main sites of action for the parathyroid hormone, through cyclic AMP. Parathormone prevents renal tubular reabsorption of phosphate. When plasma phosphate concentration is low, phosphate is released from the bone and calcium is released into the blood by the degradation of hydroxyapatite. If the plasma phosphate concentration is high, bone accretion occurs and the plasma calcium concentration is lowered. Other factors affect bone resorption and accretion, such as the pH of the mineral environment.

3. Physico-Chemical Factors

Calcification is not unique to any particular material. It has been observed with glutaraldehyde-treated tissue valves (xenografts) and with polymers of diverse chemical structures (6). Growing children as well as calves are particularly predisposed to calcify implanted, blood contacting bioprotheses. The calcification of xenografts has been attributed to calcium homeostasis (7) and warfarin has been claimed to reduce the calcification in calves of segmented polyether-urethanes (8). Yet this may have been an isolated observation, because others have reported calcification with polyurethanes, silicone, and other polymers despite giving warfarin. As both warfarin and coumadin are vitamin K antagonists, it is not clear why the reported suppression of calcification with the particular polyurethane elastomer was not observed with other polymers, if indeed the postulated vitamin K-dependent protein carboxylation process were to be the only decisive factor. Consequently, other mechanisms are also involved in the calicification process.

Although calcification is not restricted to flexing materials, it is nevertheless observed to take place in experimental cardiac assist devices and artificial hearts primarily at the flexing areas of pumps. For xenografts it has been proposed that removing the outside protective layer of proteoglycan during the processing of porcine valves uncovers the epsilonamino groups in lysine and hydroxylsine, thus allowing the formation of bonds between the native collagen and phosphate groups (9). Furthermore, the loss of proteoglycan opens the collagen structure to the penetration of plasma proteins and other blood constituents, thus providing additional loci for calcification (9). Thrombi formed on the surfaces can also calcify and cause calcific deposits in distant organs. Also, the crosslinks formed with glutaraldehyde are more complex than commonly depicted, especially with respect to length and unsaturated linkages, and variations in crosslinks likely influence the long-term performance of xenografts (10). Surfactant (sodium dodecy sulfate and isopropanol) treated porcine xenografts tricuspid valves showed somewhat less severe calcification in young sheep over a period of 20 weeks, but little difference or even increased calcification was observed with similarly treated bovine pericardial valves (11).

In earlier publications (6, 12) I proposed that absorption and permeation (in addition to surface adsorption) of plasma components by polymeric materials in contact with blood may be important in calcification, in addition to biochemical and physiological factors. This prediction was validated by the recent finding of calcium deposits, not only on the surfaces of polyurethane blood pump diaphrams of artificial hearts, but also deep inside the polymeric diaphram (13, 14). Occasionally, among a large number of experimental animals (mostly calves) there are some in which calcification of identically prepared protheses is minimal (13). This may, however, be associated with the relative health of the animals. Just as some people are less prone to atherosclerosis, so one could expect a similar pattern in experimental animals. Unfortunately appropiate studies are lacking. It may therefore be unrealistic to point to a few statistically insignificant experiments that have led to premature expectations in man.

As I noted elsewhere (12), repeated flexing and other so-called "macroscopic deformation" of elastomers may cause microscopic deformations of their structural elements, giving rise to "opening" and "closing" of pores. This may then allow the entry by absorption and penetration of blood components or drugs, depending on factors such as thermodynamic parameters, domain structures, surface defects, and porosity. These factors in turn are functions of the molecular and supramolecular structures of the polymer systems. To achieve optimum and reproducible properties of a given polymer system is a formidable task and poses special problems with biomaterials.

It should be noted that the ultra-smooth, dense, pyrolitic carbon used in many mechanical heart valves does not calcify. Should calcification occur, it is restricted to the fabric covering of sewing rings. The lack of calcification of this carbon may be due not only to its good blood compatibility and thromboresistance (12) but also to its impermeability to blood components and administered drugs.

We have recently shown that poly(dimethyl siloxane) [Silastic^{R}], and polyether-urethane/poly(dimethyl siloxane) copolymer [Avcothane^R, Cardiothane^R] are permeable to various drugs (15-17). In as much as large quantities of various drugs are usually given during cardiac implantation surgery, these drugs and the plasma commponents could find their way inside the polymeric materials of the protheses and serve as loci for calcification (17). Once calcification begins, it will eventually cause cracks and weaken the polymer. This in turn will promote further calcification and thrombosis, not only within the polymer but also in distant organs.

As it was recently noted by Rahimtoola (18), single or few events are often prematurely exaggerated. He and others (18-20) have also called attention for the need to reemphasize broad-based health-care priorities rather than medical spectaculars.

We have recently developed a proprietary polymer system that shows significantly less calcification than other polymers, including polyurethanes. However, it should be noted that calcification is just one of many problems involved with flexing blood-contacting polymers. Showers of blood clots may detach from "clean" surfaces to cause strokes and damage in organs. Thus, even though a surface may appear "clean" (without the presence of visible clots), the material itself may not be blood compatible. This phenomenom has often produced misleading results and premature optimism. Hence, the calcification and associated problems indicate the need for re-evaluating its clinical effect and what ought to remain in the laboratory. Similar sentiments have been expressed by several respected authorities (21).

CONCLUSIONS

During the past 30 years, the emphasis on surface phenomena in biomaterials research has given rise to some dogmatic views. Yet recent findings (13, 14) and ongoing animal work with proprietary materials clearly indicate the contribution of my previously predicted absorption and permeation phenomena (6) (in addition to biochemical and surface characteristics) in the calcification of polymeric compounds of flexing implanted devices. While perhaps this process could be slowed, I doubt that it can be entirely prevented, considering our meager understanding of calcification, both in natural tissues and synthetic materials. For this and other reasons, the enthusiasm for long-term clinical use of totally implantable, ambulatory ventricular assist devices, artificial hearts, and associated devices must be tempered with reality (21). Considering the current state-of-art, these devices may find some limited uses for short-term, interim periods until organ transplantation can be effected.

REFERENCES

1. J. W. Suttie and C. M. Jackson, Physiological Reviews, 57, 1-70 (1977).
2. C. T. Esmon, Blood, 62, 1155-1158 (1983).
3. M. Colucci, J. M. Stassen and D. Collen, J. Clin. Invest., 74, 200-204 (1984).
4. T. Drakenberg, P. Fernlund, P. Roepstorff and J. Stenflo, Proc. Natl. Acad. Sci., U.S.A., 80, 1802-1806 (1983).
5. G. E. Mortimore, in: "Best & Taylor's Physiological Basis of Medical Practice", 10th edition, J. R. Brobeck, ed., Williams & Wilkins, Baltimore, 1979, chapter 5.
6. S. D. Bruck, Biomaterials, 2, 14-18 (1981).

7. M. M. Silver, J. Pollock, M. D. Silver, W. G. Williams and G. A. Trusler, Am. J. Cardiol., 45, 685-696 (1980).
8. W. S. Pierce, J. H. Donachy, G. Rosenberg and R. E. Baier, Science, 208, 601-603 (1980).
9. V. J. Ferrans, S. W. Boyce, M. E. Billingham, M. Jones, T. Ishihara and W. C. Roberts, Am. J. Cardiol., 46, 721-734 (1980).
10. S. D. Bruck, Int. J. Artif. Org., 2, 223-225 (1979).
11. E. Arbustini, M. Jones, R. D. Moses, E. E. Eidbo, R. J. Carroll and V. J. Ferrans, Am. J. Cardiol., 53, 1388-1396 (1984).
12. S. D. Bruck, "Properties of Biomaterials in the Physiological Envirnment", CRC Press, Boca Raton, Florida, 1980.
13. Jaromir Vasku, J. Cerny, M. Dostal, P. Urbanek, Jan Vasku, P. Guba, T. Sladek, V. Trbusek, Anna Vasku, B. Hartmannova, E. Urbanek, S. Dolezel, V. Pavlicek and P. Svoboda, Proceedings of the XI Annual Meeting of the European Society for Artificial Organs, Alpbach-Innsbruck, Austria, September 9-12, 1984, published in : Life Support Systems, vol. 2, Suppl. 1, p 242-244 (1984).
14. Jaromir Vasku, private communication.
15. S. D. Bruck, M. Kojima, Y. Kadoma and E. Masuhara, Med. Progr. Technol., 10, 161-169 (1983/1984).
16. S. D. Bruck: Rate-controlled drug release system. U. S. Patent No. 4,559,054, December 17, 1985.
17. S. D. Bruck, Proceedings of the XI Annual Meeting of the European Society for Artificial Organs, Alpbach-Innsbruck, Austria, September 9-12, 1984, published in: Life Support Systems, vol. 2, Suppl. 1, p 95-97 (1984).
18. S. H. Rahimtoola, Circulation, 71, 1-3 (1985).
19. H. Hutten, Med. Prog. Technol., 10, 133-134 (1983/1984).
20. S. D. Bruck, Biomat., Med. Dev., Art. Org., 11, 271-280 (1983/1984).
21. J. G. Copeland, R. W. Emery, M. M. Levinson, J. Copeland, M. J. McAleer and J. E. Riley, Circulation (Part II), 72, II-7 -II-12 (1985).

METAL ION COMPLEXATION OF POLY(ETHER)URETHANES

R. J. Thoma[a], T. Q. Hung[a], E. Nyilas[a*], A. D. Haubold[a], and R. E. Phillips[b]

(a) CarboMedics, Inc.
1300-D East Anderson Lane
Austin, TX 78752
(b) Intermedics, Inc.
P.O. Box 617
Freeport, TX 77541

The interaction of low molecular weight models of poly(ether)urethanes (PEUs), used in various biomedical applications, with metal ions was studied. Interaction of calcium ion and silver ion with urethane and urea models of the hard segment were observed using protein nuclear magnetic resonance (^{1}H NMR). Soft segment models, poly(tetramethylene ether)glycols and poly(propylene ether)glycols, were shown by atomic emission spectroscopy (AES) to extract metal ions, quantitatively and selectively, from aqueous solutions. Among the polyethers, those having a molecular weight of 1000 showed an enhanced capacity for extracting metal ions. These models studies support other data which indicate that PEUs extract metal ions from aqueous media, over a wide range of pHs, including physiologic pH, and the hypothesis that the ionic interaction or complexation and the mineralization of PEU may be interrelated.

INTRODUCTION

Segmented poly(ether)urethane (PEU) elastomers have been used as coatings on pacemaker leads for the past seven years (1-3). Reports, in 1981, of surface fissuring in insulation of some specific explanted leads have resulted in work to identify the crack initiation mechanism and the severity of the surface fissuring, propagation problem (4).

Two closely related mechanisms have been identified which may result in PEU fissuring. First, it is now understood that PEUs are susceptible to environmental stress cracking (5,6): a crack initiating and propagating process which requires that the polymer be in contact with an active chemical environment and be under tensile stress (7). More recently, specific metal ions were found which interact in some manner with segmented PEUs resulting in changes in key mechanical properties (8). This

* Deceased

ion/PEU interaction may result in an even greater susceptibility of the polymer to stress cracking. The discovery of metal ion/PEU interactions in blood contact applications, combined with reports of the calcification or mineralization of PEUs in artificial heart bladders in areas of high flex stress (4), have led to the hypothesis that the ionic interaction, or complexation, and calcification of PEUs may be interrelated. The work reported herein tests this hypothesis and further elucidates the effect that the polymer backbone chemistry has on environmental stress cracking and/or mineralization.

Segmented PEUs are block copolymers. The so-called soft block is a low molecular weight polyether. The hard block is made from the reaction of a diisocyanate with a diamine or diol resulting in a urea or urethane linkage. The elastomeric behavior of segmented PEUs has been attributed to the formation of microphase separated domain structures consisting of hard-segment rich and soft-segment rich domains. Since these two blocks are phase separated, the approach to the problem of polyurethane cracking, whether as a result of metal ions in the cracking of specific models of pacemaker lead coatings (8), or as a result of calcium or other ion in the mineralization of mechanical heart bladders, is based on a study of compounds which are low molecular weight models of the hard and soft segments which make up the PEUs of interest.

Pellethane^{R}-80A and -55D, Biomer^{R}, Tecoflex^{R}, and Cardiothane^{R} -51 are PEUs currently used commercially in implant applications. All of these are composed of a soft-segment poly(tetramethylene ether)glycol, and an aliphatic or aromatic, chain extended, diisocyanate hard-segment. Cardiothane^{R}-51 also contains approximately 10 % by weight of polydimethylsiloxane.

In studying the metal ion complexing or coordinating capability of PEU fragments or domains, a great deal of information can be obtained through a study of the coordinating or complexing ability of models of the various fragments. The following preliminary data are the result of a study of the metal ion (M^{+n}) coordinating ability of the hard segment, using protein (^{1}H) Nuclear Magnetic Resonance (NMR), and of the soft segment [the poly(ether)glycols], using Atomic Spectroscopy (AES).

EXPERIMENTAL

Small molecular weight analogs of the hard and soft segments were purchased from commercial sources and were used without further purification, or were prepared in our laboratory by standard synthetic procedures and recrystallized one time. The metal ion salts, $AgNO_3$, $CaCl_2$, $Ca(NO_3)_2$, and $Fe(NO_3)_3$ were purchased from commercial sources and used without further purification.

Proton NMR were recorded on a Varian EM-390 NMR spectrometer; AES measurements performed on a Perkin-Elmer model 5500 Inductively Coupled Plasma unit. Infrared spectra were recorded on a Perkin-Elmer IR-297 IR spectrometer.

The hard-segment model compounds were dissolved in d_6-dimethylsulfoxide (d_6-DMSO). The hard-segment models gave solutions of 3-5 %/wt in d_6-DMSO. Saturated solutions of $AgNO_3$ and $CaCl_2$ in d_6-DMSO were also prepared.

D_6-Dimethylsulfoxide was used because it is one of few solvents in which the models, $AgNO_3$, and $CaCl_2$ are soluble, and because it is comparable to water in its solvating ability (9). Use of a solvent which is compatible with both the models and the salts reduces the probability of introducing changes in chemical shifts by a variation in the solvent system upon mixing.

Tetramethylsilane (TMS) was the internal reference in all cases. It is only moderately soluble in d_6-DMSO. Its solubility is reduced by the presence of silver nitrate, however, adequate locks were obtained on the TMS reference for all samples.

A ^1H NMR spectrum was taken of the model, the sample tube removed from the probe and an approximately equal volume of M^{+n} solutions added. The tube was well shaken and the solution thoroughly mixed. A ^1H NMR spectrum was taken of the "reaction mixture". The chemical shifts were determined relative to TMS and the change in chemical shift resulting from addition of the metal salt was determined. (Tables I and II)

Metal ion abstraction, by the soft segments, was determined using a technique like that used to evaluate the distribution equilibria of salt complexes of cyclic polyethers (10). In this procedure, 20 % aqueous solutions of $Ca(NO_3)_2$, $AgNO_3$, and $Fe(NO_3)_2$, and 20 % toluenic solutions of poly(tetramethylene ether)diol (MW = 650, 1000, and 2000) and poly-

Table I. The Change in the ^1H NMR Chemical Shift (δ) of the N-H proton of Urethane Models upon Complexation with Ca^{++} and Ag^+ (a).

	δ	δ+Ca^{++}	δ+Ag^+	$\Delta\delta Ca^{++}$	$\Delta\delta Ag^+$
Et-O-C(=O)-NH_2	6.36	6.45	6.34	0.09	-0.02
Et-C(=O)-N-Et	6.93	7.09	6.93	0.16	0.00
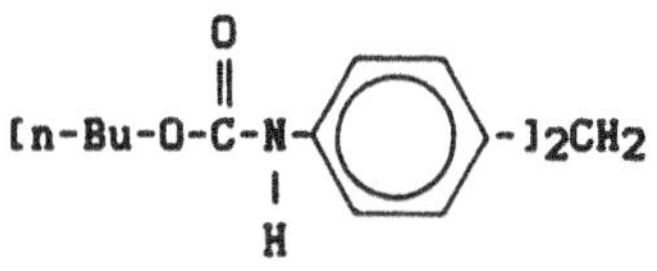	9.56	9.73	9.55	0.17	-0.01
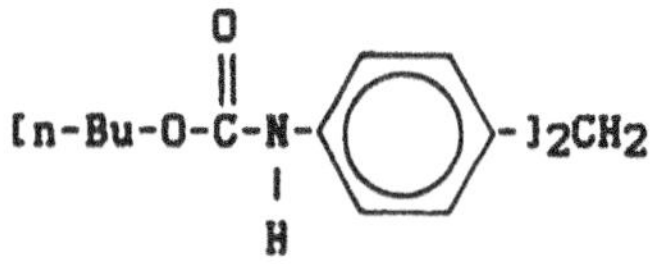	6.90	6.95	6.90	0.05	0.00

(a) Chemical shifts in ppm, are downfield from zero, with tetramethylsilane as the internal standard set at 0.00 ppm.

Table II. The Change in the ^1H NMR Chemical Shift (δ) of the N-H Proton of the Urea Models upon Complexation with Ca^{++} and Ag^+ (a).

	δ	$\delta + Ca^{++}$	$\delta + Ag^+$	$\Delta\delta Ca^{++}$	$\Delta\delta Ag^+$
Et-N(H)-C(=O)-N(H)-Et	5.75	6.36	6.34	0.61	0.59
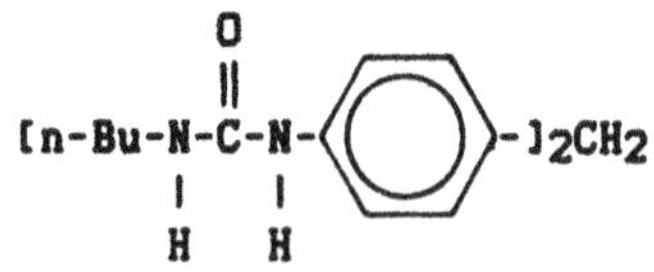					
(n-Bu-NH-)	6.06	6.74	6.25	0.68	0.69
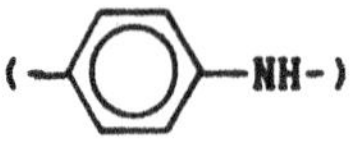	8.33	9.20	8.46	0.87	0.13
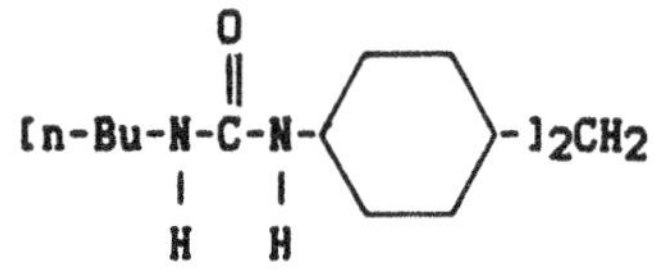					
(n-Bu-NH-)	6.06	6.40	6.12	0.34	0.06
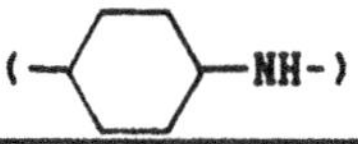	5.98	6.20	6.02	0.22	0.04

(a) Chemical shifts in ppm are downfield from zero, with tetramethylsilane as the internal standard set at 0.000 ppm.

(propylene ether)diol (MW = 425, 1000, and 4000) were prepared. For each metal ion species and each polyether, the solutions were mixed such that the metal-ion:ether-oxygen ratio was 1:1. The two phase mixture was placed on a platform shaker for 15 minutes. The mixture was then transferred to a separatory funnel and the toluene layer extracted. The toluene layer was dried for 15 minutes over Na_2SO_4, filtered and diluted to 25 ml with toluene. The toluene solution was analyzed by AES by aspirating the solution directly into the ionizing plasma.

An extraction following the same procedure, except that there were no dissolved polyethers, served as the control. Each extraction was repeated five times, and for each sample three ion readings were made. A reading of less than 0.001 ppm indicated that contained metal ion was below the detection limit of the instrument (Figures 1 and 2, Tables III and IV).

Table III. Metal Ion Complexation - PPM (±S.D.) of Metal Ion Extracted from Aqueous Solution Due to Complexation with Poly(tetramethylene ether)glycols (PTMEG) of Three Molecular Weights.

	Ag^{+}(ppm)	Ca^{+2}(ppm)	Fe^{+2}(ppm)
PTMEG 650	0.0826 (±0.1152)	0.0024 (±0.0015)	NDA[a]
PTMEG 1000	0.9628 (±0.5676)	7.2056 (±0.5542)	1.212[b]
PTMEG 2000	2.436 (±2.042)	0.3288 (±0.0176)	0.5152 (±0.2200)
PTMEG Control[c]	NDA[a]	NDA[a]	NDA[a]

(a) NDA: non-detectable amount (<0.001 ppm)
(b) Average of four extractions
(c) 0 % polyether

Table IV. Metal Ion Complexation - PPM (±S.D.) of Metal Ion Extraction from Aqueous Solution due to Complexation with Poly(propylene ether) glycols (PPEG) of three molecular weights.

	Ag^{+}(ppm)	Ca^{+2}(ppm)	Fe^{+2}(ppm)
PPEG 425	a	0.0216 (±0.0150)	NDA[b]
PPEG 1000	3.904 (±3.6609)	0.0032 (±0.0020)	NDA[b]
PPEG 4000	0.6948 (±0.4374)	0.0105 (±0.0049)	NDA[b]
PPEG control[c]	NDA[b]	NDA[b]	NDA[b]

(a) Data unavailable
(b) NDA: non-detectable amount
(c) 0 % polyether

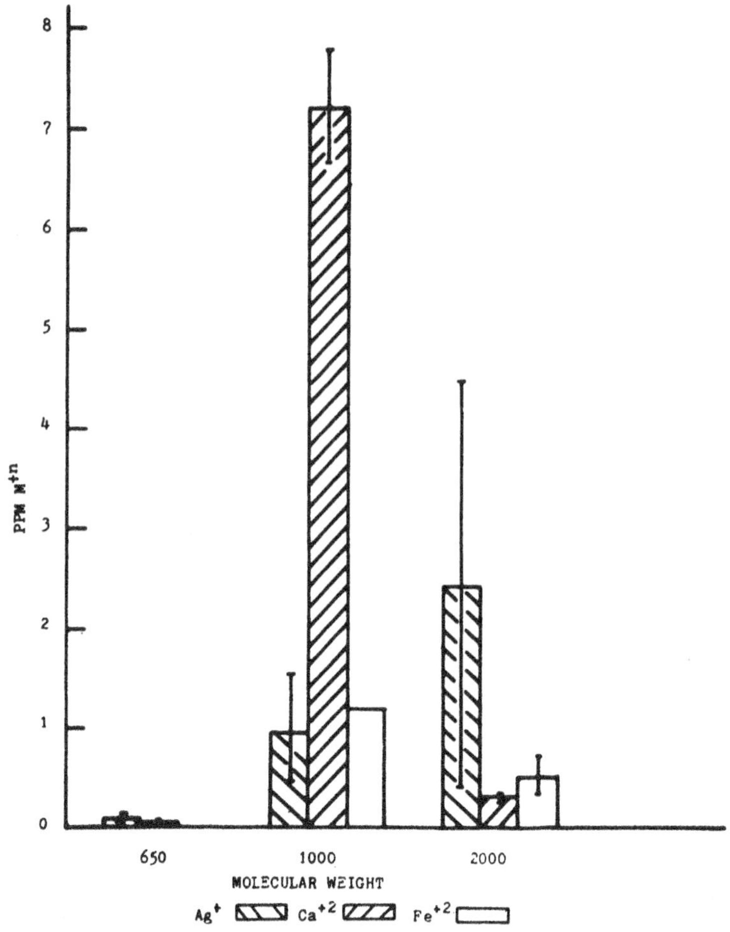

Figure 1. Metal ion complexation - PPM of metal ion extracted from aqueous solution due to complexation with poly(tetramethylene ether)glycols of three molecular weights.

RESULTS AND DISCUSSION

Low molecular weight models of polyurethane and polyurea hard segments are capable of chelating metal ions (M^{+n}) as a monodentate ligand through the carbonyl oxygen, the ester oxygen, or the amide nitrogen (Diagram 1), and as a bidentate ligand through various combinations of these coordination sites (Diagram 2).

Proton NMR has been used, in a qualitative sense, to establish whether coordination can occur in the PEU hard segment. Since this analytical technique is a measure of the magnetic environment about a proton, any change in that magnetic environment, due to complexation of M^{+n} by the urethane or urea linkage, is indicated in the ^{1}H NMR spectra of the model compounds by a change in the frequency at which the amide proton resonates (11).

Tables I and II show that there is a change in the ^{1}H NMR chemical shift of the N-H protons of the urethane and urea models. Some metal ions' interactions with the urea analogs cause larger changes in the

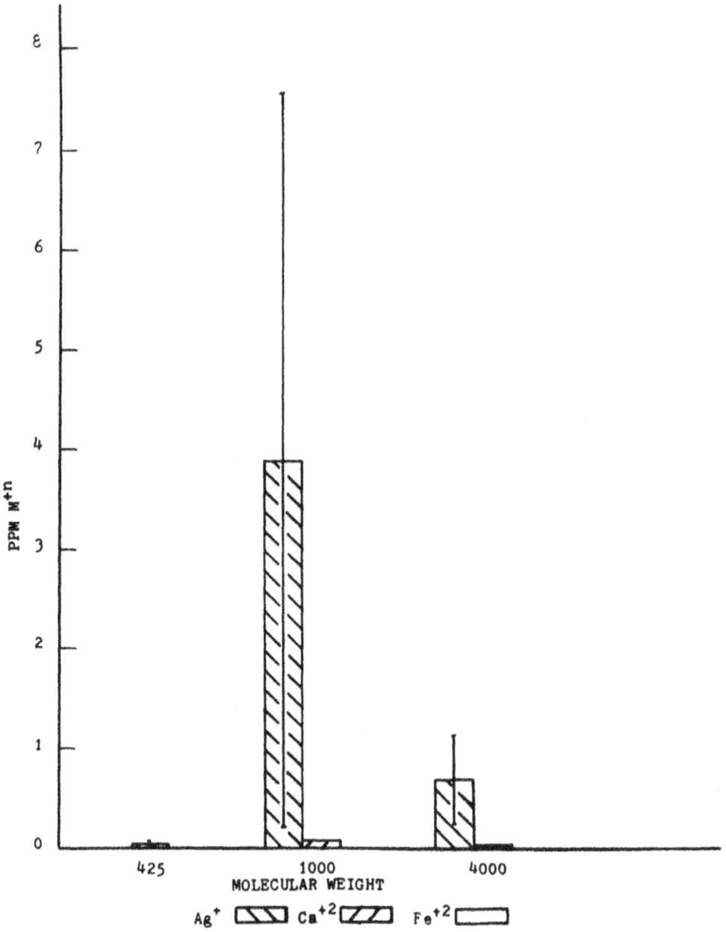

Figure 2. Metal ion complexation - PPM of metal ion extracted from aqueous solution due to complexation with poly(propylene ether)glycols of three molecular weights.

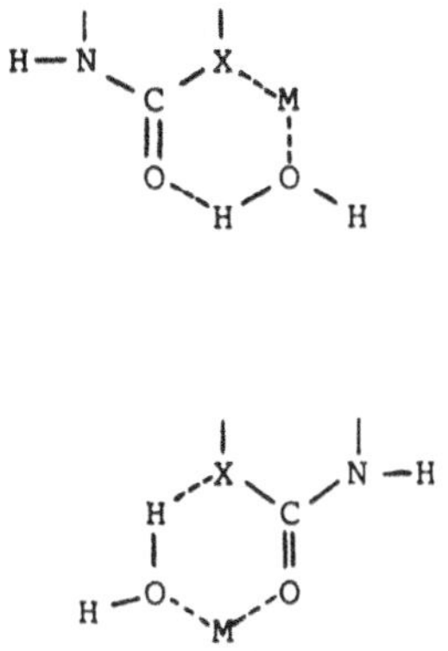

Diagram 1. Monodentate coordination by the urethane hard-segment ($X = O, NH$; $M = M^{+n}$).

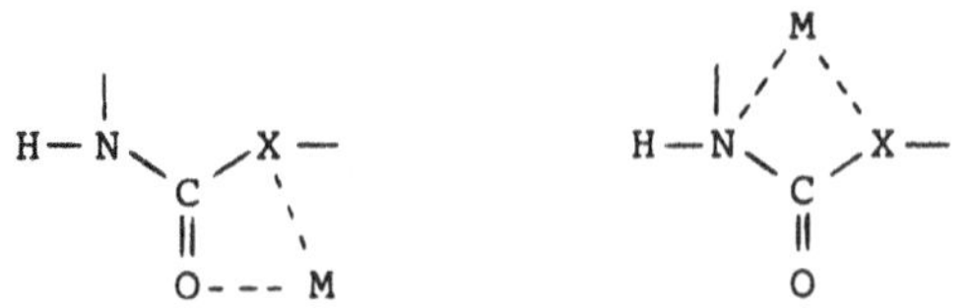

Diagram 2. Bidentate coordination by the urethane hard-segment ($X = O, NH$; $M = M^{+n}$).

chemical shift of the amide proton than do metal ion interactions with the urethane analogs. Calcium ion causes larger changes in the chemical shifts of both the urethane and urea models than does silver ion. The data indicate that there is complexation or interaction of the models with M^{+n}.

Studies of metal ion complexation by the soft segment are based on similar work with cyclic polyethers called crown-ethers (Diagram 3). Crown-ethers are well known for their chelating ability and their selectivity. The conditions necessary for the formation and factors influencing the stability of the complexes include: (a) the relative size of the ion and the hole in the polyether ring, (b) the number of oxygen atoms in the polyether ring, (c) the coplanarity of the oxygen atoms, (d) the symmetrical placement of the oxygen atoms, (e) the basicity of the oxygen atoms, (f) steric hindrance in the polyether ring, (g) the tendency of the ion to associate with the solvent, and (h) the electrical charge on the ion (12).

Low molecular weight acyclic polyethers have also been shown, by single crystal x-ray analysis, to form metal ion complexes from metal salts in either a relatively random or "S" configuration (Diagram 4) (13), or in a relatively ordered or coiled configuration (Diagram 5) (14). The criteria for formation and stability hold equally for acyclic polyether complexes with metal ions as for cyclic analogs.

Chow and coworkers (15,16) have used various PEUs to extract metal ions, quantitatively, from aqueous solutions. The mechanism of extraction has been termed the "Cation-Chelation Mechanism" and is based on the complexation of cations, metal and ammonium cations, at the center of a coil composed of the polyether soft segment of the polyurethanes (Diagram 6) (16). These urethanes, like the crown-ethers, are also selective with respect to size, charge, and solvent association of the cations.

The soft-segment models were studied such that the only driving force for M^{+n} extraction is interaction between polyether oxygen atoms and M^{+n}.

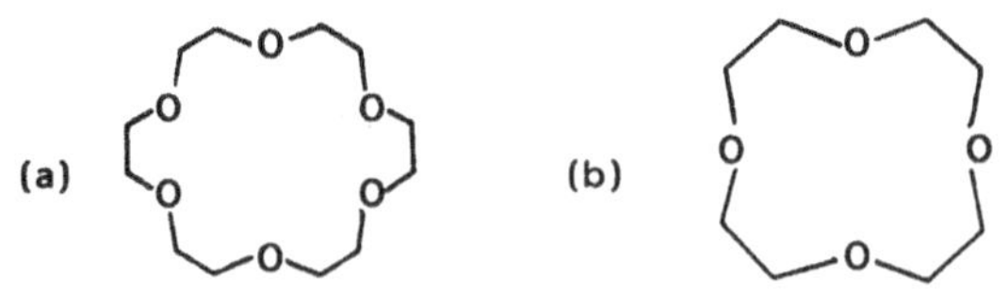

Diagram 3. Cyclic polyethers (crown-ethers); (a) 18-crown-6, (b) 12-crown-4.

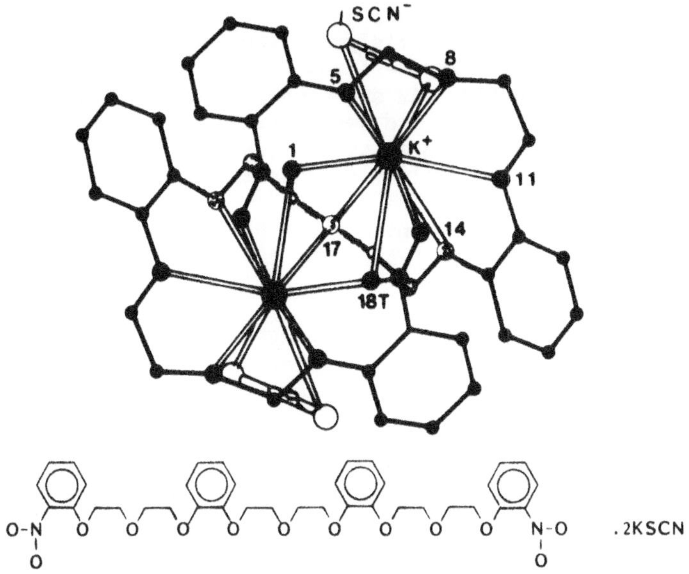

Diagram 4. S-shaped binuclear cation complex of a linear polyether.

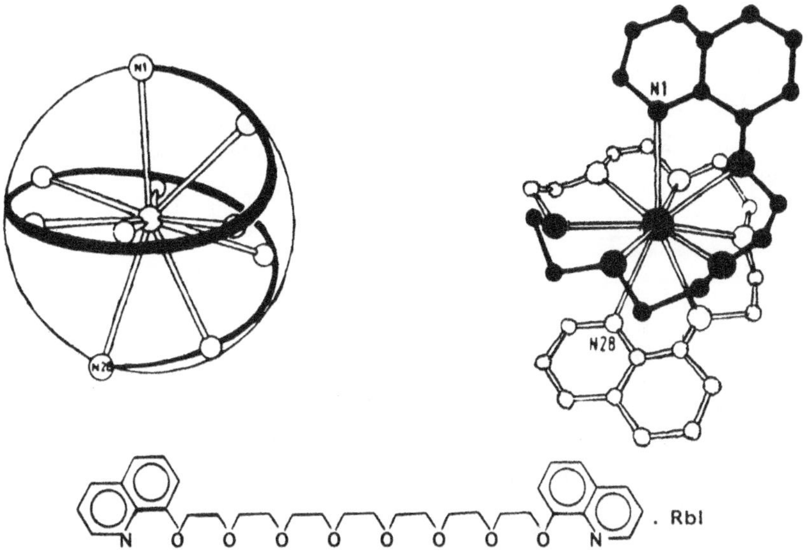

Diagram 5. Spherical wrappings of a linear polyether around cation.

Extraction data for calcium (Ca), silver (Ag), and iron (II) ions (Fe) from aqueous solutions using poly(tetramethylene ether)glycols (PTMEGs) and poly(propylene ether)glycols (PPEGs) of varying molecular weights (Tables III and IV, Figures 1 and 2) indicate that the polyethers are capable of extracting M^{+n}. As expected , from the selectivity exhibited by the crown-ethers (12) and demonstrated in polyurethanes by Chow (15), these polyethers are also selective toward specific cations, and this selectivity is molecular weight dependent.

The PTMEGs extracted Ca far better than PPEGs. Only trace amounts of Ca were extracted by any of the PPEGs. Of the PTMEGs, Ca was selectively extracted by PTMEG-1000.

Silver ion was extracted by all the PTMEGs and PPEGs. Selectivity was greatest for PPEG-1000; PTMEG-2000 extracted more Ag than PTMEG-1000.

Iron selectivity was very similar to Ca selectivity, with PTMEG-1000 preferred. No Fe extraction is observed by any of the PPEGs. This does not mean, however, that there is no interaction between the polyether and Fe; only that it is too slight to be detected by the particular experimental technique.

When the criteria for the formation and stability of polyether-M^{+n} complexes are considered, as mentioned above, the selectivities shown by the various polyethers and metal ions can be easily explained. The criteria which are of most importance in these systems are those which relate to the relative size of the ion and the space available in the polyether ring, the number of oxygen atoms in the polyether ring, and coplanarity of the oxygen atoms, the symmetrical placement of the oxygen atoms, and the steric hindrance of the polyether ring.

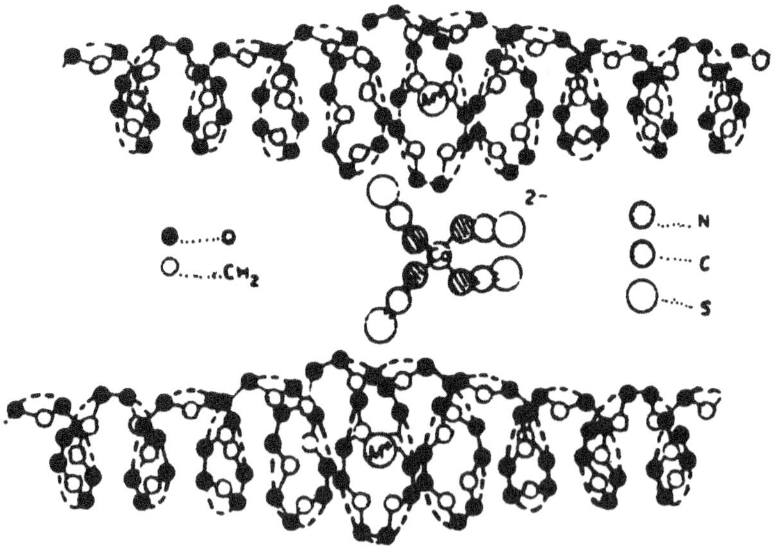

Diagram 6. Proposed helical structure (not to scale) of sorbed $M_2Co(NCS)_4$-polyether complex suggested by the cation chelation mechanism.

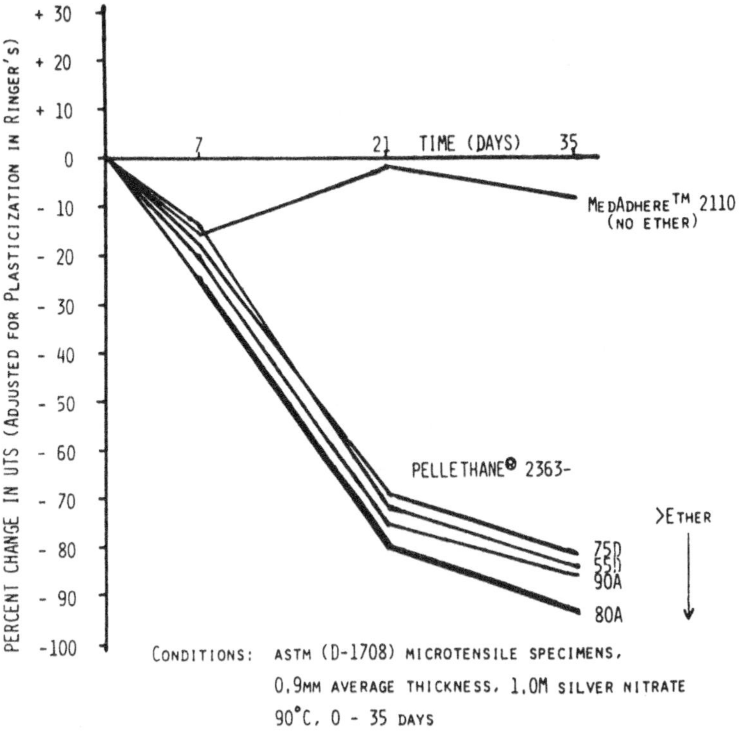

Figure 3. Aging of polyurethanes with various ether contents in silver nitrate.

By using molecular models (scales: 1 Å = 1 cm) of the systems, it can be shown that the steric constraints of the PPEG system do not allow a polyether ring having six oxygen atoms (like the 18-crown-6 above) to achieve a small enough diameter, while still retaining the coplanarity and symmetrical nature required of the oxygen ring for formation of a stable complex, to interact substantially with small cations like Ca and Fe (ca. 1.98 Å and 1.52 Å, respectively). The diameter is adequate, however, to readily accommodate the relatively larger Ag (ca. 2.52 Å) thus forming a stable complex. Models of smaller rings of PPEG have internal diameters which are too small to accommodate any of the above mentioned ions. Poly(propylene ether)glycols, and PPEG-containing PEU, can only be expected to complex M^{+n} which are of appropriate size. The selectivity demonstrated by these PPEGs, among alkali-metal cations, is comparable to the selectivity shown by the crown-ether, 18-crown-6. The crown-ether, 18-crown-6, exhibits a definite order of preference: K^+ (ca. 2.66 Å) > Rb^+ (ca. 2.96 Å) > Cs^+ (ca. 3.39 Å) > Na^+ (ca. 1.90 Å) > Li^+ (ca. 1.20 Å) (17). Preference by the crown-ether for the intermediate sized cation (K^+) is analogous to the selectivity shown by PPEG for Ag; the smaller cations are not readily extracted by either system.

Poly(tetramethylene ether)glycols, due to inherent flexibility of the longer carbon chain between oxygen atoms, can easily achieve the smaller ring sizes (as demonstrated by Fischer models). The flexibility of the PTMEG allows the ring to expand to accommodate the larger ions as well. A molecular model of a PTMEG having only four oxygen atoms in the internal ring shows that the internal diameter can vary from 1 - 4 Å. Thus PTMEGs are expected to form stable complexes with cations having a wider range of ionic diameters than are PPEGs. This is evident from the data above.

The molecular weight selectivity shown by both PTMEG and PPEG is significant, especially when considering the fact that most PEU currently commercially available for biomedical applications are composed of a polyether with molecular weight of 1000. Indeed, polyethers in this study which demonstrated the greatest capacity for M^{+n} extraction are those, with one exception, having MW = 1000.

Several workers have shown the affect of M^{+n} complexation on the mechanical properties of PEUs. Coury et al. (18) have demonstrated that as the concentration of polyether soft-segment increases, in a family of PEUs called Pellethane^R, the ultimate tensile strength, due to Ag interaction, decreases (Figure 3). It is also suggested that the changes in the ultimate tensile strength of some pacemaker lead coatings are a result of oxidation of the soft segment by silver salts generated as a result of interactions of physiological fluids with the conducting lead (18). Oxidation by Ag does not, however, account for the presence of surface microfissures, "frosting", which appear on the outer surface of all explanted pacemaker leads. It should also be pointed out that such an interaction must have been present in the lead before the Ag could have been electrolytically generated. Silver ion participation, therefore, whether by chelation or by oxidation, could not have been the cause of the initial cracking or crazing in PEU coatings, but may have enhanced it.

Benson et al. (19) have shown that LiBr causes changes in the morphology of PEUs resulting in an increase in water absorption and decrease tensile strength. Dynamic mechanical measurements were able to show that although no gross changes in the morphology of the PEUs occurred, as compared to PEUs without LiBr salts, sufficient minor changes caused an apparent increase in the stiffening of the polyether segment.

In the current study, the counter-anion has been held constant. Hamon et al. (16) have shown, however, that PEUs are anion selective as well as cation selective. It is an important consideration that in order to maintain charge neutrality of M^{+n} polyether or PEU complexes, an equivalent number of anions must accompany any chelated cations. Anion extractability depends mainly on the hydrophobicity of the ion as dictated by size, charge density, hydrogen bonding ability, and other factors. Ionic metal complexes which are large, have low charge densities, and lack the ability to form strong hydrogen bonds with water (as might be expected, in an _in vivo_ environment, from complexes of lipoproteins and phosphates) are expected to be highly extractable. Evidence for this is apparent in work by Coury et al. (18) in studies of the aging of Pellethane^R-80A in oxidants, organic species, and their combinations (Figure 4). Silver nitrate alone does not have a significant influence on the ultimate tensile strength of the urethane. It is not substantially greater than that of the two organic acids, as might be expected if Ag alone were causing the oxidation of the polyether. Larger changes in the ultimate tensile strength of the urethane result from complexation of silver salts of organic acids due to the larger, more hydrophobic counter ions, than with either silver nitrate, lactic acids, or acetic acid by themselves. This is observed in the case of silver acetate, which is a silver salt of an organic acid, and in the case of a mixture of silver nitrate and lactic acid. It is important to realize, though not originally indicated (18), that the effect of the silver nitrate/lactic acid mixture is due to PEU complexation of silver lactate. Lactic acid, in aqueous media, is partially dissociated (pKa = 3.98). Since the lactate anion is larger and more hydrophobic than the nitrate anion, it is more easily extracted by the PEUs, causing a greater decrease in the ultimate tensile strength.

The phenomena of M^{+n} complexation has been shown to be essentially

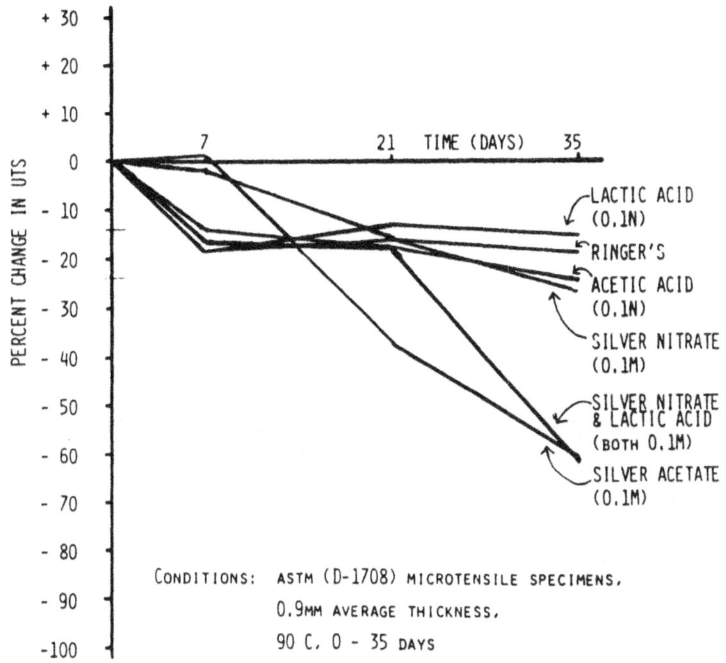

Figure 4. Aging of Pellethane[R] 2363-80A in oxidants, organic species and combination.

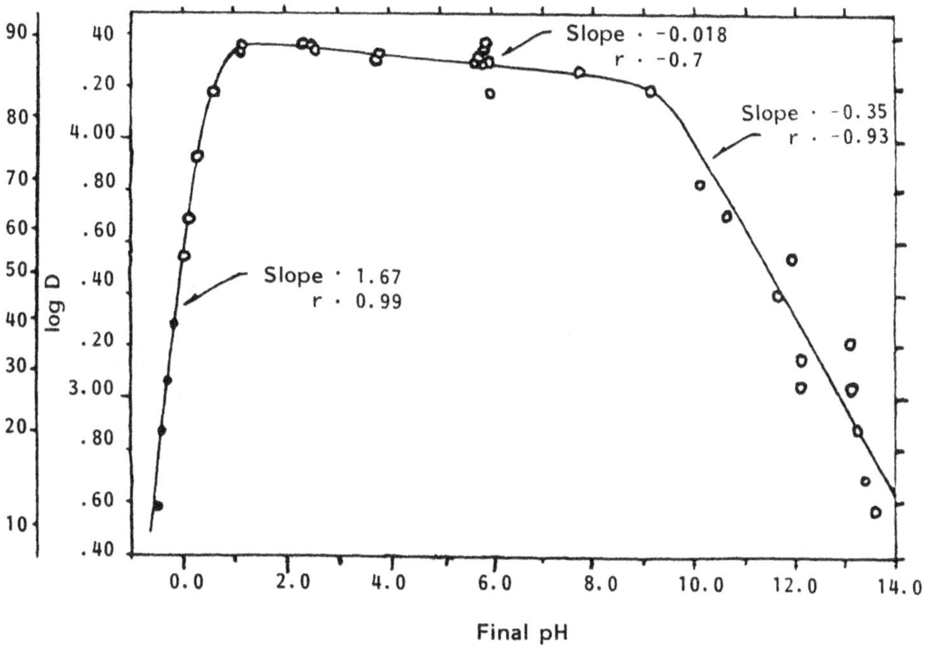

Figure 5. Effect of pH on Co sorption from aqueous solution.

independent of pH over a very wide range, pH 1 - 9 (Figure 5) (16). Extraction in the physiological environment by polyethers or PEUs at physiological pH, is expected to be comparable to extraction in the in vitro systems mentioned.

CONCLUSIONS

The data presented above, have shown that both the hard and soft segments of PEUs are capable of complexing M^{+n}. It has been shown that M^{+n} complexation by PEUs is not unique to models of the hard and soft segments. Metal ion complexation is also independent of pH, and is likely to occur in the in vivo environment, as well as in the model systems employed. The in vivo environment, as a result of many combinations of cations, metal and nitrogen containing, and anions, chloride, phosphate, and the like, is likely to be more susceptible to complex formation by PEUs than in vitro environment which have been used for various studies of M^{+n} coordination or biomaterials compatibility.

It appears, upon considering data presented by others, regarding the effect of salts on PEU mechanical properties (19), the effect of reduced polyether content on the reduction in the mechanical properties (18), the fact that PEUs are composed of a much greater proportion of polyether and that the surfaces of some PEUs, which have been used for biomedical applications, are composed primarily of the soft-segment polyether (20), that complexation by the soft segment is of far greater consequence than that by the hard segment.

Soft-segment polyethers have been shown, under some conditions, to be susceptible to oxidation, via free radical initiated autooxidation (21). It has also been reported, from Attenuated Total Internal Reflectance Infrared (ATIR-IR) studies of some PEU based pacemaker leads, that the PEU surface may show the presence of oxidative by-products (22). It is not surprising then that many manufacturers of polyethers and PEUs include in their formulations antioxidant additives (21). These antioxidants, however, can be inadvertently removed during a manufacturing step such as solvent swelling extruded PEU tubes for insertion of the coil during pacemaker lead fabrication. Antioxidant removal would make the fabricated lead susceptible to autooxidation. Unless the antioxidants are removed, either through a manufacturing process or through an (as yet) unidentified, in vivo leaching process, soft segment autooxidation should not be a factor in the cracking or crazing of PEUs.

Enhanced environmental stress cracking by M^{+n} chelation, which is not precluded by premature removal of antioxidants, can be viewed in terms of a cation complexation mechanism, and is easily imagined as resulting from induced linear compression of the polyether chains in which cations are chelated along the axis (16). In such an arrangement a chelated cation acts as a source for localized stress which can be relieved by cracking or fissuring. Metal ion complexation may also be an important factor in the formation of calcific deposits on PEUs or other blood contacting surfaces as a result of a build-up of a local calcium ion concentration at the surface. High local Ca concentration may result in precipitation of hydroxyapatite leading to stiffening and consequential susceptibility to fatigue, or degradation of mechanical integrity.

REFERENCES

1. J. W. Boretos and W. S. Pierce, Science, 158, 1481 (1968).
2. J. W. Boretos, W. S. Pierce, R. E. Baier, A. F. Leroy and H. J.

Donachy, J. Biomed. Mater. Res., 9, 327 (1975).
3. K. Stokes, Medtronic News, 12 (1981).
4. D. J. Parins, K. M. Black, K. D. McCoy and N. J. Horvath, Cardiac Pacemakers, Inc., St Paul, MN (1981).
5. P. Termin, Pacing Concept Paper, 4 (1982).
6. K. Stokes, W. A. Frazer and E. A. Carter, Prtesented at the SPE ANTEC Meeting, New Orleans, LA, May (1984).
7. L. Engle, H. Klingele, G. W. Ehrenstein and H. Schaper, "Atlas of Polymer Damage", Prentice-Hall, Inc., Englewood Cliffs, NJ (1981).
8. A. J. Coury, P. T. Cahalan, E. L. Schultz and K. B. Stokes, Transactions of the Second World Congress on Biomaterials, 7, 252 (1984).
9. P. M. Henrichs, J. J. H. Ackerman and G. E. Maciel, J. Amer. Chem. Soc., 99, 2544 (1977).
10. H. K. Frensdorf, J. Amer. Chem. Soc., 93, 4684 (1971).
11. R. S. Drago, "Physical Methods in Chemistry", W. B. Saunders Co., Philadelphia, PA (1977).
12. C. J. Pederson, J. Amer. Chem. Soc., 89, 7017 (1967).
13. G. Weber and W. Saenger, Angew. Chem., Int. Ed., 12, 227 (1979).
14. G. Weber, W. Saenger, F. Vogtle and H. Sieger, Angew. Chem., Int. Ed. Engl., 18, 226 (1979).
15. A. Chow and S. L. Ginsberg, Talanta, 30, 620 (1983).
16. R. F. Hamon, A. S. Khan and A. Chow, Talanta, 29, 313 (1982).
17. I. M. Kotthoff, Anal. Chem., 51, 1R (1979).
18. A. J. Coury, P. T. Cahalan, P. C. Slaiken and K. B. Stokes, Trans. Society for Biomaterials, 7, 134 (1985).
19. R. Benson, S. Yoshikawa, K. Knutson and D. J. Lyman, Adv. in Chem. Ser.-199, American Chemical Society, Washington, D.C., 133 (1982).
20. C. S. Paik-Sung and C. B. Hu, J. Biomed. Mater. Res., 13, 161 (1979).
21. (a) Polymeg Polyols Applications and Properties Data, The Quaker Oats Company, Chicago, IL (1976).
(b) DuPont Glycols; Properties, Uses, Storage and Handling, E. I. DuPont de Nemours & Co., Wilmington, DE (1983).
22. M. Szycher, D. Dempsey and V. L. Poirier, Second World Congress on Biomaterials, 7, 24 (1984).

ENVIRONMENTAL STRESS CRACKING IN IMPLANTED POLYURETHANE DEVICES

K. B. Stokes and M. W. Davis

Medtronic, Inc.
Fridley, MN 55432

The tissue contacting surfaces of many early polyurethane insulated cardiac pacing leads have experienced microscopic surface crazing and cracking. Some models have also experienced insulation failures. The failures were model dependent, demonstrating that this was a problem with some early polyurethane devices as opposed to a problem with polyurethane per se. Extensive analysis of explants demonstrates no significantly measurable chemical changes in cracked polymer, with the exception of some slight oxidation found by Fourier transform infrared spectroscopy. The failure mechanism appears to require a critical strain. The time to failure then depends on the amount of strain in excess of a critical value. An <u>in vivo</u> test has been developed to study the phenomenon using fixed strain as an accelerator. Four hypotheses exist to explain this cracking, true environmental stress cracking, solvent stress cracking, enzymatic stress cracking, and oxidative stress cracking. In all cases, however, residual molecular stress/strain is at the root of the mechanism. Thus, elimination of such strain through optimized extrusion techniques, annealing, etc. can prevent stress cracking failures. Certain polyether polyurethanes (optimally processed) are safe and reliable in implanted devices as has been demonstrated both by many early models and a new series of leads with excellent long-term <u>in vivo</u> performance.

INTRODUCTION

Neurologic and cardiac pacing leads insulated with polyether polyurethanes have been implanted in humans since 1975 and 1977, respectively. These elastomers have been proven to be biostable in long-term animal studies (1,2). Nonetheless, while most lead models have demonstrated excellent long-term performance, a few have not (3). Two mechanism have been discovered that can cause failure of polyurethane insulation, a form of stress cracking and metal catalyzed oxidation. This report will summerize the results of our investigation into cardiovascular lead insulation failure due to stress cracking only, since oxidation processes have been discussed separately (4). Basically, stress cracking appears to be primarily a mechanical phenomenon where residual polymer stress/strain interacts with tissues to develop crazed cracks.

MATERIALS

The large majority of implanted polyurethane insulated pacing leads use tubing extruded from Pellethanes 2363-80A (P80A) or 2363-55D (P55). These polyether polyurethanes (from Upjohn, Inc., CPR Division) are composed of polytetramethylene ether oxide (PTMO) soft segments and methylene di(isocyanatobenzene)/propanediol hard segments. The softer material (P80A) has a higher percentage of soft segments than the harder (P55D) polymer. The resin contains several additives including an antioxidant and an extrusion lubricant. Polyurethane cardiac pacing leads are typically constructed of one or two (coaxial) insulated conductor coils. Typically, conductor coils are inserted in extruded tubing. The tubing ends must then be expanded over electrode and terminal assemblies to make leak resistant joints. The details of manufacturing, however, can vary considerably from model to model.

Four early P80A insulated transvenous cardiovascular lead models (Medtronic, Inc.) will be used as examples. Model 6972 has thin (0.13 mm) insulation over 2 (coaxial) conductors (Table I). Models 6971 and 6957 have thicker insulation over single conductor coils. Model 6991U has thin (0.13 mm) insulation over a "memory coil" which forms the distal portion of the lead into a J shape for placement in the right atrial appendage. The majority of the lead body, however, has much thicker (0.64 mm) insulation.

METHODS

All explanted and control specimens were evaluated by optical microscopy at 30-70X. Specimens selected for scanning electron microscopy (SEM) were mounted on aluminum stubs and gold-coated prior to examination on an ISI-40 SEM with a Princeton Gamma-Tec electron dispersive analysis by x-ray (EDAX) attachment. A Perkin-Elmer Model 398 grating spectrophotometer was used for bulk and surface infrared analysis (IR). Surface studies employed a frustrated multiple internal reflectance (FMIR) attachment. In some studies, cracked surfaces were rinsed for about 10 seconds in a drop of tetrahydrofuran on a salt plate. This removed a very thin layer as verified by SEM, including almost all cracked material, for infrared analysis. Fourier transform surface IR (FTIR) has been used for limited surface studies. Molecular weights have been determined by gel permeation chromotography and high pressure liquid chromotography. Additional analytical techniques have been described elsewhere (6-8).

Specimens for analysis included leads explanted from humans and returned to the manufacturer, or from canines. Not all leads can be removed from humans and not all leads that are removed are returned to the manufacturer. Thus, actuarial survival statistics were determined from telephone monitoring data (covering over 18,300 leads) by Cardio-Care, Inc.

Canine explants (always 1 atrial and 1 ventricular lead per dog, inserted through the same vein) were removed atrumatically. These were usually washed free of biologic debris in water, then air dried. Ten leads with $\geq$ 2 years implant each were removed intact with all surrounding tissues from 12 dogs. Optical microscopy was then done to assess the location of cracks in relationship to tissues, etc.

Unprocessed polymer pellets were extracted in chloroform for 48 hours in a soxlet extractor using a 5 liter thimble. About 3-4% by volume was extracted, containing primarily low molecular weight polymer, extrusion lubricant and antioxidant (by infrared analysis) (6). Some tubing was extruded from extracted pellets. Some had an antioxidant (AO) added, some had extrusion lubricant (EL) added, and some had both (AO/EL). The same extrusion lubricant and antioxidant were added in approximately the same concentrations contained in the original, unextracted polymer.

The effects of these and other processes were evaluated in vivo as a function of fixed strain. Tubing was extruded using a 2.5 cm Killian extruder with a 24.1 length to diameter screw. Prior to extrusion, all materials were thoroughly dried. A 0.50 mm diameter steel wire was bent into staple or figure 8 shapes, 12.5 mm long. Extruded tubing 2.0-2.25 mm diameter was placed over the mandrels, then strained to elongations up to 500% in increments of 100%. Firm ligatures were placed over the tubing at each end of the mandrel to fix strain. Specimens were tied in strings of 3 or 4, colored-coded with glass beads and implanted subcutaneously in rabbits or dogs under general anesthesia (see Figures 1 and 2). Explants from rabbits were done weekly for 12 weeks. Canine explants have been done after 9-10 months and are continuing for long-term studies. All explants were evaluated by optical microscopy. Selected explants were studied by SEM.

RESULTS

Most of the results of our investigation of canine and human explanted leads have been presented elsewhere (4,6,8). Thus, these data will not be discussed in detail here. In summary, however, shallow, usually microscopic surface cracking (to remarkably uniform depths) occurs at various places on the lead, frequently at strained sites near fixation ligatures, expansion joints, kinks, sharp bends, or in surface imperfections such as die lines, etc. (Figure 3). Microscopic surface cracks also appear to occur at random spots on explanted leads. On devices evaluated with all tissues intact, however, such "frosted" areas in the vascular system are usually (not always) found near the ends of slowly growing chronic fibrous sheaths. Not all areas of a chronically implanted lead are encapsulated in fibrous tissue. Usually, polyurethane leads in canines are at least partially free of capsule in the ventricle and atrium. They are encapsulated with very thin fibrous tissue only at the most distal tip (by design to prevent dislodgment) and from the superior vena cava/atrial junction, proximally. These sheaths are not necessarily continuous. Cracked surfaces, however, are always encapsulated with thin fibrous tissue, even when they are located in otherwise "clear" areas. Stress cracking failures have always been associated with severe strains (Figure 4) (7,8).

Stress cracking failure is model dependent, as is shown in Table I. An estimated half of the Model 6972 failures are believed to have failed

Figure 1. A 0.50 mm diameter wire is bent into a figure 8 shape, 12.5 mm long. This is inserted in 2.0-2.55 mm diameter tubing which is strained. The strain is fixed with 2.0 polyester ligatures. Specimens with combinations of variables (usually different strains) are tied in strings of 3 or 4 and color-coded with a glass bead.

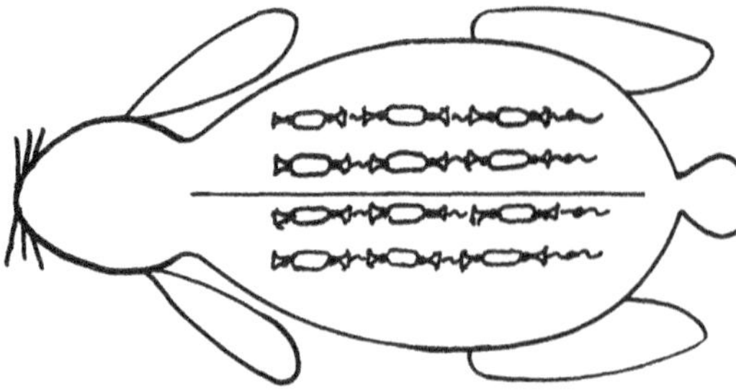

Figure 2. Four sterile strings (see Figure 1) are implanted subcutaneously in rabbits or canines under general anesthesia. Weekly rabbit explants are done for 12 weeks while canine studies continue for extended periods.

because of stress cracking, the remainder due to oxidative processes. Also, 23 out of 24 model 6971 failures and all Model 6991U failures were due to stress cracking. In December 1981, Model 6991U manufacturing processes were changed to eliminate (axial) stretching of insulation and the use of swelling solvents during coil/tubing assembly. Such solvents allowed the occasional accidental use of tubing with an inside diameter smaller than the outside diameter of the memory coil. While microscopic surface cracks are still found, no Model 6991U leads made since December 1981 are known to have failed as a result of stress cracking (as of January 1985).

Based on extensive long-term animal tests, shallow microscopic surface cracks typically appear 3-6 months postimplant (7,10). These may

Figure 3. Shallow microscopic cracks in explanted lead surface. A ligature has been tied tightly just to the left of the micrograph to prevent slippage. Cracks decrease in severity as a function of distance from the ligature, 1010x.

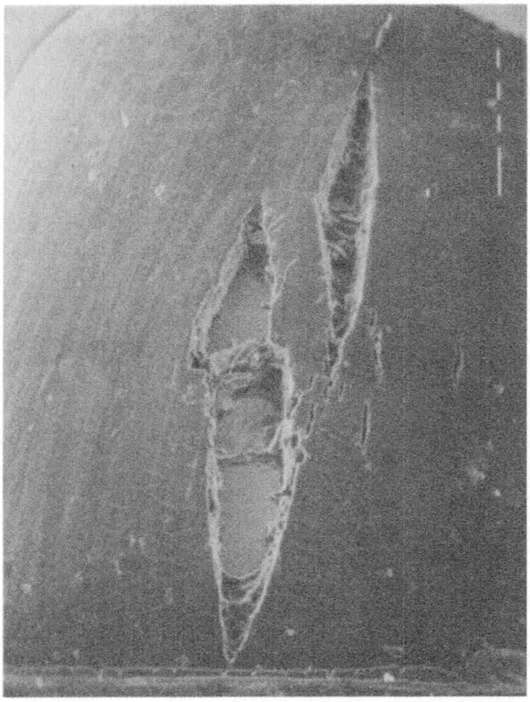

Figure 4. Insulation failure where tubing was expanded over metallic electrode shank. Note orientation of cracks normal to strain and evidence of crazing, 70x.

Table I. Experience With Medtronic Polyurethane Insulated Leads (U.S. Only) As of January 1985 (3).

Model No.	Insulation Thickness (mm)	Number of U.S. Implants	Returned Leads* With Insulation Failure N	%	Actuarial** Survival (% at X Mo.)
6957/57J	0.58	24,900	0	0	99.7/24
6971	0.25	56,700	24	0.04	99.3/48
6991U	0.13/0.64	5,800	19	0.33	--
6972	0.13	43,300	440	1.0	89.8/48
All Others***	--	63,300	62	0.10	--
		194,000	545	0.28	

* Based on analysis of returned products, includes both SC and MO.
** Based on telephone monitoring of pacemaker patients.
*** Includes 5 different models with no known failures.

spread, covering larger surface areas, but do not appear to propagate deeper as a function of time unless strain exceeds a critical value (7,9). Thus, the time to failure appears to depend upon the amount of strain in excess of the critical value. Stress cracks form from crazes (Figure 5) (4,7,8). Propagating crack tips are always crazed whereas cracks that have stabilized no longer have crazes at their tips (Figure 6) (10).

Typically, nothing untoward is found even in severely cracked insulation by chemical and physical analyses. Others have found no changes in cracked surfaces by electron spectroscopy for chemical analysis or Auger spectroscopy (11). No chemical change has been found by bulk IR, FMIR, or stripped surface infrared spectroscopy (4,7,10). By FTIR surface spectroscopy some evidence of oxidation has been seen in a few severely cracked P80A surfaces, but not in others. Oxidation of ether soft segment can be detected primarily by a loss of PTMO ether at 1110 cm^{-1}, compared to urethane ether (1080 cm^{-1}) and amide (1220 cm^{-1}). Molecular weight analyses have revealed no significant changes in cracked versus uncracked leads as is shown in Table II. Cracked material retains essentially the same stress/strain characteristics as uncracked material (4,7,10). Usually nothing is found by EDAX.

Specimens made from reformulated extracted polymer showed very definite differences in the <u>in vivo</u> fixed strain test. "As is" and E1 tubing developed deeper, more intense cracks as a function of fixed strain and implant time (6). Both A0 and A0/E1 tubing developed significantly shallower, less extensive surface cracking and no failures at elongations less than 200%.

The effects of extrusion parameters on P80A have also been reported previously (4). P80A tubing extruded into a relatively hot cooling tank exhibited substantially better SC resistance than the same tubing extruded into a very cold tank (Tables III and IV). Specimens strained, then annealed to remove all residual stress showed no cracking at all in

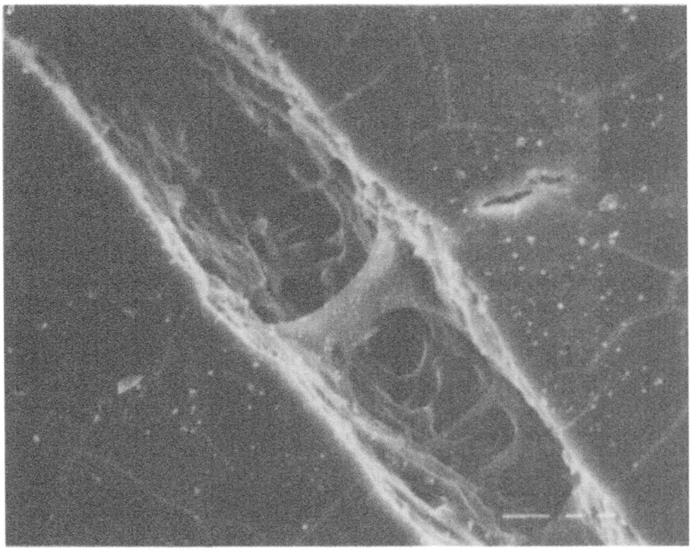

Figure 5. A stress crack in the outer surface of an explanted lead. Some craze tie bundles still remain, 700x.

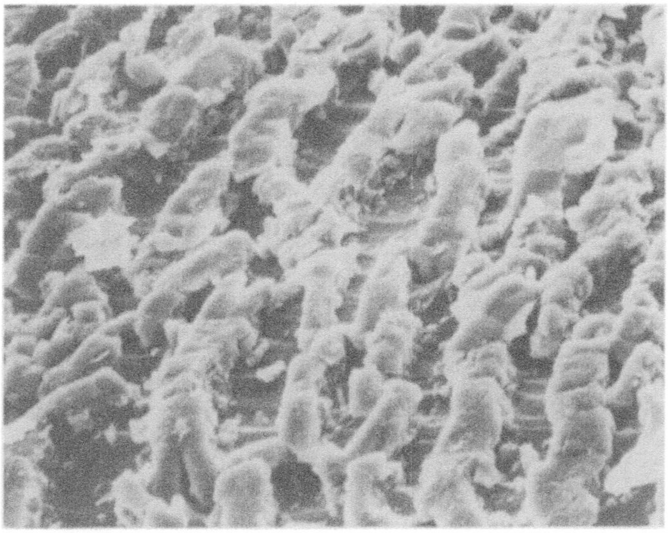

Figure 6. A cracked surface similar to that in Figure 3 was mounted on the SEM stub under fixed strain (510x). That separated cracks, revealing what looks like an undamaged substrate. No crack propagation occured. No evidence of crazing was observed at crack bases at 5,000x.

this test (Table V). Specimens removed from dogs nine months postimplant so far confirm that even unannealed P80A is acceptable at low fixed strains and that the stress cracking resistance of P55D is excellent at strains up to 400%.

DISCUSSION

Two mechanisms are known that can cause cracking of polyurethane insulation. Transition metal catalyzed oxidation of ether soft segment inside a pacing lead has been identified (12). Polymer so affected shows clear evidence of destructive chemical changes. However, even in extreme cases little evidence has been presented that chemical reactions are involved in the P80A and P55D stress cracking phenomenon that has occured on tissue contacting surfaces. The term environmental stress cracking (ESC) has been used in its broadest sense to describe the phenomenon, but this may not be strictly correct. To be specific, at least three other possibilities exist, oxidative stress cracking (OSC), some form of *in vivo* solvent stress cracking (SSC), and enzymatic stress cracking (ZSC) (13). While metal catalyzed oxidation has been duplicated in accelerated tests at 37°C in 3% hydrogen peroxide, no evidence of stress cracking has been found *in vitro*. So far, no stress cracking has been observed in 37°C water, saline, 3% hydrogen peroxide, 1M glycolic acid solution, albumin solution, blood, solutions of various metal ions, etc. at fixed strains up to 500% elongation. Baier, et al., has demonstrated that the tissues adjacent to low surface energy materials are lipid rich (14). Nonetheless we have been unable to demonstrate stress cracking in 37°C synthetic plasma or phospholipid emulsion.

While FTIR has revealed oxidation degradation on some cracked surfaces, it has not on others. Bulk IR and surface FMIR on very severely

Table II. Some Typical molecular Weights of P80A Leads (High Pressure Liquid Chromatography).

Sample source (condition)	N	$M_w \times 10^3$	$M_n \times 10^3$	d
Leads from Sterile Package (OK)	4	253 ± 14	145 ± 13	1.75 ± 0.06
Cracked Leads from Human Explants (SC)	3	253 ± 28	149 ± 16	1.70 ± 0.08

Table III. <u>In Vivo</u> Fixed Strain Test P80A Tubing Extruded Into Very Cold Tank (a).

Implant Time (Weeks)	70x Optical Appearance at Various Fixed Elongations (%)					
	0	100	200	300	400	500
1	O,O	O,O	O,O	O,O	O,O	O,O
2	O,O	O,O	F,O	O,O	F,O	F,O
3	O,O	O,S	F,F	S,F	F,O	F,F
4	S,S	F,VS	F,F	O,F	O,F	F,F
5	O,S	O,O	F,F	F,L	F,L	L,L
6	VS,VS	F,F	F,F	F,F	F,F	F,F
7	S,F	F,S	F,F	F,F	F,F	F,F
8	VS,S	S,VS	F,F	F,F	O,F	F,F
9	S,S	S,S	F,S	F,F	F,F	F,F
10	S,S	F,F	F,F	F,F	F,F	F,F
11	S,S	F,F	F,F	F,F	F,F	F,F
12	S,VS	F,S	F,F	F,S	F,F	F,F

O = No defect
VS = Barely visible defect
S = Severe cracking
F = Cracked to failure
L = Artifact (ligature cut-through)

(a) In this experiment failures occured at 0-100% elongation with a 1-3 week induction period.

Table IV. In Vivo Fixed Strain Test P80A Tubing Extruded Into Hot Tank (a).

Implant Time (Weeks)	70x Optical Appearance of Various Fixed Elongations (%) 0	100	200	300	400	500
1	O	O	O	O	O	H
2	O,O	O,O	O,O	O,O	O,O	H,F
3	O,O	O,O	O,O	O,O	O,O	H,O
4	O,O	O,O	O,O	O,O	O,O	F,O
5	O,O	O,O	O,O	F,O	F,F	H
6	O,O	O,O	O,O	F,F	F,F	H,F
7	O,O	O,O	C,O	F,O	F,F	H,F
8	O,O	O,O	O,O	F,O	F,F	H,H
9	O,O	O,O	O,O	F,F	F	F,F
10	O,O	O,O	O,O	F,F	F,F	F,F
11	O,O	O,O	O,O	F,F	F,F	F,F
12	VS,VS	O,O	O,O	F,F	F,F	F,F

O = No defect
VS = Barely visible defect
F = Cracked to failure
H/C = Artifacts (hole and cut)

(a) Note that a 200-300% elongation appears to be necessary for cracking in this experiment with a 4-5 week induction period.

cracked material have shown no such evidence. Thus oxidized polymer must be present in such thin layers that it is not always detectable by these techniques. Slight oxidation could be either a cause or result of cracking. P55D has been shown to be superior to unannealed P80A in stress cracking resistance, yet it appears to be about equally susceptible to oxidative degradation (15,16). Thus, it does not seem reasonable that autooxidation alone is the source of stress cracking.

If stress cracking is due to absorptive processes (SSC), then why would removal of antioxidant exacerbate the problem? A craze is a reordering of polymer molecules under tension into bundles separated by space (Figure 5). This occurs as absorbed material plasticizes a strained surface, reducing its "yield point" to that of any residual stress. Crazing relieves some stress, but now the smaller diameter craze tie bundles must support larger loads. Molecules break, forming two free radicals per break. Due to tension, the broken molecules' ends retract so that the free radicals cannot recombine. Thus, they "migrate" by disproportionation across other strained molecules until enough tension has been relieved that they can recombine (17). Of course, oxygen is ubiquitous, but in very low partial pressures in the venous system and capillary bed (<<100 mm Hg). The diffusion of oxygen into the polymer is probably relatively slow compared to the rate at which strain generated free radicals can form. The breaking of relatively large numbers of molecules can deplete absorbed oxygen resulting in the formation of hydroperoxides. This could allow further disproportionation to progress

Table V. <u>In Vivo</u> Fixed Strain Test P80A Tubing Extruded Into Ambient Cooling Tank Assembled Specimens Annealed to Provide 100% Stress Relaxation (a).

Implant Time Weeks	70x Optical Appearance of Various Fixed Elongations (%)					
	0	100	200	300	400	500
1	0,0	0,0	0,0	0,0	0,0	0,0
2	0,0	0,0	0,0	0,0	0,0	0,0
3	0,0	0,0	0,0	0,0	0,0	0,0
4	0,0	0,0	0,0	0,0	0,0	0,0
5	0,0	0,0	0,0	0,0	0,0	0,0
6	0,0	0,0	0,0	0,0	0,0	0,0
7	0,VS	0,0	0,0	0,0	0,0	0,0
8	0,0	0,0	0,0	0,0	0,0	0,0
9	0,0	0,0	0,0	0,0	0,0	0,0
10	0,0	0,0	0,0	0,0	0,0	0,0
11	0,0	0,0	0,0	0,0	0,0	0,0
12	VS,VS	0,0	0,0	0,0	0,0	0,0

0 = No defect
VS = Barely visible defect

(a) Note significant improvement in stress cracking resistance compared to Tables 3 and 4.

relatively unincumbered, unless there was a sufficient concentration of appropriate antioxidant present to react with the free radicals.

When a lead is implanted, thrombus forms due to blood flow perturbations as well as material interactions. This resolves into a thin translucent fibrous sheath (as a result of acute inflammatory processes) in about 6 weeks. The sheath itself perturbs flow so that at its very end, microscopic thrombi continue to form and resolve into fibrous tissue, causing it to thicken and grow slowly in length (about 1 cm per year in canines, in our experience). A newly cracked rough surface on a chronic lead is no longer biocompatible in the sense that it must provoke inflammation (resulting in encapsulation or thickening of existing capsule). Normal inflammatory reactions result in the evolution of oxidants, especially hydrogen peroxide to further exacerbate autooxidation (18).

Autooxidative processes depend on permeation of oxygen/oxidant into the bulk (19). It is expected, therefore, that as acute inflammation subsides, oxygen/oxidant concentration should drop to normal, very low physiologic values. Thus, stress generated autooxidation should slow and for all practical purposes stabilize (that is, the rate of degradation should be so slow as to be of no further significance within the expected lifetime of the device). Oxidized polymer would be evident only as a thin layer over cracked P80A and P55D surfaces, not always detectable by IR surface techniques. Other polymers may exhibit greater damage depending on their susceptibility to autooxidation.

According to this hypothesis, oxidation is a result of stress cracking, not necessarily the cause. But the presence or absence of an appropriate antioxidant is critical to either mechanism, terminating or allowing destructive free radicals caused by mechanochemical processes to continue. The strains necessary for these processes come from extrusion/molding, device assembly and implant procedures. Under certain conditions, tubing and molded parts can have skin-core morphologies where skin molecules are in tension and core molecules are in compression (or vice versa), with neutral axes in-between (4,7,8,10). The magnitude of skin-core stresses are not necessarily uniform along a length of tubing since they can vary with small changes in extrusion draw ratio, small temperature fluctuations, variations in flow profile through the die (caused by raw material variations) and even postextrusion processes. Strain induced orientation (SIO) can eliminate neutral axes. Rapidly quenched thin wall tubing may be completely amorphous just postextrusion. Subsequent phase separation almost certainly must result in molecular morphologies different from annealed samples. The concentration of antioxidant is also not necessarily uniform because certain device manufacturing or biologic processes can extract it. The antioxidant can be excluded from crystalline regions, concentrating it in amorphous regions, thus, varying its concentration as molecular morphology varies (19).

Another hypothesis is that of enzymatic stress cracking. Gladhill has demonstrated that in unstrained polymer, a few urethane bonds can be broken by enzymatic catalyzed reactions resulting from inflammation (20). Such bond cleavage is limited, however, and does not appear to progress very extensively. The degree of damage is so low that there would be no affect on the polymer's physical properties or macroscopic biostability. Theoretically, however, strained polymer surfaces would be expected to suffer more extensive enzyme catalyzed damage which could nucleate or initiate craze formation.

In all of the above mechanisms, the propensity for lead insulation to crack will depend upon the polymers molecular morphology (in turn, dependent upon extrusion processes, annealing, etc.), the amount of antioxidant present (dependent upon extraction, morphology, and/or consumption during annealing processes), and device configuration (that is, thicker tubing will be less strained by a given stress, etc.). Excessive skin-core strains can result in cracking to a neutral axis where stress is relieved and cracking stops. Of course, elimination of neutral axes will allow craze/cracks to propagate to failure. On the other hand, complete stress relaxation through proper annealing can completely eliminate cracking in accelerated <u>in vivo</u> tests. Since early lead models varied considerably in these factors, it is not surprising that the incidence of failures varied dramatically from model to model.

CONCLUSIONS

While certain polyether polyurethanes have been shown to be biostable in themselves, they can develop microscopic surface cracks <u>in vivo</u> under certain conditions. The most likely mechanism appears to be absorption of body components by strained polymer, or emzymatic degradation of strained polyurethane linkages, resulting in craze formation. The degradation of crazes into cracks involves the formation of free radicals which, in the absence of sufficient absorbed oxygen or appropriate antioxidant, can propagate by disproportionation. This can initiate autooxidation, even in the low oxygen, <u>in vivo</u> environment. The propagation rate of autooxidation will then depend on polymer composition, subsequent inflammatory process, diffusion of <u>in vivo</u> oxygen and other factors. The depth of cracking (hence, device failure) will depend upon the degree of residual strain in the polymer/device. These issues must be addressed in the development of any implantable device made from polyurethanes (or any other polymer for that matter). The mechanism, once understood, can be controlled as is demonstrated by the excellent long/term performance of most polyurethane insulated cardiovascular and neurological pacing leads. A new series of transvenous leads was introduced in 1983, employing many process improvements based on our understanding of the above factors. As of June 1985, none of the more than 80,000 of these devices implanted worldwide were known to have failed as a result of stress cracking.

REFERENCES

1. J. W. Boretos, Biomed. Mater. Res., 6, 473 (1972).
2. K. Stokes, K. Cobian, Biomaterials, 3, 225-231 (1982).
3. Medtronic product performance report, February 1985 (842589a).
4. K. B. Stokes, W. A. Frazer and E. A. Carter, Proceedings, ANTEC 84, 1073 (1984).
5. J. E. Upton, in "Proceedings of the VI World Symposium on Cardiac Pacing", C. Meere, ed., Pacesymp, Montreal, 1979, pp 29-36.
6. M. W. Davis and K. B. Stokes, 11th Ann. Meeting of Soc. Biomater., San Diego, April, 1985 p xxx.
7. K. Stokes, Stimucoeur Medical, 10, 205-215 (1982).
8. C. L. Byrd, W. McAuthur and K. Stokes, Pace., 6, 868 (1983).
9. F. Beyersdorf and J. Kreuzer, International Symposium on Pacemaker Leads, Leuven, Belgium, 1984, p 78.
10. K. Stokes, in "Modern Cardiac Pacing", S.S. Barold, ed., Futura, Mount Kisco, 1985, pp 172-197.
11. G. C. Timmis, D. C. Westveer and R. O. Martin, Pace, 6, 845-857 (1983).
12. K. B. Stokes, W. A. Berthelson and M. W. Davis, "Metal Catalyzed Oxidative Degradation of Implanted Polyurethane Devices", Proceedings of the ACS Division of Polymeric Materials: Science and Engineering, 53, 5-10 (1985).
13. N. Bikales, "Mechanical Properties of Polymers", Wiley-Interscience, New York, 1971, pp 73-103.
14. R. E. Baier, A. E. Meyer and J. R. Natiella, J Biomed. Mater. Res., 18, 337-357 (1984).
15. A. J. Coury, P. T. Cahalan and P. C. Slaikeu, 11th Ann. Socv. Biomater., San Diego, 1985, 134.
16. A. J. Coury, P. T. Cahalan and E. L. Schultz, 10th Ann. Soc. Biomater., Washington, D>C> 1984, 254.
17. A. Casale and R. S. Porter, "Polymer Stress RTeactions", Acadamic Press, New York, 1978.
18. J. M. Anderson and K. M. Miller, Biomaterials, 5, 5-10 (1984).
19. W. L. Hawkins, "Polymer Stabilization", Wiley-Interscience, New York, 1972.
20. K. W. Gladhill, "In vitro studies of biodegradation of Polyurethanes in Enzymatic Solutions", Masters Thesis, Department of Chemical Engineering, 1985.

METAL CATALYZED OXIDATIVE DEGRADATION OF IMPLANTED POLYURETHANE DEVICES

K. B. Stokes, W. A. Berthelson and M. W. Davis

Medtronic, Inc.
Fridley, MN 55432

While most early models of implanted polyether polyurethane insulated cardiac pacing leads have demonstrated excellent long-term performance, a few have experienced a relatively high rate of failure. In these cases, soft or brittle cracked polymer exhibits evidence of polyether soft segment oxidation. These reactions are catalyzed by corrosion products from internal silver-MP35N composite wire conductor coils. It is believed that unacceptable corrosion rates occur secondary to implant damage and/or leakage of body fluids into the device. It is also believed that minute quantities of hydrogen peroxide released as part of normal inflammatory reactions can permeate very thin insulation, seep through loose joints, etc. Hydrogen peroxide decomposes on transition metal surfaces to form free radicals which can initiate and propagate autooxidative chain reactions. Metal catalyzed oxidation can be controlled by assuring that conductor corrosion rates and insulation/device permeability to biologic oxidants are minimized. A second generation of polyurethane insulated leads has been developed using principals learned from prior experience. Their long-term biostability has been excellent.

INTRODUCTION

Certain polyether polyurethane elastomers have been proven to be biostable in long-term animal implants (1,2). Neurological and cardiac pacing leads insulated with polyether polyurethanes have been implanted in humans since 1975 and 1977, respectively. Most models have demonstrated excellent long-term performance but a few have not (3). Two mechanisms have been discovered that can cause failure of polyurethane insulation; a form of stress cracking and metal catalyzed oxidation. This report will summarize the results of our investigation into cardiovascular lead insulation failures due to oxidation only, since stress cracking has been discussed separately (4).

MATERIALS

Cardiac pacing leads are insulated conductor coils with one or more electrodes and a terminal for connection to an implanted pulse generator. Insulation tubing is extruded from segmented polyether polyurethanes composed of polytetramethylene ether oxide (PTMO) soft segments and methylene di(isocyanatobenzene)/butanediol hard segments. Several different hardness versions of the same polymer are used in the large majority of polyurethane insulated pacing leads; primarily Pellethanes 2363-80A (P80A) and 2363-55D (P55D) from Upjohn, CPR Division. This chapter will concentrate on four early P80A insulated devices manufactured by Medtronic, Inc. Models 6972 and 6990U are coaxial bipolar leads (two conductors and electrodes) while Models 6971 and 6957 are unipolar (one conductor and electrode). All four models have Pt-Ir alloy electrodes, passivated 304 stainless steel terminal pins, and silicone rubber terminal sealing assemblies. Models 6957 and 6990U have MP35N (Ni,Co,Cr,Mo) conductor coils. Models 6972 and 6971 have conductor coils made from a silver-MP35N composite wire called DBS. Six MP35N and one silver wire are drawn through a die and brazed. The finished wire has 6 MP35N elements separated by silver, surrounding a silver core. Silver also coats the outer wire surface (see Figure 1).

METHODS

Some leads explanted from humans are returned to the manufacturer for analysis. Not all chronic leads can be removed and not all explanted leads are returned to the manufacturer. Thus, data from Medtronic's CardioCare telephone monitoring service (covering over 18,300 leads) have been used to develop clinical actuarial survival statistics (3).

Returned leads are checked electrically, then examined by light microscopy at 30-70X for defects. Samples for scanning electron microscopy (SEM) are mounted on aluminum stubs with conductive adhesive. After application of a thin layer of gold, the specimens are studied on an ISI-40 SEM with a Princeton-Gamma Tec electron dispersive analysis by x-ray (EDAX) attachment. Films for infrared spectroscopy are cast on salt plates from tetrahydrofuran solution. A Perkin-Elmer Model 398 grating instrument is used. Some polymer solutions are filtered to remove any particulates, then are evaporated to dryness. These samples are ashed or extracted with nitric acid for analysis by atomic absorption using a Perkin-Elmer Model 4000 spectrophotometer.

Similar analyses have been done on leads removed from canines after implant periods ranging from 12 weeks to 3.5 years and on leads tested _in vitro_. Results from one such _in vitro_ test will be briefly mentioned. Whole leads were immersed at 37°C, in 3% hydrogen peroxide contained in polyethylene bags. The peroxide was changed three times weekly. The leads were examined about weekly at 30-70X. At the end of the experiment, leads were analyzed as above.

RESULTS

As of January 1985, Medtronic had sold over 194,000 P80A insulated cardiac pacing leads in the United States alone, representing 14 models. A total of 545 with insulation failure have been explanted and returned to the manufacturer as shown in Table I. However, 440 of these (81%) represented one model (6972). An estimated 50% of these (0.51% of sales) failed by metal catalyzed oxidation, the remainder by stress cracking. Only four model 6990U leads failed due to oxidative processes (0.07% of sales). Model 6971 has experienced only one known similar oxidative failure (0.0018% of sales). Model 6957 has experienced no insulation failures whatsoever due to either mechanism. Clearly, therefore, there is

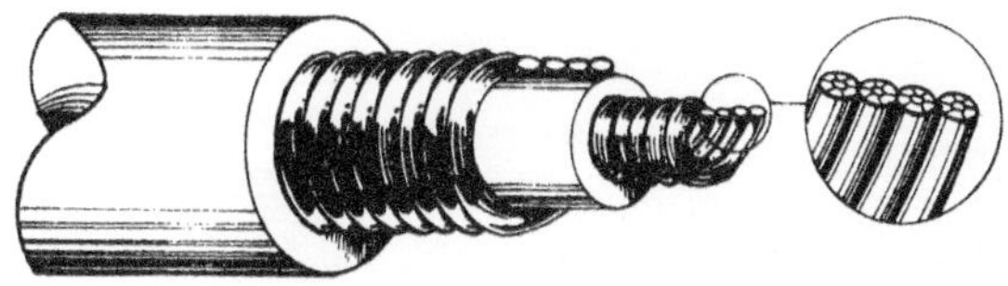

Figure 1. Section of a coaxial lead conductor assembly using DBS (drawn-brazed-strand) wire. The wire is made by drawing one strand of silver and six strands of MP35N alloy through a die while brazing. The resulting composite wire has the desirable low electrical resistance of silver and excellent MP35N derived mechanical properties.

a model dependency which shows that one lead has a > 99% survival after four years implant, while another made from exactly the same materials has a < 90% survival in the same time frame.

Infrared spectra of degraded material typically shows a loss of PTMO ether at 110 cm^{-1}, relative to the 1220 cm^{-1} amide and 1080 cm^{-1} urethane ether bands. A loss of PTMO CH_2 (sym. bend) occurs at 2860 cm^{-1} relative to the sym. bend CH_2 peak at about 2940 cm^{-1}. In extremely degraded cases, an unbonded OH peak is found at 3650 cm^{-1}. It is difficult to see much change in the 1690-1750 cm^{-1} carbonyl bands, although there is typically an increase in an ester band at 768 cm^{-1}.

Visually, oxidative degradation is manifested as either discolored (amber) hard brittle, or soft gummy, cracked polymer inside a lead. The extent of physical damage ranges from none at all to complete loss of inner insulation over a length of 5-10 cm. Cracks on inside surfaces are always smooth-faced, brittle fractures as opposed to stress cracks (on outer surfaces) which are always crazed (Figure 2). Inner insulation

Table I. Experience With Medtronic Polyurethane Insulated Leads (U.S. Only) as of January 1985 (3)

Model No.	Insulation Thickness (mils)	Conductor Material	Number of U.S. Implants	Returned Leads* With Insulation Failure N	%	Actuarial** Survival (% at X Mo.)
6957/57J	23	MP35N	24,900	0	0	99.7/24
6971	10	DBS	56,700	24	0.04	99.3/48
6990U	5	MP35N	5,660	36	0.6	---
6972	5	DBS	43,300	440	1.0	89.8/48
All Others	--	---	63,800	45	0.07	---
			194,000	545	0.28	

* Based on analysis of returned products, includes both SC and MO.
** Based on telephone monitoring of pacemaker patients.

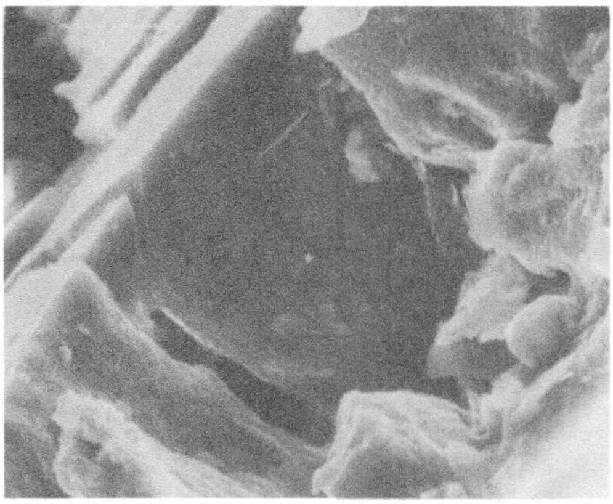 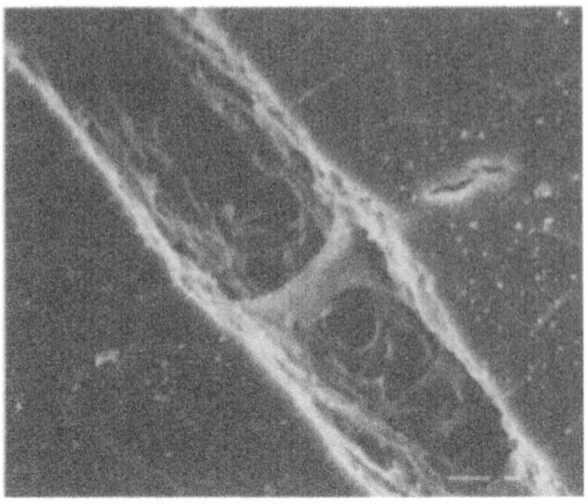

Figure 2. Right: A stress crack still retaining some craze tie bundles on the outer surface of an explanted polyurethane lead, 700x. Left: The fracture faces of cracks inside an explanted lead due to metal catalyzed oxidation at 830x. (The lumen surface is in the upper left hand corner).

cracks usually, but not always, have a random mud-cracked appearence. Usually the cracks appear to initiate at points of implant damage or are associated with metal parts (Figure 3). The pattern of failure with regard to time seems random. Some leads show no sign of cracking after 3-4 years implant, while others do in as little as 6 months. Almost all explanted devices with polymer degradation appear to have suffered some form of leakage such as lost bonds, implant damage, stress cracking or connector leakage.

Frequently, a grey to black discoloration associated with cracked polymer is observed on DBS conductor coils (not on MP35N coils). At

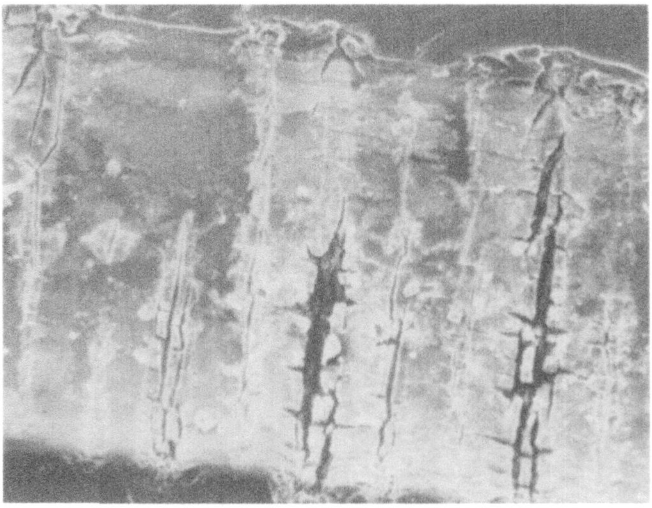

Figure 3. The lumen of polyurethane tubing from a lead explant. Note the extensive random cracks at the places where metallic conductor coils touch the polymer, 55x.

higher magnifications these blackened areas resolve into approximately cubic crystals (Figure 4). EDAX shows that these crystals are composed of Ag/AgCl where Cl concentration ranges from below detectable limits to large quantities consistent with Ag/Cl per se. Silver also commonly delaminates from the DBS surfaces, progressing to a splitting of the wire into its separate MP35N elements. In extreme cases, the silver is completely gone (Figure 5). Nickel crystals (tightly adherent to insulation surfaces) have also been found (Figure 6). Insulation lumen surfaces in and near cracks (containing no crystals) typically contain Ni and Fe, although Co, Cr, Mo, and Aq are also occasionally present (detection limit is about 0.1%). While EDAX detects elements within a few microns of the surface, atomic absorption measures bulk concentrations. As is shown in Table II, certain metals (Ni and Fe in particular) appear to permeate insulation in small amounts even when the device is stored on the shelf. When implanted, however, the migration of certain metals into the insulation increases substantially. The DBS lead data show variable, but high concentrations of transition metals in degraded polymer. The concentration of metals are relatively high even in insulation not yet exhibiting outward signs of degradation. The MP35N lead data show relatively consistent concentrations of Ni, Co, Cr, Mo and Ag, but increasing concentrations of Fe with implant time.

Numerous 37°C in vitro test environments have been used in unsuccessful attempts to duplicate the above phenomena. Some examples include synthetic plasma, phospholipid emulsion, glycolic acid solution, saline, blood (changed several times weekly), albumen solution (also changed frequently), solutions of various metal salts, etc. Similar phenomena have as yet been duplicated only in leads immersed in 3% hydrogen peroxide at 37°C. In these tests, cracks began to be visible in Model 6972 leads at the conductor coil/tubing contact points in as early as 30 days.

Table II. Metals in P80A Lead Tubing (Typical Examples)

Model No.	Conductor Metal	Tubing Condition	Implant Time (Mo.)	Metal Concentration (ppm) in tubing Ni	Co	Cr	Mo	Fe	Ag
New Tubing		OK	0	0.3	0.4	0.6	--	1	0.1
6972	DBS	OK	0*	3	0	0	--	8	4
6972	DBS	OK	23	12	2	0	--	83	1,600
6972	DBS	Brittle	29	280	320	0	--	79	16,000
6972	DBS	OK	39	7	0.2	0	--	15	180
6972	DBS	Gummy	?	9	1	8	1	620	66
6990U	MP35N	OK	0*	5	0.1	0	--	29	1
6990U	MP35N	OK	14	0.4	5	2	1	11	0.1
6990U	MP35N	OK	25	3	2	1	1	19	0.1
6990U	MP35N	OK	36	7	9	2	1	84	0.6

* Two years shelf storage in sterile packages.

Figure 4. Crystals of silver on DBS conductor coils adjacent to a small hole between coaxial conductors. Note the rough corroded appearence of the underlying silver surface, 1000x.

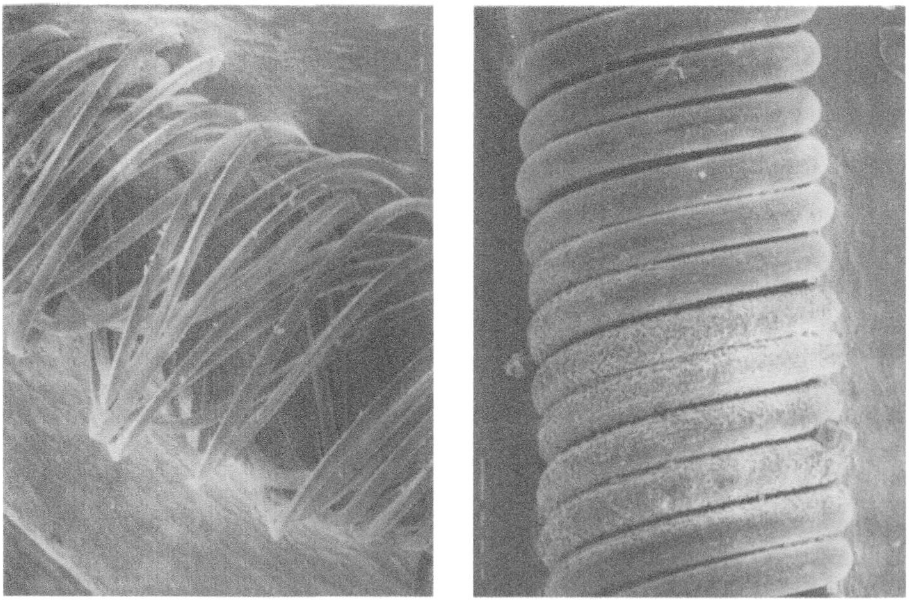

Figure 5. Left: Completely delaminated DBS wire associated with severe degradation of polyurethane. There is no silver left, 70x. Coils should look more like those on the right, also at 70x. Some silver crystals are visible in the right hand micrograph, similar to those in Figure 4.

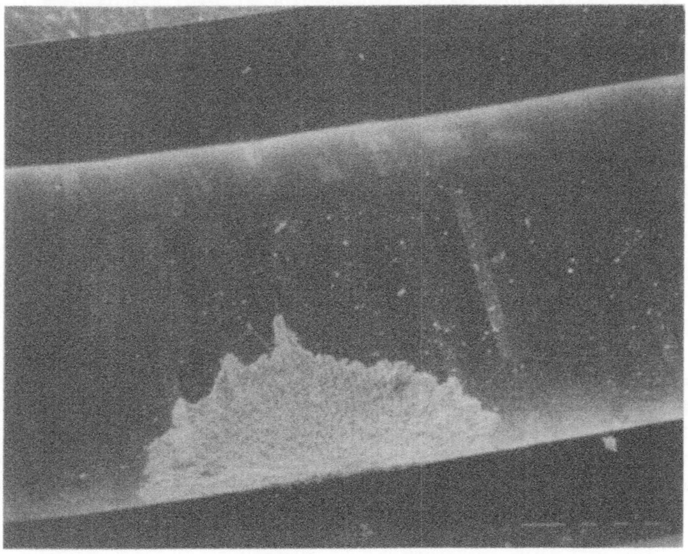

Figure 6. Longitudinal section of lead insulation, lumen view. Nickel crystals (identified by EDAX) adhered to the surface. The conductor was DBS, 70x.

Cracking was severe in 60 days. No such cracking was seen in leads containing MP35N conductors. Infrared spectra, optical and electron microscopy, EDAX, atomic absorption, etc. gave exactly the same results as were obtained with returned human and canine explants.

DISCUSSION

The general mechanisms for polymer autooxidation are well known (5,6). These are complex series of chain reactions which are usually categorized in 3 steps; initiation, propagation (sometimes with kinetic chain branching) and termination. Initiation occurs either through hydrogen abstraction or chain scission to form free radicals (Equation 1, where R is polymer).

$$RH \longrightarrow R\cdot + H\cdot \qquad \text{(Equation 1)}$$

Significant initiation can occur during secondary processes (such as extrusion or injection molding) where heat is combined with shear. Initiation is most likely to occur at the weakest bond in the polymer (5). In this case, the ether linkage per se and/or the H-C bond alpha to the ether oxygen appears to be the most probable site. Oxygen is ubiquitous in the environment (even within an extruder). The combination of polymer free radicals with oxygen is so fast that the concentration of peroxy free radicals is >> polymer free radicals. Thus, hydroperoxides are ubiquitous in polymers usually at concentrations below the detection limits of most analytical techniques.

$$R\cdot + O_2 \longrightarrow ROO\cdot \qquad \text{(Equation 2)}$$

$$ROO\cdot + RH \longrightarrow ROOH + R\cdot \qquad \text{(Equation 3)}$$

Propagation occurs due to decomposition of hydroperoxides into reactive free radicals.

$$POOH \dashrightarrow P\cdot + \cdot OOH \text{ or } PO\cdot + \cdot OH \text{ or } POO\cdot + \cdot H \qquad \text{(Equation 4)}$$

Catalysts for this step include ultraviolet light, heat and transition metals. Chain branching increases the kenetic chain length prior to termination. That is, the reaction can take many paths depending upon the rate of reaction, catalysts, interactions between free radicals, etc. This helps to explain why degraded lead insulation is sometimes gummy, sometimes brittle. Termination reactions are those that result in inactive products.

Generally speaking, autooxidation reactions are expected to occur at relatively slow rates in vivo, especially given the lack of U. V. light and excessive heat. The partial pressure of oxygen in the in vivo environment experience by pacing leads is relatively low (usually << 100 mmHg). Why then the relatively rapid degradation in some leads and not in others? Based on analysis of explanted leads, metals from the device itself appear to be implicated in this relatively rapid degradation. Coury et al., has established that all polyurethanes tested containing ether functionality are susceptible to metal catalyzed oxidation in accelerated (90°C) in vitro tests (7,8).

The normal events following implantation of a biomedical device include acute and chronic inflammatory processes. These processes are responsible for the formation of fibrotic encapsulation. Such biological reactions involve phagocytic cell activity which can result in the generation of enzymes and hydrogen peroxide (9). It is difficult to envision enzymes permeating very far into intact solid polyurethane insulation, but moisture does and hydrogen peroxide can (especially if insulation is very thin). When hydrogen peroxide contacts certain transition metal surfaces it decomposes through free radical intermediates (10) (Equation 5-9).

$$M^{n+} + H_2O_2 \dashrightarrow M^{(n+1)+} + HO^- - HO\cdot \qquad \text{(Equation 5)}$$

$$M^{n+} + HO\cdot \dashrightarrow M^{(n+1)+} + HO^- \qquad \text{(Equation 6)}$$

$$HO\cdot + H_2O_2 \dashrightarrow H_2O + HOO\cdot \qquad \text{(Equation 7)}$$

$$HOO\cdot + H_2O_2 \dashrightarrow H_2O + O_2 + HO\cdot \qquad \text{(Equation 8)}$$

$$HOO\cdot \dashrightarrow O_2^{\dot{-}} + H^+ \qquad \text{(Equation 9)}$$

Where the polymer and metal are in intimate contact, these free radicals can cause initiation as well as propagation of polymer autooxidation. Peroxide permeation may virtually cease within months of implant as acute inflammation subsides, but it may have been sufficient to get self sustaining chain reactions going.

The corrosion of metallic parts inside pacing leads has generally been thought to occur at very slow (acceptable) rates. In the presence of traces of chloride ion and hydrogen peroxide, however, corrosion may occur at faster rates than previously thought (11). Transition metal corrosion products such as Mo and Co are capable of catalyzing both initiation and propagation steps of polymer autooxidation (Equations 10-12).

$M^{(n+1)+} + RH \dashrightarrow M^{n+} + R\cdot + H^+$ (Equation 10)

$M^{n+} + ROOH \dashrightarrow M^{(n+1)+} + RO\cdot + OH^-$ (Equation 11)

$M^{(n+1)+} + ROOH \Longleftrightarrow M^{n+} + ROO\cdot + H^+$ (Equation 12)

In addition, the termination reactions from Equations 11 and 12 produce oxygen which is necessary to sustain autooxidation (Equations 13 and 14).

$2\ ROO\cdot \dashrightarrow 2RO\cdot + O_2$ (Equation 13)

$2\ ROO\cdot \dashrightarrow ROOR + O_2$ (Equation 14)

Oxygen, whether from this source or from the surrounding environment can form destructive free radicals to initiate and propagate oxidation (Equations 15-17) (10).

$M^{n+} + O_2 \dashrightarrow M^{(n+1)+} + O_2^{\dot{-}}$ (Equation 15)

$O_2^{\dot{-}} + H^+ \dashrightarrow HOO\cdot$ (Equation 16)

$O_2^{\dot{-}} + RH \dashrightarrow RO\cdot + HO^-$ (Equation 17)

In an intact sealed lead, the above reactions depend strongly on metallic corrosion rates and the permeability of reactants/catalysts into the polymer. The permeability of the polymer can be variable, depending on its molecular morphology. The appearent randomness of insulation failure with respect to time may be due simply to variability in these factors.

If the lead's inner insulation is breached, conductor corrosion can be greatly accelerated. For example, a thin stylet wire is passed (temporarily) through the leads conductor coil lumen to aid in implant procedures. If stylet perforation of inner (coaxial) insulation occurs, a very small hole will result which can constitute an electrical leak path. Corrosion of DBS silver or damaged stainless steel surfaces will then be accelerated by the pacing pulse. Moisture does permeate polyurethane insulation within weeks. But the aqueous solubility of silver is very low. The pacing pulse is basically a pulsed DC waveform, usually biphasic but not equally so. During the anodic phase, corrosion can occur and silver ion will drift. During the cathodic phase the metal ion will reverse direction raising its concentration above its solubility limit, forcing its precipitation and the formation of miroscopic (black) crystals.

Breaches of outer insulation do not necessarily affect pacing performance, and thus may go unnoticed. But protein and electrolyte containing body fluids can enter at rates proportional to the size of the defect to form soluble metal complexes. We have found, for example, that the solubility of silver chloride in blood plasma is 1,000 times greater than it is in distilled water. Body proteins and electrolytes can also accelerate the corrosion of metallic parts, especially Ag, in the presence of an electrical leak path. Any allied corrosion of the substrate MP35N is similarly accelerated.

In addition to body fluid components, minute amounts of low molecular weight polymer, additives, contaminants and oxidative degradation products can act as ligands to form complexes with metallic corrosion products. Metal complexes can be even stronger oxidents than their uncomplexed forms (12). If these complexes are compatible with the polymer, they can migrate into the bulk. Thus, especially in the presence of an insulation leak, MO reactions can proceed at relatively rapid rates, allowing damage to proceed to clinically significant failures in months rather than decades.

Why have various models of leads demonstrated such a wide range of performance? Model 6972 leads had thin insulation (0.13 mm thick) which was relatively easily breached by stress cracking, scalpel cuts, stylets, etc. Some early versions also had problems with joint integrity which allowed body fluid leakage. Corrosion of DBS silver under these circumstances resulted in a high degree of device failure. Model 6990U has the same thin insulation, but no significant joint integrity problems and a substantially heavier conductor coil that better resisted stylet perforation. Lacking DBS wire, it is believed that the four known oxidative failures were related to possible insulation breaches and corrosion of stainless steel parts. Model 6971, while containing DBS wire, has thicker insulation (0.25 mm) which much more effectively resists stylet damage, stress cracking, cuts, etc. Thus, the excellent performance of this device demonstrates the very large differences in degradation rates due to diffusion controlled processes alone, versus leads with a higher probability of insulation leaks. Model 6957 has much thicker insulation (0.5 mm), no DBS wire, and only one stainless steel part (the terminal pin sealed in the generator connector).

Thus, while polyether polyurethanes can be susceptible to oxidative degradation reactions under certain conditions, implantable devices can be designed and manufactured by processes that control or prevent the problem. The immediate means of controlling oxidation include leak resistant designs, thicker, more stress cracking resistant insulation, and control of transition metal corrosion (preferably elimination of suspect metals). In addition, maximizing crystalline phase separation by annealing should help reduce permeability (13,14). A new series of four transvenous polyurethane leads that made use of these principles was introduced in 1983. So far with over 39,000 leads implanted in the United States (over 80,000 worldwide as of June 1985) there has been no evidence of metal catalyzed oxidation and not one single failure.

REFERENCES

1. J. W. Boretos, J. Biomed. Mater. Res., 6, 473 (1972).
2. K. Stokes and K. Cobian, Biomaterials, 3, 225-231 (1982).
3. Metronic product performances report, February 1985 (842589a).
4. K. B. Stokes, W. A. Frazer and E. A. Carter, Proceedings, ANTEC 84, 1073 (1984).
5. B. Ranby and J. E. Rabek, "Photodegradation, Photooxidation and Photostabilization of Polymers", Wiley, London, 1975.

6. J. R. Shelton, in: "Polymer Stabilization", W. L. Hawkins, ed., Wiley-Interscience, New York, 1972, pp 29-117.
7. A. J. Coury, P. T. Cahalan and P. C. Slaikeu, 11th. Ann. Soc. Biomater., San Diego, 1985, p 134.
8. A. J. Coury, P. T. Cahalan and E. L. Schultz, 10th Ann Soc. Biomater., Washington DC, 1984, 252.
9. J. M. Anderson and K. M. Miller, Biomaterials, 5, 5-10 (1984).
10. J. F. Rabek, in: "Comprehensive Chemical Kinetics", C. H. Bamford and C. F. Tipper, eds., Elsevier, Oxford, 1974, pp14,265.
11. M. G. Fontana and N. D. Greene, "Corrosion Engineering", McGraw-Hill, New York, 1978, pp 18-22 and 54.
12. G. Scott, "Atmospheric Oxidation and Antioxidants", Elsevier, Amsterdam, 1965.
13. W. L. Hawkins, in: "Polymer Stabilization", W. L. Hawkins, ed., Wiley-Interscience, New York, 1972, Chapter 1.
14. B. D. Gesner, in: "Polymer Stabilization", W. L. Hawkins, ed., Wiley-Interscience, New York, 1972, pp 353-376.

A COLLAGEN-DACRON COMPOSITE VASCULAR GRAFT FOR ARTERIAL RECONSTRUCTIONS

Shu-Tung Li

Meadox Medicals, Inc.
Oakland, N.J. 07436

The advancement of medical technology in vascular reconstructions has prompted the need for the development of a non-porous Dacron graft which does not require intra-operative preclotting while not compromising the healing characteristics of the porous Dacron substrate. We have developed such a Dacron prosthesis that can be implanted without pretreatment. This has been achieved through the use of a bioresorbable bovine skin collagen as an impregnating and coating material for a Collagen/Dacron composite graft fabrication. *In vitro* studies were conducted to ensure that the collagen was strongly attached to the graft and that the extent of cross-linking was maintained for optimal *in vivo* resorption. Canine implantation demonstrated that the Collagen/Dacron composite graft was safe and highly effective for promoting graft healing. Thus the composite graft has the following features: 1) sustain long term *in vivo* stability; 2) does not require preclotting and reduces blood loss during implantation; 3) enhances post-operative healing and 4) does not induce adverse side effects. A careful consideration of the physical, chemical and biological properties of collagen suggests that an ideal Collagen/Dacron graft can be designed to provide surface and mechanical properties that are suitable for small vessel replacement.

INTRODUCTION

Poly(ethylene terephthalate), commonly known as Dacron, whose repeat unit, as shown below, has been widely accepted as a biomaterial for arterial vessel reconstruction, mainly due to ease of fabrication, high mechanical strength and long term *in vivo* stability. Various textile engineering techniques have been applied to construct the prosthesis with different porosity or surface textures in order to meet particular surgical needs. Due to the porous nature of the construction it has been necessary to preclot the prosthesis before implantation in order to prevent blood loss. The preclotting procedure involves the formation of host blood clot on the surface and in the interstitial spaces for an effective seal of the pores of the vessel prosthesis. Although this has become a routine intra-operative procedure, the effectiveness of preclotting depends on the blood chemistry of the patient, the pore size of the graft, and the time period between preclotting and graft implantation. For patients with blood disorders, the administration of a clotting factor is needed during the preclotting. Additionally, the risk of blood loss increases in thoracic aortic surgery. It has been recommended to use the specially designed low porosity woven grafts for these procedures even at the sacrifice of some healing characteristics.

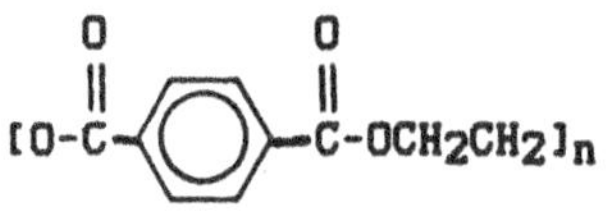

This and other reasons have prompted the need for the development of a vascular prosthesis that has the following features: 1) sustained long term *in vivo* stability; 2) does not require preclotting and reduces blood loss during implantation; 3) enhances post-operative healing and; 4) does not induce adverse side effects. A composite utilizing the existing Dacron prosthesis and a suitable biodegradable material has the potential of fulfilling these requirements. A number of approaches have been recently tried. These include the use of gelatin, soluble collagen, and albumin as the resorbable surface coating material on the Dacron graft. However, none of these materials has proven sufficiently effective as far as the above requirements are concerned. We have, in the past several years, employed purified, insoluble, bovine skin collagen as a biodegradable polymer for a Collagen/Dacron composite graft fabrication. Insoluble type I collagen has a number of properties that are most suitable for this application. The material is easily accessible in large quantities from a number of animal tissues in highly purified forms. The chemistry and biology of type I collagen have been intensively investigated in the past several decades and a great deal of information has been accumulated (1-3). A Collagen/Dacron composite graft that has the features described above is presented and discussed.

MATERIALS AND METHODS

Collagen Material

Purified, insoluble, bovine skin type I collagen was supplied by Bioproducts, Inc., Tucson, Az. The material was well characterized before being shipped to the manufacturer for graft production. The purity of the collagen was studied by an analysis of the amino acid composition from a Beckman autoanalyzer, a determination of the total hydroxyproline content by the method of Stegemann (4), the total hexosamine content by the method of Elson and Morgen (5), and the trace metal content by an atomic absorption spectrophotometer. The extent of molecular aggregation was estimated by differential centrifugation. Collagen was suspended in a physiological saline solution pH 7.4, and centrifuged at different velocities. The supernatant and the precipitate were analyzed for hydroxyproline content.

The thermal stability of the collagen was determined by optical means. Collagen was first equilibrated with phosphate buffered saline solution, pH 7.4, and subsequently inserted into a capillary tube. The sample was then placed into a heating chamber equipped with a polarized microscope at a heating rate of 2°C/minute. The shrinkage temperature was defined as that temperature at which the majority of collagen was transformed from a polar anisotropic fibrillar state to a random coiled isotropic state. Toxicity and pyrogenicity tests were performed according to the standard USP test methods at the established testing laboratories.

The extent of swelling of the collagen material at pH 4 was measured by equilibrating a known amount of collagen with a known volume of an aqueous solution in a centrifuge tube. After a period of equilibration, the collagen was centrifuged at 770 g for 10 minutes. The amount of solution associated with the collagen matrix was operationally defined as the swelling ability of a given collagen matrix. The fibrillar state of the material was also characterized by transmission electron microscopy.

Dacron Vascular Prosthesis

Microvel® (Meadox Medicals, Inc., Oakland, N.J.) was used for the composite graft fabrication. Microvel is a knitted Dacron vascular graft with a velour surface on both sides and a porosity specifically designed to enhance graft healing. It has been proven clinically safe and highly effective for promoting tissue ingrowth and healing. In its present form, intra-operative preclotting is a necessary requirement.

Collagen/Dacron Composite Vascular Prosthesis (Microvel® with Hemashield™)

Collagen was processed to provide the proper concentration and uniform dispersion prior to the vessel prosthesis application. The interstitial pores were impregnated repeatedly with the processed material. The lumen and adventitia surfaces were also coated uniformly by a controlled mechanical means. The extent of collagen cross-linking was carefully controlled by a cross-linking agent for optimal in vivo resorption. Surface and morphological analyses indicated an essentially non-porous structure for the graft (Figures 1-4).

In Vitro Studies

The safety and effectiveness of the Collagen/Dacron grafts were tested by a determination of the water porosity, the shrinkage temperature of the collagen, and the mechanical stability of the collagen under pulsatile pumping conditions at physiological pressure (120 mmHg). The water porosity was measured as the amount of water leak per unit area and time under physiological pressure. The shrinkage temperature of the collagen was determined by the optical method described above. In this case, the grafts were first equilibrated with a phosphate buffered saline solution. Subsequently, the collagen fibrils were scraped from the graft for shrinkage temperature measurement. In the mechanical stability studies of collagen, the composite grafts were attached to a loop with an in-line filter. A physiological solution was circulated through the grafts under pulsatile pumping at physiological pressure for various time periods. The filter papers were collected and assayed for the collagen content.

Figure 1. The appearance of a knitted, double velour Dacron vascular prosthesis.

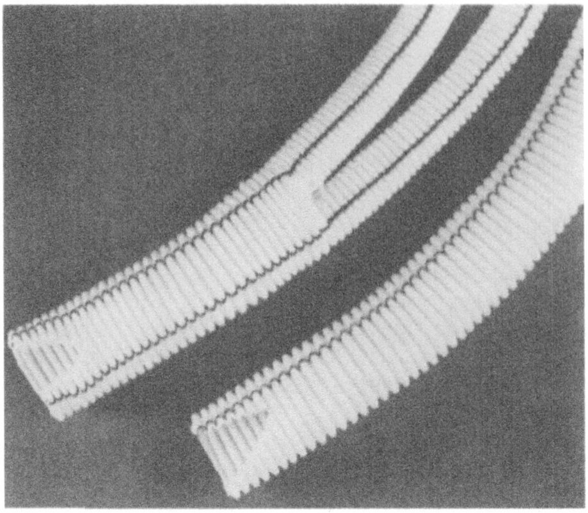

Figure 2. The appearance of a Collagen/Dacron composite vascular prosthesis.

In Vivo Canine Studies

Canine studies were conducted at several research centers aimed to test the performance of the composite grafts in vivo. In the first study, 6 cm x 6 mm Collagen/Dacron composite grafts were implanted in the abdominal aorta of canines by an end-to-end anastomosis and in the iliac artery by an end-to-side anastomosis. At various time intervals grafts were explanted for surface and healing characterizations. Non-impregnated Microvel grafts served as controls.

In the second canine study, 30 cm x 8 mm Collagen/Dacron grafts were implanted in the decending thoracic aorta in a loop configuration, employing an end-to-end anastomosis. Non-impregnated Microvel grafts were used as controls. The objective of this study was to investigate, in a greater detail, the effect of collagen on post-operative graft healing. Specifically, the extent and uniformity of endothelialization and the extent of neovascularization were examined. The 30 cm graft implantation eliminated the complication of data interpertation due to pannus ingrowth from the anastomotic sites. The grafts were harvested at different time intervals and examined for surface morphology and extent of neovascularization. The surface was examined by light and scanning electron microscopes. Histology was examined by standard techniques. Tissue slices were stained with either Masson trichrome or hemotoxylin and eosin. The grafts were cut into 3 cm long sections and each section was further sliced along the longitudinal or flow direction, allowing a histological examination and an estimation of the extent of microvessel formation over the whole length of the graft.

RESULTS

The amino acid composition of the purified insoluble bovine skin collagen was consistant with that observed by other investigators (Table I). The low content of tyrosine, the absence of cystine residues, and close to 33% of glycine indicated a minimum contamination of non-collagenous proteins in the preparation. A minimum 95% of the dry weight of

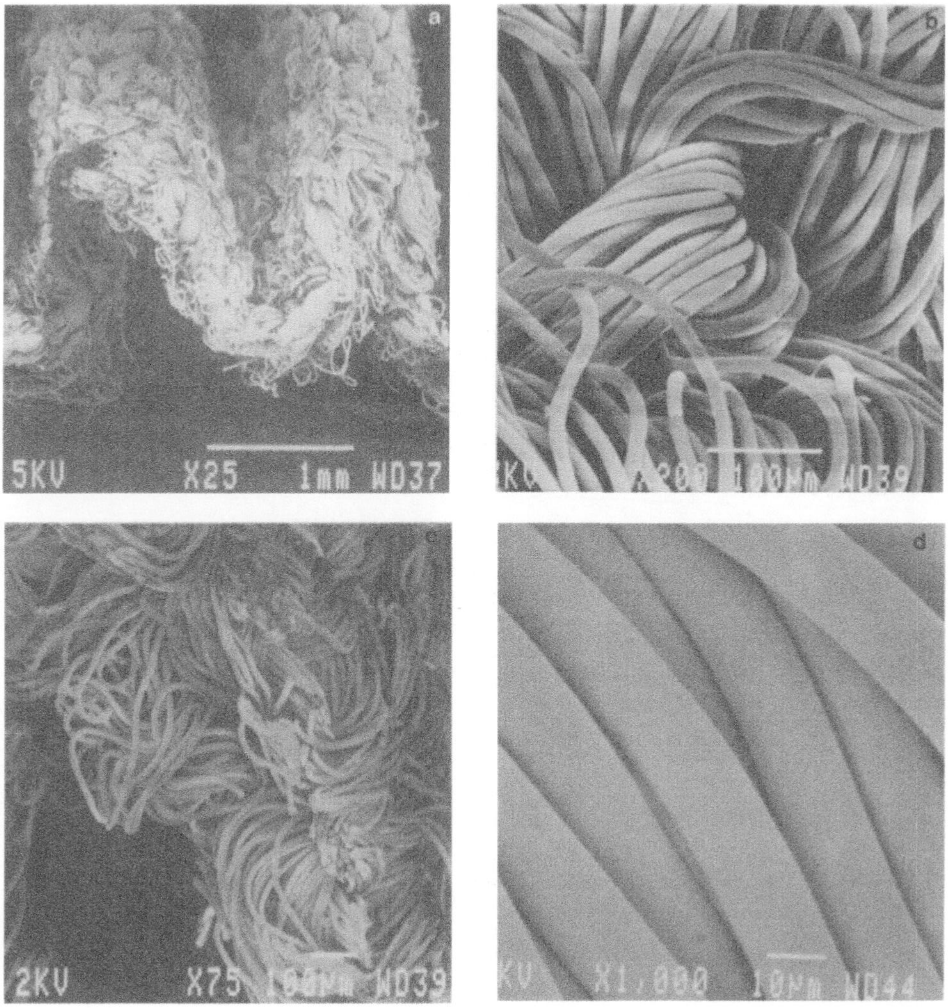

Figure 3. Scanning elecron micrographs of a knitted, double velour Dacron vascular prosthesis: (a) cross-sectional view at 25x; (b) cross-sectional view at 75x; (c) surface view at 200x; (d) surface view at 1000x.

the purified material could be accounted for as collagen by the hydroxyproline determinations. No hexosamine was detectable in the preparation, indicating a lack of glycoprotein and glycosaminoglycan moieties. The purified collagen contained no toxic metals, such as Cd, Hg, As and Pb, in the preparation. Additional biological tests showed that the purified collagen did not contain any pathogens and toxins. The insoluble material remained reasonably intact after an extensive purification procedure, as seen by an electron micrograph, showing the typical banding pattern observed for the native fibrillar structure (Figure 5). A broad temperature of shrinkage was observed for the material, suggesting that the collagen fibrils were heterogeneous in nature. The temperature of transition between the onset of change of the optical property to the completion of change usually spanned a range of 5-10°C. Differential centrifugation studies showed that most of the collagen existed in the form of insoluble higher aggregates. Only approximately 1% of the collagen re-

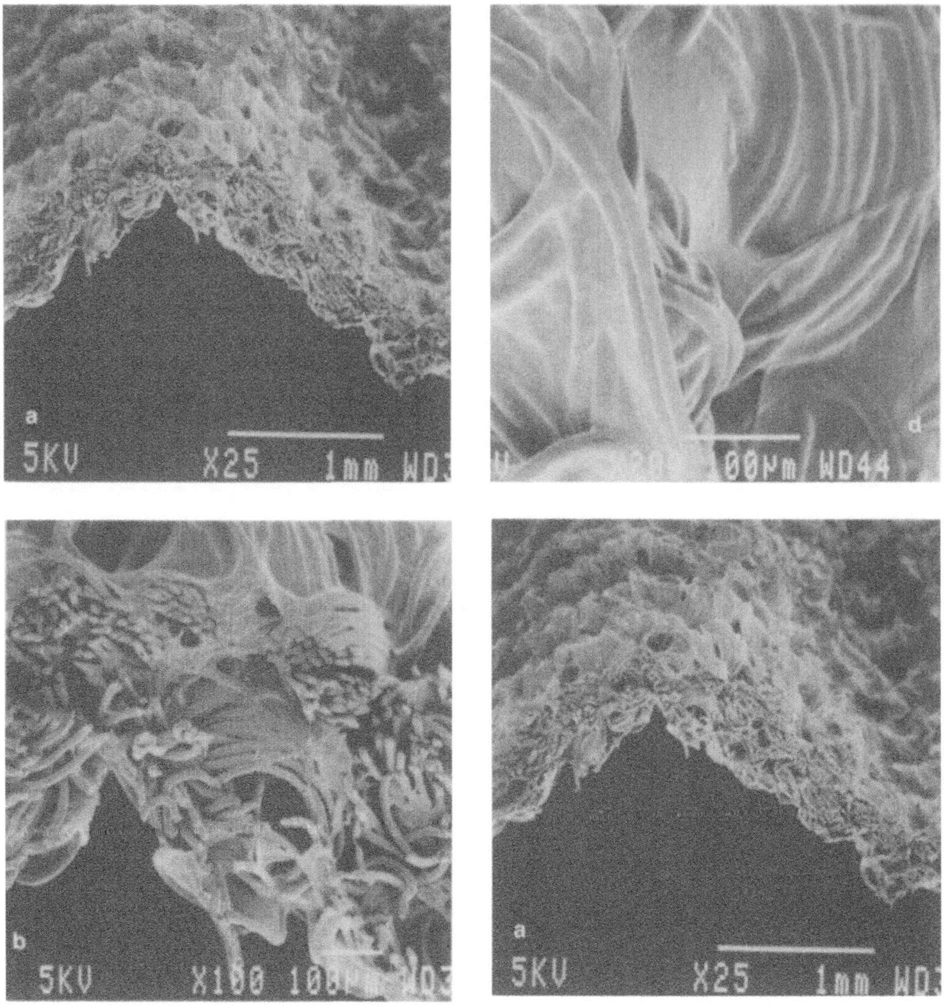

Figure 4. Scanning elecron micrographs of a Collagen/Dacron composite vascular prosthesis: (a) cross-sectional view at 25x; (b) cross-sectional view at 100x; (c) surface view at 200x; (d) surface view at 1000x

mained in solution at a centrifugal force of 200 g. Under experimental conditions the collagen slurry retained 4-6 g of water per mg of collagen, indicating a highly swollen nature of the material.

Table II summarizes the results of <u>in vitro</u> studies of the Collagen/Dacron composite grafts. The porosity test of the composite grafts showed that the grafts were essentially impervious to water under physiological pressure. The porosity of microvel for the corresponding test condition was 1900 mls/min/cm^2, as seen by the highly porous structure of the graft (Figure 3). The thermal stability analysis showed that the collagen on the graft was heterogeneous, as revealed by a broad thermal transition, in contrast to a sharp thermal transition usually seen in intact tissues. This resulted from the mechanical breakdown of the material into fine fibrils during material processing. The range of shrinkage temperature that was defined for the canine trial was between 55-65°C. Under simulated pulsatile pumping conditions, collagen was found to be strongly anchored to the Dacron substrate. No detectable collagen was released from the graft for a continuous pumping period of 15 minutes at flow rates of 250-750 ml/min. Continuous soaking of Collagen/Dacron graft in saline under mild agitation up to 72 hours did not detach any collagen from the graft. These findings were consistant with the morphological observations that collagen was uniformly adhered to the surface and impregnated into the interstitial pores (Figure 4). The appearance and handling characteristics of the composite grafts were very close to that of the Microvel grafts. Collagen impregnation and coating did not affect any of the mechanical properties of the Dacron graft as determined by the manufacturer's quality control test methods, including burst, suture pull out, and tensile strength of the grafts. The values reported in Table 2 were identical to that observed for the Microvel substrate.

The first canine study indicated that the control grafts required several minutes of handling to obtain sufficient preclotting and hemostasis, in contrast to the Collagen/Dacron graft which was impervious to blood during implantation without any pre-operative treatments. It was shown that collagen has no effect on the patency of a 6 mm diameter graft during nine months implantation.

Table I. Amino Acid Composition of Purified Bovine Skin Collagen (a).

Hydroxyproline	115.0 (b)	90.3 (c)	94 (d)
Aspartic Acid	46.0	42.1	46
Threonine	16.0	14.0	17
Serine	34.0	30.1	36
Glutamic Acid	70.0	67.9	72
Proline	121.0	119	127
Glycine	329.0	342	326
Alanine	107	106	112
Half-cystine	0	0.5	--
Valine	18.0	19.3	23
Methionine	6.4	3.6	6
Isoleucine	10.5	11.1	11
Leucine	23.0	21.8	25
Tyrosine	1.8	1.9	3
Phenylalanine	11.7	11.5	13
Hydroxylysine	6.8	4.0	6
Lysine	28.0	23.5	27
Histidine	4.2	3.7	5
Arginine	48.0	44.8	51

(a) Data expressed as residues/1000 residues.
(b) Material used for Collagen/Dacron graft fabrication.
(c) Reported in reference 11.
(d) Reported in reference 12.

Figure 5. An electron micrograph of the purified insoluble bovine skin collagen.

Table II. Results of *In Vitro* Studies of Collagen/Dacron Grafts (a).

Porosity (ml/min/cm^2)	2.8 $\pm$ 1.7 (b)
Thermal Shrinkage Temperature of Collagen (^{o}C)	60.3 $\pm$ 3 (c)
Mechanical pumping for Collagen Stability	Non-detectable collagen release from grafts
Morphology by SEM	Uniform surface coating and impregnation (d)
Burst Strength (lbs/in^2)	124 $\pm$ 15 (e)
Tensile Strength (lbs)	41 $\pm$ 2 (e)
Suture Pull Out Strength (lbs)	5.0 $\pm$ 0.7 (e)

(a) Value reported as the mean $\pm$ S.D.

(b) The porosity for the knitted Dacron Microvel graft is 1900 mls/min/cm^2.

(c) Acceptable shrinkage temperature range for *in vivo* implantation is 60 $\pm$ 5°C.

(d) The SEM of knitted Dacron Microvel graft indicated a highly porous structure.

(e) The mechanical strengths of Collagen/Dacron grafts are identical to that observed for the Microvel grafts.

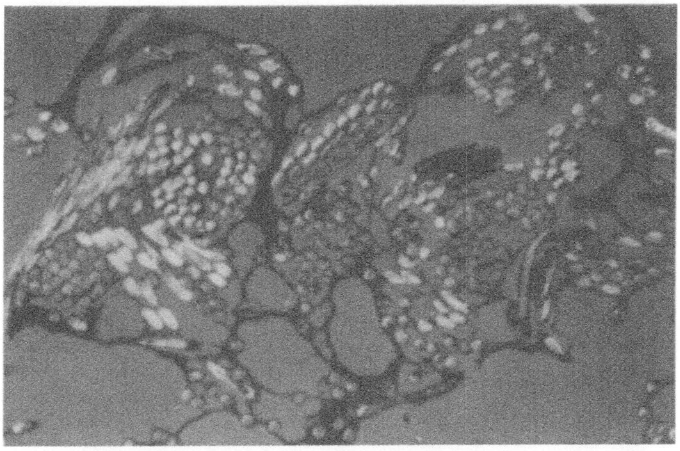

Figure 6. Histological staining of bovine collagen with Picro sirius Red F3BA, showing intense deep red staining of bovine collagen.

The rate of collagen resorption was estimated by a histological method. The histological sections of the explants were stained with Picro sirius Red F3BA in saturated aqueous picric acid and observed under polarized microscope. The exogenous bovine collagen showed an intense deep red color under polarized microscope as seen in Figure 6. The method provided a qualitive way of estimating the amount of bovine collagen remaining at each graft explantation. It was observed that approximately 40% of the collagen was resorbed in the first two weeks of implantation. At the end of four to six weeks, most of the original bovine collagen has been resorbed and substituted by an ingrowth of the host connective tissue between the Dacron fibers. A recent study of Li, et al, employing a ^{14}C radiolabeled collagen as a means to follow the biodegradation of collagen in vivo, indicated that 70-80% of collagen was resorbed in the first week of implantation accompanied by a rapid healing of the graft (6). A complete resorption was observed at the end of three months, accompanied by a full tissue ingrowth with a smooth neointima formation (Figure 7). Figure 8 shows the histology of a six month explant. Fully developed neointima has completely replaced the initial collagen matrix. The excellent tissue ingrowth associated with highly porous knitted Microvel grafts was not affected by the presence of collagen. The final healing of the Collagen/Dacron graft, revealed at the light microscopic level, also appeared to be similar to the control grafts. A smooth neointima formation was found on the healed surface of both grafts.

In the second canine study, careful ultrastructural and morphological anaylses of the explants demonstrated that the Collagen/Dacron grafts had not only a more uniform coverage of endothelial cells but, more significantly, a several fold increase in the extent of microvessel formation than that of the preclotted control grafts. A typical endothelialized surface of a Collagen/Dacron composite graft is shown in Figure 9.

DISCUSSION

Collagen is a fibrous protein which accounts for about 30% of the total proteins in the body. At present there are five major types of collagen that have been extensively characterized and a great deal of information is known about them. In addition, several new types of colla-

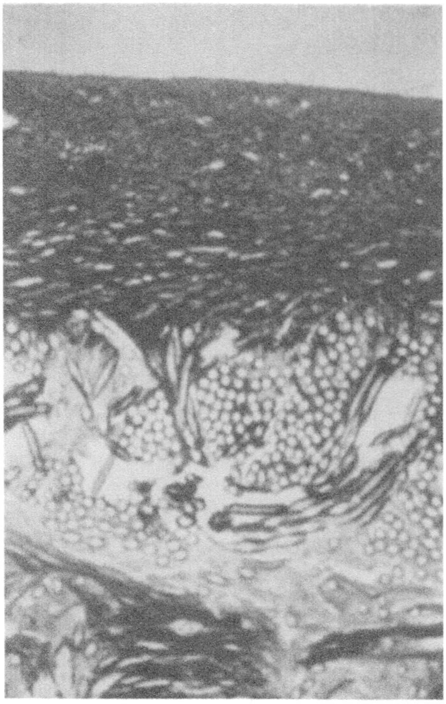

Figure 7. Histology of a three months canine explant of a Collagen/Dacron prosthesis, showing full tissue ingrowth and a well developed neointima. Stained with hematoxylin and eosin.

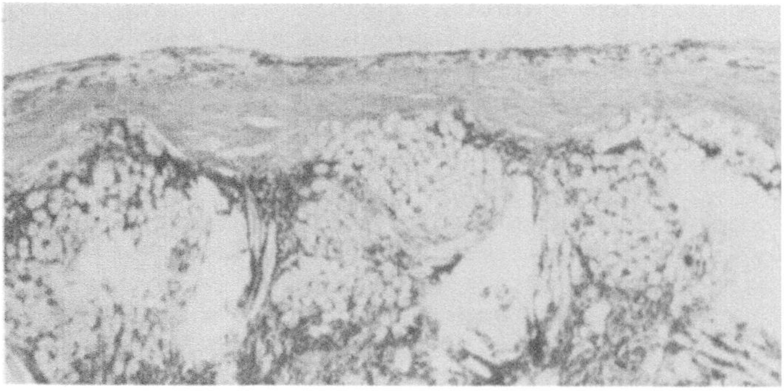

Figure 8. Histology of a six months canine explant of a Collagen/Dacron prosthesis. Fully developed neointima with well organized collagenous connective tissue and cellular activity. Stained with Masson trichrome.

gen have recently been discovered and are being investigated. Type I collagen, which is used for the Collagen/Dacron graft production, is the major component of bone, skin and tendon. Due to it's accessibility in purified forms, the type I collagen is the most well characterized collagen of all. In the tissue, type I collagen exists mostly as insoluble fibrils and fibers that are intermolecularly cross-linked after extracellular assembly of the molecules. A type I collagen molecule consists of three polypeptide (α) chains, each having about 1000 amino acid residues. Due to the unique amino acid sequence of the α-chains (every third residue is a glycine), the chains are intertwined forming a triple helical coiled-coil molecular structure, having a rigid rod-like conformation of 300 nm long and 1.5 nm in diameter. Each molecule contains approximately 240 acidic side chains of glutamic and aspartic acids and 250 basic side chains of lysines, hydroxylysines, and arginines. Under physiological conditions the acidic and basic side chains are charged. As a result of the unique molecular packing of collagen in fibrils (67 nm or multiple integral of 67 nm intermolecular staggering), most of the charged side chains are internally neutralized, forming salt-linkage of the type, protein-$NH_3P^{+-}OOC$-protein (7). The charged side chains of collagen, however, can be specifically modified in vitro, resulting in a variety of new materials having diverse physico-chemical and biological properties (8). Depending on the specific needs, the properties of collagen can be judiciously controlled to meet the requirements of particular biomedical applications. For example, one of the major desirable properties for the Collagen/Dacron graft is the rate of in vivo collagen resorption. The rate of resorption of collagen after implantation depends upon the extent of intermolecular cross-linking, a parameter which is related to it's in vitro thermal stability. Since collagen molecules contain a number of side chains that are in close proximity for chemical cross-linking, the rate of in vivo resorption can be controlled by chemical means, monitored by the thermal stability measurements. Thus the thermal stability of collagen in the final product was defined in order to ensure the safety level of the in vivo rate of resorption. The concept of controlled resorption has also been effectively utilized to produce a number of collagen based medical products such as resorbable collagen hemostats, sutures and drug delivery systems.

Another important property of collagen is its surface reactivity towards ions, macromolecules and cells. Collagen can be specifically modified into a positively charged, a negatively charged or into a totally non-charged moiety. The modified collagen matrices have drastically different surface electrostatic properties, resulting in a different reactivity toward ions, macromolecules, cells and platelets (9). These properties have important biomedical applications in both implantable devices as well as in drug delivery systems.

Collagen is the major component of the connective tissue, and forms the essential framework of the tissues and organs. It has such diverse biological functions ranging from controlling cell differentiation in morphogenesis and development to the structural supporting functions of bone, tendon and tooth. Many cells are found either resting on the collagen-containing basal lamina or within the collagenous matrix. It is well established that the collagenous surface promotes cell growth and differentiation in in vitro tissue culture experiments. The collagen-coated and impregnated Dacron graft, therefore, should provide an ideal substrate for the in vivo regeneration of endothelial cells. In deed, the results of canine studies support this notion.

Type I collagen possesses another key property for the biomedical application which should be briefly discussed here. It is a protein of minimum species specific differences in the primary as well as higher

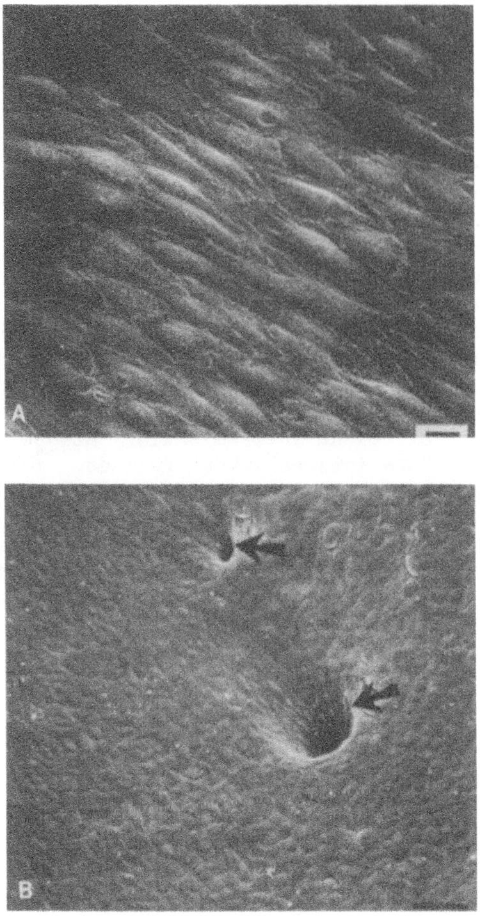

Figure 9. Scanning electron micrographs of the inner surface of a fully healed Collagen/Dacron prosthesis. The luminal surface is completely endothelialized. (a) Near the proximal anastomotic line. The bar indicates 10 μ thick, 300x (b) Near center of the prosthesis. Two openings with endothelial cells of newly formed microvessels are observed. The bar indicates 100 μ, 100x.

order structure. The biodegradable products are natural metabolites of the body. The minor structural differences of type I collagen between species that could elicit adverse biological reactions, such as antigenicity, is primarily associated with the non-helical segments of the molecule, which are commonly referred to as N- and C- terminal telopeptides. The telopeptides accounts for only approximately 5% of the total collagen molecule.

Numerous studies on the antigenicity of collagen have been conducted at a number of institutions. It has been concluded that pure collagen in its soluble form is a poor antigen (10). The antigenicity decreases further, when the solubility of collagen decreases, or when the rate of biodegradation of collagen slows down in vivo. Since the type I collagen used for Collagen/Dacron composite graft production is insoluble and its in vivo rate of resorption is carefully controlled, the collagen on the graft should not elicit any detectable antigenic reactions, as was noted in the passive hemagglutination test which showed a non-detectable level of antibody formation.

CONCLUSION

In this paper, the advantages of employing collagen for a medium to large diameter vascular application were presented. Since it is a well known fact that native collagen induces platelet adhesion and aggregation, the extension of collagenous surface to small caliber (<6 mm internal diameter) vascular graft applications requires careful consideration of both the physico-chemical and biological properties of collagen. These applications can only be fully appreciated when the properties of collagen are considered as a whole. In this regard, an ideal Collagen/Dacron graft can be designed to provide a surface and mechanical properties that are suitable for vessel replacement.

ACKNOWLEDGEMENTS

I wish to thank Dr. R. Kennedy for his encouragement and support of this work and Dr. M. Sunwoo, Mr. H. Hoffman, Ms. P. Feigel, Ms. S. Daum and Dr. T. Freund for technical assistance and helpful discussions. I also thank Dr. M. Chvapil for sharing his canine results before publication, and Drs. W. Moore and D. Ruhland for the *in vivo* canine evaluation of the Collagen/Dacron grafts.

REFERENCES

1. J. H. Fessler and L. I. Fessler, Ann. Rev. Biochemistry, 47, 129 (1978).
2. P. Bornstein and W. Traub, in: "The Proteins", H. Neurath and R. L. Hill, eds., 3rd ed., Vol. 4, Academic Press, New York, 1979, pp 411-632.
3. E. J. Miller in "The Chemistry and Biology of Mineralized Tissues", W. T. Butler, ed., EBSCO Media, Birmingham, 1985, pp 80-93.
4. H. Z. Stegemann, Physiol. Chem., 187, 41 (1958).
5. L. A. Elson and W. T. J. Morgan, Biochem. J., 27, 1824 (1933).
6. S. T. Li, P. Feigel, M. Chvapil and A. Pinder, Manuscript in preparation.
7. S. T. Li, E. Golub and E. B. Katz, J. Mol. Biol., 98, 835 (1975).
8. S. T. Li and E. P. Katz, Biopolymers, 15, 1439 (1976).
9. E. P. Katz, N. Wax and S. T. Li, in: "Electrical Properties of Bone and Cartilage", C. T. Brighton, J. Black and S. Pollack, eds., Grune and Stratton, Inc. 1979, pp 131-140.
10. R. Timpl, in: "Biochemistry of Collagen", G. N. Ramachandran and A. H. Reddi, eds., Plenum, New York, 1976, p 319.
11. I. Oneson, D. Fletcher, J. Olivo, J. Nichols and R. Kronenthal, J. Am. Leather Chemists Assoc., 65, 440 (1970).
12. D. Volpin and A. Veis, Biochemistry, 12, 1452 (1973).

DESIGN AND IN VITRO TESTING OF NEWLY MADE BICOMPONENT KNITTED FABRICS FOR VASCULAR SURGERY

C. C. Chu and L. E. Lecaroz

Department of Textiles and Apparel
Martha Van Rensselaer Hall, Cornell University
Ithaca, NY 14853-4401

The purpose of this study is to advance an alternative design of knitted vascular grafts in order to achieve low bleeding porosity at implantation without the expense of high healing porosity for full-wall healing. The design is based on the utilization of both biodegradable (polyglycolic acid) and nonbiodegradable (Dacron) synthetic fibers blended at various composition ratios. The fabric construction is of single jersey weft knitted tubular structure. The experimental part of this study included the fabrication of the knitted specimens at various composition ratios of PGA to Dacron fibers, characterization of the knitted specimens, and in the *in vitro* physical, mechanical and morphological testing of the specimens after various periods of immersion in phosphate buffer of pH 7.4. The overall properties of these bicomponent fabrics were determined by the relative magnitudes of the two opposite factors: the absorption of PGA component and the inherent relaxation shrinkage of the weft knitted fabrics. The most important observation of the study was the achievement of increasing water porosity with the duration of immersion without the significant expenses of structural integrity and strength of the specimens. This unique property has not been found in commercial vascular graft fabrics of similar construction.

INTRODUCTION

Vascular diseases have always been associated with human beings as recorded history shows. They frequently result in either the loss of lives or limbs. The development of vascular grafts of either synthetic or natural origin has made significant progress and has allowed the reconstruction of obstructed or injured blood vessels with remarkable success. However, some serious problems still remain unsolved (1). While homografts are still considered to be the preferred arterial replacement, they are used in limited quantity and are insufficient to meet the increasing needs for vascular replacement, due mainly to an inadequate supply, nonuniform properties, and difficulty in preparation (1,2).

In the long search for an ideal graft replacement, various synthetic materials had been experimented with. They include rigid nonporous tubes made of gold, silver, aluminum, glass and polyethylene. However, these materials completely failed due to the lack of porosity and compliance. These two important requirements of an ideal vascular graft have been partially fulfilled by the incorporation of fibrous polymers and textile structure. The compliance of vascular grafts has recently been advanced further by the using of elastomeric fibers, such as Spandex (3). In contrast to a solid tube, the inherent porous nature of textile materials has provided some available areas for the fibrous connective tissue to grow into to achieve full-wall healing (defined by Sauvage as "incorporation of the entire graft within a fibrous tissue matrix whose flow surface is covered with endothelium") (4).

However, the two basic fabric structures, woven and knitted, which have been used for synthetic vascular graft construction are quite different from each other in terms of porosity and other properties.

The woven structure provides strength with high dimensional stability, low permeability to blood, and is less prone to kinking. Its main disadvantages are low healing porosity leading to poor full-wall healing, difficulty in suturing, fraying of cut edges, and poor compliance. The knitted structure has a much higher porosity and hence has better full-wall healing properties than woven grafts. However, due to the high porosity, they are difficult to preclot during the time of implantation and therefore impose the problem of blood leakage.

Because all of the existing commercial vascular grafts are constructed from a single type of non-biodegradable fiber (e.g., Dacron, Teflon), their porosity is fixed and does not change as a function of time. It is thus impossible to vary the porosity of these single component fabrics to make them very tight during implantation to prevent the occurrence of blood leakage, and make them very porous during healing to promote fibrous tissue ingrowth for full-walled healing.

The purpose of this study is to provide an alternative design for a vascular graft to achieve these two contradictory requirements. The design is based on the utilization of both biodegradable (polyglycolic acid) and nonbiodegradable (Dacron) synthetic fibers blended at various composition ratios.

EXPERIMENTAL METHOD

1. Fabrication of Specimen

The Model LZ80 two-feed circular knitting machine from Lamb Knitting Machine Corporation was utilized for the manufacturing of the single jersey weft knit tubular fabric specimens. The machine is capable of knitting two types of yarn simultaneously. It has 48 hosiery type needles of 72 gauge and the cylinder size is 7/8". The fabric specimens were constructed at varying composition ratios of Dacron to polyglycolic acid fibers. Variable yarn tensions were incorporated to control strength, elasticity and porosity. Knitted fabrics made from 100% Dacron served as the control. The various design composition ratios are summarized in Table I. Basically, there were two groups of bicomponent fabrics depending on the denier of Dacron yarns (i.e., 250 and 200 deniers). The PGA composition ratio ranged from 24% to 57%, by weight.

2. Sample Conditioning

Prior to testing the knitted tubular specimens were immersed in

Table I. Composition of Eight Single Jersey Bicomponent Tubular Knitted Fabrics

Sample Code	Yarn Component	Composition % PGA by weight	% Dacron
K1*	250 denier Dacron	0	100
K2	250 denier Dacron/1+ -90 denier PGA	24	76
K3	250 denier Dacron/2+ -90 denier PGA	40	60
K4	250 denier Dacron/3+ -90 denier PGA	49	51
K5*	200 denier Dacron	0	100
K6	200 denier Dacron/1+ -90 denier PGA	31	69
K7	200 denier Dacron/2+ -90 denier PGA	47	53
K8	200 denier Dacron/3+ -90 denier PGA	57	43

(*) control fabrics
(+) # of PGA yarn strands

phosphate buffered solution (pH = 7.4) for 0, 14, 30, 60, 90 and 120 days at 37 ± 1°C. All of the unimmersed samples (0 day) were washed in 5% sodium bicarbonate solution for five minutes to remove lubricants and excess impurities introduced during fiber and yarn production and knitting.

After immersion the samples were gently washed in distilled water to remove excess salt deposits. The fabrics were dried and placed in a desiccator with Drierite to prevent the PGA component from premature undesirable degradation. Prior to testing the knitted tubular specimens were cut 2 inches in length along the axial direction of the tubes, and were placed in a conditioning room at 21 ± 1°C and 65% ± 2% relative humidity for twenty four hours.

3. Fabric Characterization

The knitted fabrics in each design were characterized in terms of number of wales and courses per cm, stitch density, loop length, thickness, weight per unit area, density and packing factor. Changes in fabric properties were later correlated to the changes in fabric characterizations.

The courses and wales per cm were determined using a pick glass. The wales per cm were measured by counting the loops in the course direction and the course per cm were measured by counting the loops in the vale direction. Stitch density is the total number of stitches per unit area, usually a square cm. It is the product of courses per cm and wales per cm and is expressed in stitches/cm^2. The loop length was obtained by raveling a length of yarn from the fabric and counting the number of loops and dividing it by the length of the same yarn in the straightened form.

ASTM D1777 Standard Method for Measuring Thickness of Textile Materials was used to measure the thickness of the knitted tubular specimens (5). The test was conducted on an Ames Thickness Testing instrument using a 2.54 cm diameter presser foot with a pressure of 0.50 psi. The diameters of the eight knitted tubular fabrics were measured to determine the changes in fabric dimensions as immersion progressed. The packing factor indicates the tightness of the structure and is calculated from Equation 1-3, where Dp is 1.68 g/denier and D_d is 1.38 g/denier.

Packing Factor = Fabric density x 100/ Fiber density (Equation 1)

Fabric Density (g/cm^3) = weight/(area x thickness) (Equation 2)

Fiber Density = (% PGA x Dp) + (% Dacron x D_d) (Equation 3)

4 Gross Morphological Study

The changes in gross morphology of the fabric specimens were observed by the JEOL Model-35CF Scanning Electron Microscope. The knitted specimens were cut and placed on SEM metal stubs with a double stick scotch tape and sputter-coated with gold-palladium. Both the exterior and interior sides of the knitted specimens were examined. Observations in fabric structure were recorded by polaroid micrographs at a magnification of 39x.

5 Fabric Physical and Mechanical Properties

Stiffness

ASTM D4032 Stiffness of Fabrics by the Circular Bend Procedure was used to measure the fabric stiffness of the knitted tubular specimens (6). The test was conducted by J. A. King and Company, Inc., Greensboro, North Carolina. A modified version of the King Fabric Stiffness Tester machine was used in the determination of the fabric stiffness because of the limited size of the specimens. Due to the small dimensions of the knitted tubular specimens, a 5.3 mm diameter flat tip plunger was utilized to force the fabric through a 8.75 mm diameter hole. Four replicates were taken for each knitted specimen. The stiffness of the fabric (resistance to bending) was indicated by the amount of force in pounds necessary to push the fabric through the oriface.

Weight Loss Measurement

The knitted specimens of 1" x 0.5" were conditioned in a conditioning room for 24 hours before weighing. The amount of weight loss per unit area of the specimen at any specified period of immersion was determined and expressed as in Equation 4 where Wo = initial weight before immersion in grams/cm^2, and W_1 = weight after immersion in grams/cm^2. An average weight was taken from three fabric specimens for each immersion period for an adequate statistical average.

% of Weight loss = $(W_o - W_1)/W_o$ (Equation 4)

Porosity Measurement

The porosity of the fabric specimens was determined by the water flow through method developed by Wesolowski (7). The measurement was done at a pressure of 120 mm Hg. The amount of distilled water flow through the fabric was collected in a beaker for one minute and measured. The data were calculated and expressed in terms of $ml/min/cm^2$.

Bursting Strength

The bursting strength of the knitted fabrics was measured by using the Instron Tensometer Table Model TM. The test was performed according to the procedures in ASTM D3787 Constant-Rate-of-Traverse (CRT) Ball Burst test (8). However, the specimen holder and ball burst attachment were modified to accommodate the smaller dimensions of the fabric.

To minimize fabric slippage, a 1.5 x 2 inch fabric was securely taped between two square shaped metal plates with an oriface diameter of 2.00 cm. The metal plates were then mounted on two pieces of square wooden blocks and clamped with four butterfly screws. The square wooden blocks were then placed on the top of a compression cell (CCTM). A metal rod measuring 5 inches in length and 3/16 inch diameter was secured on the crosshead of the Instron tensile tester. A polished hardened steel ball was placed on the fabric and was aligned with the metal rod to allow the rod to push the ball through the fabric at a constant speed of 1 inch/min until rupture. An average of three tests per sample was done. The data were expressed in terms of Kg/cm^2.

Suture Retention Strength

Suture retention properties were measured to determine the amount of force necessary for the fabric to break at the suture line, and how far the suture must be placed from the fabric edge. Three knitted tubular specimens were used for each design to measure suture retention strength using the Instron Table Model TM. A 1-0 Polypropylene monofilament suture was used because the thickness and strength of the knitted fabrics caused a finer suture to break before tearing the fabrics apart during the tests. The suture bit the fabric at a distance of either 2 mm or 5 mm from the edge of the fabric. One third of the knitted specimen was then gripped on the lower jaws of the Instron at a gauge length of 76.2 mm (3 inches) while the suture was gripped by its two ends in the upper jaws of the Instron. The crosshead was lowered at a speed of 100 mm/min until rupture of the fabric occurred.

6. Statistical Analyses

The experimental data were examined by the one-way analysis of variance and linear regression. The one-way analysis of variance was used to compare the means of each fabric and analyzed for changes in physical properties (stiffness, weight loss, porosity, bursting and suture retention strength) as a result of PGA degradation with time immersion. A 95% confidence interval was used to determine if data were statistically significant. The linear regression was utilized in each fabric design to determine if there was a correlation between the data of porosity, bursting strength and suture retention strength, and time of immersion.

RESULTS

1. Characterization of Eight Tubular Knitted Fabrics Before Immersion

The characterization data of the eight tubular fabric specimens before immersion are shown in Table II. Obviously, a wide range of fabric characteristics was observed as a result of both different fiber deniers of Dacron and PGA composition. The course and wales per cm of all fabrics ranged from 8 to 11 and 13 to 17, respectively. This variation of courses and wales per cm of the fabrics was the result of varying composition ratio of PGA to Dacron yarns. When comparing with commercial weft knitted fabrics (20 to 39 wales/cm and 26 to 43 courses/cm), these numbers are considerably lower(9). However, both control fabrics (K1 and K5) had an identical number of courses and wales per cm.

The stitch density ranged from 104 to 170 stitches/cm^2 with both control fabrics having identical and the highest stitch density. As more PGA yarns were incorporated in the fabric, stitch density decreased due to the increased thickness of the knitted loops. This is because the thick knitted loops would occupy a wider space and hence accomodate less courses and wales per cm. This range of stitch density was significantly lower than all the existing commercial weft knitted vascular grafts, which have ranges from 520 to 1,680 (9).

Fabric thickness increased as the amount of PGA yarns increased in the fabrics and ranged from 0.0504 to 0.0772 cm. K1 was thicker than K5 due to the heavier denier Dacron yarns. This range of thickness is higher than most of the commercial weft knitted vascular graft fabrics, which have thickness ranging from 0.018 to 0.058 cm.

Fabric weight ranged from 0.0106 to 0.0240 g/cm^2. This range is comparable to commercial weft knitted surgical fabrics. Fabrics containing 250 denier Dacron (K1 to K4) were heavier than fabrics containing 200

Table II A. Fabric Characterization of the Eight Knitted Specimens Before Immersion.

Sample Code	Courses per cm	Wales per cm	Stitch[a] Density (St/in^2)	Thickness (mm)	Loop Length (cm)	Diameter (cm)	Weight (g/cm^2)
K1	10	17	170	0.508 $\pm$ 0	0.1875	1.50	0.0128 $\pm$ 0.0003
K2	9	16	144	0.576 $\pm$ 0.059	0.1071	1.80	0.0179 $\pm$ 0.0018
K3	9	13	117	0.660 $\pm$ 0	0.1667	2.00	0.0202 $\pm$ 0.0001
K4	8	13	104	0.707 $\pm$ 0.006	0.1786	2.25	0.0240 $\pm$ 0.0004
K5	10	17	170	0.504 $\pm$ 0.008	0.1875	1.50	0.0106 $\pm$ 0.0003
K6	11	14	154	0.553 $\pm$ 0.011	0.1389	1.75	0.0190 $\pm$ 0.0005
K7	10	13	130	0.711 $\pm$ 0	0.1550	2.00	0.0195 $\pm$ 0.0003
K8	10	13	130	0.772 $\pm$ 0.003	0.1607	2.10	0.0223 $\pm$ 0.0004

(a) the product of course/cm and wales/cm

Table IIB. Fabric Characterization of the Eight Knitted Specimens Before Immersion

Sample Code	Fabric Density (g/cm³)	Fiber Density (g/cm³)	Packing[b] Factor (%)	Water[c] Porosity (cc/min/cm²)	Stiffness[d] (Kg)	Bursting Strength (Kg/cm²)
K1	0.2529 ± 0.0054	1.3800	18.30 ± 0.40	2800 ± 0	0.141 ± 0.058	17.21 ± 0.27
K2	0.3108 ± 0.0314	1.4520	21.40 ± 2.15	1700 ± 30	0.649 ± 0.055	19.93 ± 0.44
K3	0.3067 ± 0.0021	1.5000	20.50 ± 0.15	1800 ± 0	2.123 ± 0.608	22.75 ± 0.09
K4	0.3641 ± 0.0059	1.5270	23.80 ± 0.40	2550 ± 60	3.683 ± 0.624	22.30 ± 1.13
K5	0.2114 ± 0.0052	1.3800	15.40 ± 0.40	4000 ± 0	0.050 ± 0.015	10.76 ± 0.54
K6	0.3433 ± 0.0085	1.4700	23.30 ± 0.60	1950 ± 100	1.016 ± 0.071	20.07 ± 0.37
K7	0.2750 ± 0.0040	1.5200	18.10 ± 0.30	1800 ± 0	2.132 ± 0.595	20.31 ± 0.32
K8	0.2893 ± 0.0055	1.5500	18.70 ± 0.30	1500 ± 0	-----	25.78 ± 1.28

(b) fabric density/fiber density
(c) according to Wesoloski method at 120 mm Hg
(d) the amount of force required for a 5.3 mm diameter flat tip plunger to push the fabric through a 8.75 mm diameter hole.

denier Dacron (K5 to K8). Moreover, as the amount of PGA yarns in the fabric increased, fabric weight also increased. The diameter of the newly made knitted fabrics ranged from 1.5 to 2.25 cm, the diameter increasing with the increasing composition of PGA fibers because more yarns occupied the knitted loops.

The fundamental unit of the weft knitted structure is the loop length. It exists in the course direction of the structure and influences fabric dimension and properties. The two controls (K1 and K5) have an identical loop length (0.1875 cm). The incorporation of PGA yarns into the control shortened the loop length initially, but the loop length increased as more PGA yarns were added into the structure. This variation in loop length was due to the adjustment of tension of yarns during fabrication to accommodate the desired amount of PGA yarns in the structure.

The packing factor, an indication of the tightness of the structure of the newly made knitted fabrics, ranged from 15 to 23%. The packing factor of K1 was greater than K5 since K1 had heavier and thicker Dacron yarns. As PGA was incorporated into the structure, the packing factor increased as seen in K2 and K6. However, fabrics containing two PGA yarns

(K3 and K7) had lower packing factors than the corresponding K2 and K6 as a result of changes in the loop length. Loop length was adjusted to accommodate knitting of more yarns in the structure causing the fabrics to decrease in tightness. The commercial weft vascular grafts have packing factors ranging from 26 to 42%.

The water porosity of the eight knitted specimens ranged from 1,500 to 4,000 cc/min/cm^2 at 120 mm Hg pressure. K5 had the highest level of porosity while K8 had the lowest level of porosity. Although courses and wales per cm and stitch density of the two control fabrics (K1 and K5) were identical, their porosities were different from each other. K5 had a much higher value than K1 because of the fineness of the 200 denier Dacron yarns used in K5. In the bicomponent fabrics, the initial incorporation of PGA into the structure reduced water porosity significantly, while any further increase in PGA composition did not reduce water porosity proportionally. In the 250 denier set, this further increase in PGA composition actually increased water porosity instead of decreasing it. This range of water porosity of the newly made knitted fabrics was generally lower than most of the commercial weft knitted vascular fabrics which have water porosity ranging from 2,530 (DeBakey Standard) to as high as 5,300 (Milliknit) ml/min/cm^2 (9).

The incorporation of PGA yarns into the fabric structure made the biocomponent fabrics stiffer than the two controls, the magnitude of increment depending on the composition of the PGA component. In general, fabrics of the 250 denier series were stiffer than the 200 denier series because of the heavier denier Dacron yarns making them stiffer. The stiffest fabric was K4 (3.83 kg), while the softest one was K5 (0.050 kg), and an 80 x difference in stiffness observed.

The bursting strength of the newly made knitted fabrics exhibited a similar trend as with stiffness. Both the PGA composition and the denier of Dacron yarns affected their bursting strength. As PGA composition increased in the fabric, its bursting strength increased correspondingly. This was the result of the increased number of yarns presented in the knitted loops causing the loops to increase in strength. K8 was the strongest fabric (25.78 kg/cm^2), while K5 was the weakest (10.76 kg/cm^2). This increasing in bursting strength with PGA composition appeared to be more profound in the 200 denier series. For example, nearly a 140% increase in bursting strength was found in K8 as compared to K5 in the 200 denier series, while K4 gained only 30% more bursting strength as compared to K1 in the 250 denier series.

2. Fabric Structural Changes Upon Immersion

Since the newly made specimens had biodegradable component yarns in the structure, their characteristics must change as immersion progressed. The magnitude of change depended on the composition of PGA yarns in the structure as well as the extent of immersion, which determined the amount of PGA degradation.

As shown in Table III, all bicomponent fabrics showed an increase in courses/cm with time. This increase was most pronounced in the 250 denier series and with the highest PGA composition (i.e., K4). On the contrary, the change in wales of these bicomponent fabrics upon immersion was relatively small. Most of them had insignificant or undetectable changes (e.g., K3, K8).

As a consequence of these changes in courses and wales, their product, stitch density, also varied with time. In general, although there

Table III. Courses and Wales per Cm. of the Eight Tubular-Knitted Fabrics Upon Immersion

Sample Code	0 CPC	0 WPC	14 CPC	14 WPC	30 CPC	30 WPC	60 CPC	60 WPC	90 CPC	90 WPC	120 CPC	120 WPC
	Time (Days)											
K1	10	17	11	16	10	15	11	15	11	14	12	14
K2	9	16	10	14	11	14	10	13	11	13	12	13
K3	9	13	9	13	10	13	9	13	12	13	13	13
K4	8	12	9	12	9	11	9	12	14	12	14	13
K5	10	17	11	16	10	17	11	15	11	15	11	14
K6	11	14	15	14	14	15	13	13	13	13	13	14
K7	10	13	10	13	10	12	12	13	12	13	13	14
K8	10	13	13	13	13	13	13	13	13	13	13	13

CPC: Course/Cm
WPC: Wale/Cm

were some fluctuations in stitch density with time as shown in Table IV, all bicomponent fabrics showed an overall increase in stitch density over the entire period of the study (120 days), and the magnitude of the increment became larger as more PGA was incorporated into the fabric structure. This was particularly true with the 250 denier series (K1 -

Table IV. Stitch Density (Stitches/Cm^2) of the Eight Tubular Knitted Fabrics Upon Immersion*.

Sample Code	0	14	30	60	90	120
	Time (Days)					
K1	170	176	150	165	154	168
K2	144	140	154	130	143	156
K3	117	117	130	117	156	169
K4	96	108	99	108	168	182
K5	170	176	170	165	165	154
K6	154	210	210	169	169	182
K7	130	130	120	156	156	182
K8	130	169	169	169	169	169

(*) Stitch Density = (Course/cm) x (Wale/cm)

K4). In this series all bicomponent fabrics showed a continuous increase in stitch density after 60 days of immersion.

Because the PGA component yarns would degrade gradually upon immersion, it was expected to see an overall decrease in fabric thickness with time as shown in Table V. However, all bicomponent fabrics exhibited an initial increase in thickness between 14-30 days, then decreased thereafter with some fluctuations. The onset of the significant decrease in thickness of the bicomponent fabrics depended on the PGA composition in the fabrics. An increase in PGA yarn composition appeared to delay the onset time. For example, the onset time for K2 was at 14 days and was delayed to 30 days for K3 and 60 days for K4. This trend, however, was less pronounced in the 200 denier series.

Since loop length exists in the course direction of the fabric, any change in courses/cm would also affect loop length upon immersion. As shown generally in Table VI the loop length of bicomponent fabrics would initially increase with time as opposed to the two control fabrics which both exhibited an initial decrease in loop length. Depending on the PGA composition, the loop length would then decrease sometime during immersion, usually between 30-60 days. Except in the case of K2 and K6, which had an overall gain in loop length, the rest of the bicomponent fabrics had either a reduction in, or a relatively consistent, loop length. The two controls showed an overall decrease in loop length.

Because of the changes of stitch density per unit area, the diameter of the fabrics must also change. As shown in Table VII, both the controls (K1 and K5) exhibited a continuous and overall increase in diameter, while the bicomponent specimens fluctuated in diameter. One common phenomenon observed in both series was that, regardless of the initial wide

Table V. Fabric Thickness (Cm.) of the Eight Tubular Knitted Fabrics Upon Immersion.

Sample Code	Time (Days) 0	14	30	60	90	120
K1	0.0508 (±0)	0.0455 (±0.0004)	0.0483 (±0.0003)	0.0460 (±0.0003)	0.0456 (±0.0001)	0.0533 (±0.0003)
K2	0.0576 (±0.0059)	0.0622 (±0.0019)	0.0560 (±0.0001)	0.0459 (±0.0003)	0.0405 (±0.0005)	0.0431 (±0.0001)
K3	0.0660 (±0)	0.0684 (±0.0003)	0.0811 (±0.0002)	0.0510 (±0.0002)	0.0607 (±0.0003)	0.0508 (±0)
K4	0.0707 (±0.0006)	0.0889 (±0)	0.0861 (±0.0002)	0.0863 (±0.0003)	0.0633 (±0.0003)	0.0549 (±0)
K5	0.0504 (±0.0008)	0.0457 (±0)	0.0609 (±0.0001)	0.0533 (±0.0003)	0.0452 (±0.0004)	0.0330 (±0.0003)
K6	0.0553 (±0.0011)	0.0506 (±0.0005)	0.0632 (±0.0003)	0.0492 (±0.0003)	0.0533 (±0.0003)	0.0379 (±0.0003)
K7	0.0711 (±0)	0.0682 (±0.0003)	0.0760 (±0.0002)	0.0454 (±0.0004)	0.0352 (±0.0003)	0.0432 (±0.0003)
K8	0.0772 (±0.0003)	0.0808 (±0.0007)	0.0889 (±0.)	0.0811 (±0.0002)	0.0503 (±0.0005)	0.0457 (±0)

Table VI. Loop Length (Cm) of the Eight Knitted Fabrics Upon Immersion.

Sample Code	Time (Days) 0	14	30	60	90	120
K1	0.1875	0.1528	0.1111	0.1531	0.1406	0.1667
K2	0.1071	0.1500	0.1625	0.1750	0.1563	0.1571
K3	0.1667	0.2083	0.1875	0.2000	0.1786	0.1625
K4	0.1786	0.2188	0.2708	0.2222	0.1389	0.1429
K5	0.1875	0.1705	0.1571	0.1563	0.2000	0.1500
K6	0.1389	0.1563	0.1444	0.1563	0.1667	0.1667
K7	0.1550	0.1786	0.1778	0.2222	0.1571	0.1583
K8	0.1607	0.2083	0.1857	0.2083	0.1417	0.1625

range of diameter (at 0 day), all bicomponent fabrics narrowed their differences in diameter and reached a common value (1.80 cm) at the end of the immersion period (120 days). In addition, all of the bicomponent fabrics in the 250 denier series (K2-K4) showed a continuous increase in diameter until 60 days, decreasing sharply thereafter. This observation, however, was not obvious in the 200 denier series.

As a result of the changes in both courses and wales during immersion the fabric density would also change, which subsequently affected packing factor. As shown in Table VIII in general, the packing factor of

Table VII. Fabric Diameter (Cm) of the Eight Tubular Knitted Fabrics Upon Immersion.

Sample Code	Time (Days) 0	14	30	60	90	120
K1	1.50	1.55	1.65	1.60	1.70	1.70
K2	1.80	1.85	1.95	2.10	1.85	1.80
K3	2.00	2.05	2.05	2.20	1.80	1.80
K4	2.25	2.40	2.35	2.30	1.80	1.80
K5	1.50	1.60	1.60	1.70	1.70	1.70
K6	1.75	1.80	1.75	1.80	1.83	1.80
K7	2.00	2.10	2.00	2.00	1.80	1.80
K8	2.10	2.10	2.10	2.00	1.85	1.80

the bicomponent fabrics exhibited a minimal pattern with time, meaning it decreased initially after immersion, reached a minimum at about 30 to 60 days, and then increased thereafter. The final value may or may not be greater than the 0 day value depending on the PGA composition and Dacron denier. This change in packing factor indicated that the fabrics went through a series of changes of their tightness in structure due to the absorption of PGA yarns during immersion. The control fabrics, except K5, however, exhibited an overall decrease in packing factor. K5 showed the similar minimum pattern as the bicomponent fabrics. This clearly indicates that immersion alone would alter the fabric structure and would make them either tighter or looser with time.

3. Physical and Mechanical Property Changes Upon Immersion

Weight loss

The weight loss measurement was used to determine the amount of polyglycolic acid (PGA) degraded as a result of immersion time. The data are shown in Figures 1 and 2. Regardless of the composition ratio of PGA to Dacron yarns, the biggest drop in weight occurred between 14 to 60 days depending on the PGA composition. This was the period where the greatest degree of PGA degradation occurred. The percent of total weight loss of the knitted specimens was about from 2 to 7% at 14 days, from 18 to 24% at 30 days, from 26 to 43% at 60 days, and from 29 to 50% at 90 days. Obviously, the bicomponent fabrics with a higher PGA composition would show a higher percent of weight loss than those with a lower PGA

Table VIII. Packing Factor of the Eight Tubular Knitted Specimens Upon Immersion.

Sample Code	Time (Days) 0	14	30	60	90
K1	18.34 (±0.3857)	19.96 (±0.7289)	17.38 (±0.2248)	18.02 (±0.9784)	11.67 (±0.0866)
K2	21.40 (±2.16)	18.98 (±0.4957)	19.37 (±1.1561)	18.25 (±0.5774)	23.80 (±1.7551)
K3	20.45 (±0.1386)	17.94 (±0.6619)	13.30 (±0.0961)	16.97 (±0.1000)	15.27 (±0.2252)
K4	23.84 (±0.3855)	15.68 (±0.4601)	14.70 (±0.1779)	10.60 (±0.1905)	16.64 (±0.3843)
K5	15.37 (±0.3821)	14.50 (±0.5851)	11.27 (±0.2829)	11.86 (±0.0693)	16.07 (±1.0681)
K6	23.31 (±0.5797)	23.76 (±0.7716)	15.73 (±0.1002)	17.10 (±0.4388)	14.55 (±0.4965)
K7	18.09 (±0.2656)	16.98 (±0.2610)	13.23 (±0.3537)	17.62 (±0.4844)	23.20 (±0.5095)
K8	18.70 (±0.3118)	18.06 (±0.5823)	13.34 (±0.0635)	11.44 (±0.5122)	17.64 (±0.3630)

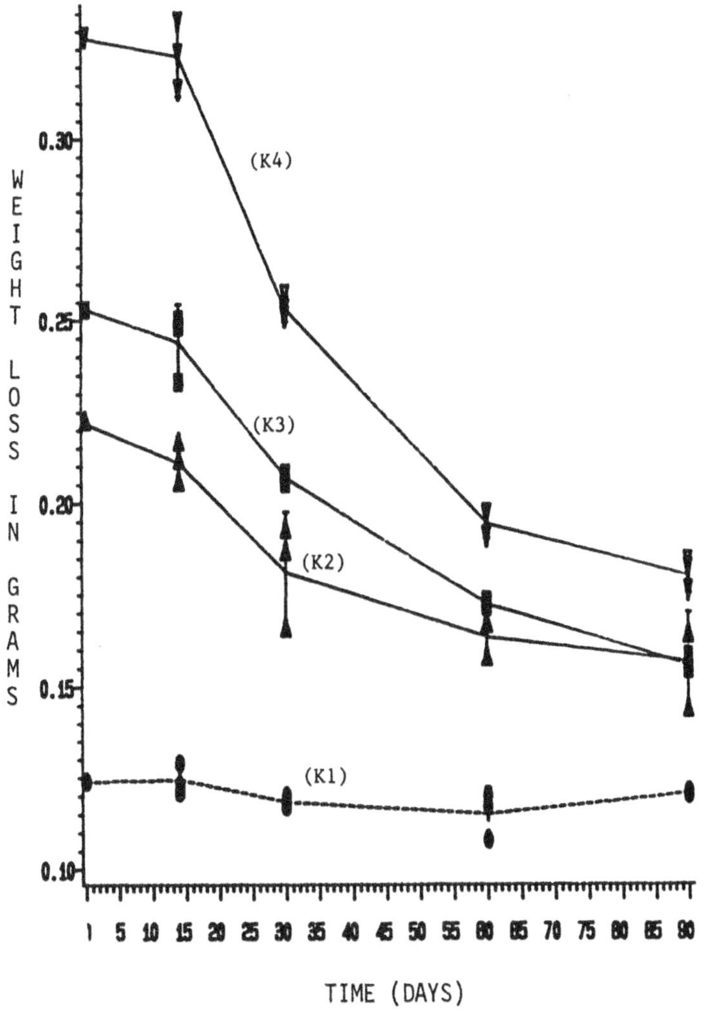

Figure 1. The changes of weight of the bicomponent knitted fabrics upon immersion in phosphate buffer of pH = 7.4, 37°C. (●)(----) K1; (▲) K2; (■) K3; (▼) K4.

composition. At 60 days PGA was almost totally degraded in K2 and K6, as evident by the smaller slope of the curves and by the comparison of the initial percent of PGA (Table I) and the total percent of weight loss data, while the bicomponent fabrics having a higher PGA composition (i.e. K3, K4, K7, K8) still had a slightly higher slope indicating there were still small amounts of PGA yarns in the structure. Control fabrics (K1 and K5) had slight changes in weight and showed a relatively constant weight during the period of study.

Water Porosity

The ability to change porosity with time is the most important advantage of these newly made bicomponent fabrics. As shown in Figures 3 and4, all the bicomponent fabrics, except K4, exhibited a net gain in water porosity from 22% to 117% at the end of immersion period (120 days). It was, however, important to understand that an increasing PGA composition

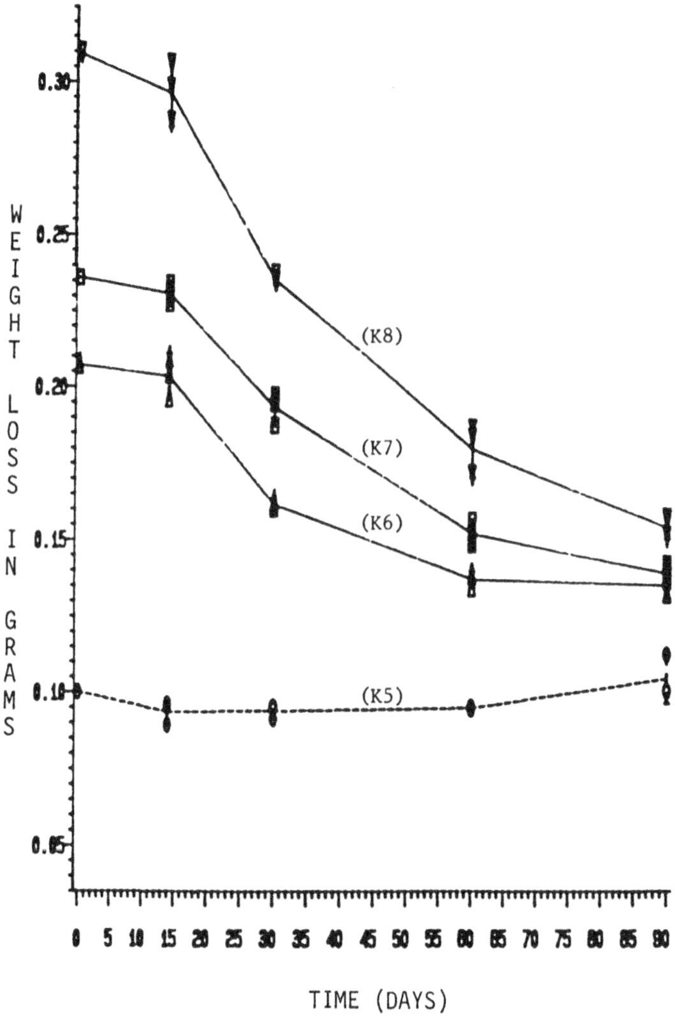

Figure 2. The changes of weight of the bicomponent knitted fabrics upon immersion in phosphate buffer of pH = 7.4, 37°C. (○)(----) K5; (△) K6; (□) K7; (▽) K8.

in the structure did not necessarily result in a higher water porosity as one would expect. In other words, the most significant change in water porosity occurred in bicomponent fabrics having the smallest PGA composition, such as K2 and K6. This indicates that no clear advantage would be found to incorporate too many biodegradable yarns into the bicomponent structure in order to increase its water porosity later. A typical example is specimen K4. For reasons unclear at this time, K4 showed a lower water porosity than 0 day specimen at all 5 measure periods (14, 30, 60, 90 and 120 days). On one occasion (90 days), the percent of porosity loss was as high as 44% of its original value.

Among all the bicomponent fabrics only K2, K6, K7 and K8 specimens showed a continuous increase in water porosity with time to 90 days, with a subsequent slight decrease to 120 days. This means their water porosities reached a maximum at 90 days. The highest percent increase in water

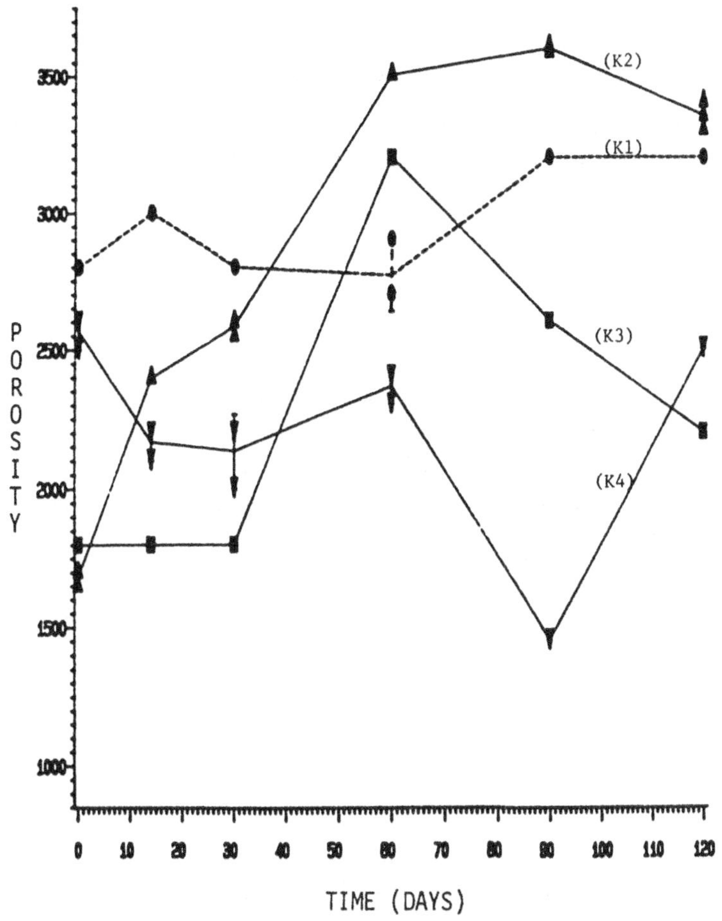

Figure 3. The changes of water porosity of the bicomponent knitted fabrics upon immersion in phosphate buffer of pH = 7.4, 37°C. (●) (----) K1; (▲) K2; (■) K3; (▼) K4.

porosity at this time interval (90 days) was found in K6 (128%) followed by K7 (122%), K2 (113%), and K8 (13%). This information also indicated that the 200 denier series fabrics had a more consistent performance in water porosity than the ones in the 250 denier series. Other bicomponent fabrics (K3 and K4) exhibited fluctuating water porosities during immersion.

It is interesting to note that the two controls also changed their porosities with time even though they did not have the biodegradable component in their structure. They also behaved quite differently from each other and from the bicomponent fabrics. K1 showed an overall gain in water porosity (14%), while K5 showed a continuous decrease in water porosity to 90 days (33% loss), and then recovered the majority of its loss at 120 days to make a net loss of about 8%.

One-way analysis of variance indicated all the above changes in water porosity with time were statistically significant.

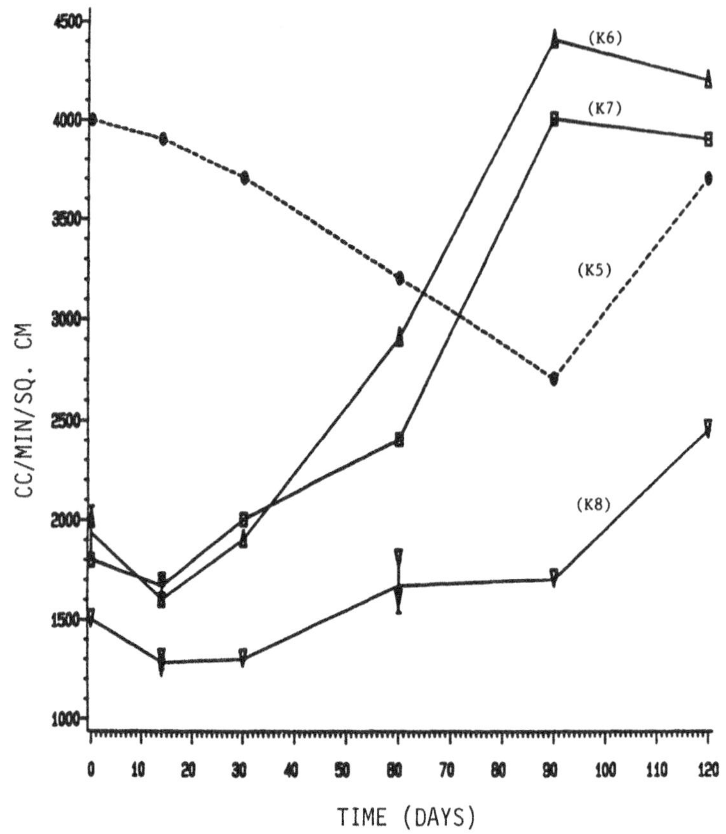

Figure 4. The changes of water porosity of the bicomponent knitted fabrics upon immersion in phosphate buffer of pH = 7.4, 37°C. (○) (----) K5; (△) K6; (□) K7; (▽) K8.

<u>Fabric Stiffness</u>

The loss of the PGA component in the fabrics upon immersion would also change their stiffness, as shown in Table IX. Generally speaking the stiffness of all bicomponent fabrics increased at the early periods of immersion, and then all decreased drastically as immersion proceeded further, usually around 30 days depending on the Dacron denier (i.e., 200 or 250 series) and the PGA composition. They recovered part of their losses in stiffness at 60 days (not in the case of 200 deniers which showed a continuous decrease in stiffness after 30 days). Thus, these bicomponent fabrics showed the highest stiffness at either 14 or 30 days. After most or all of their PGA component degraded, they had stiffness values significantly lower than their original 0 day values. For example, at 120 days, K6 and K7 had only 7.1 and 4.5% of their corresponding original stiffness. The 250 denier series bicomponent fabrics usually had much higher stiffness than the 200 denier series at 120 days. The two control fabrics showed relatively constant stiffness with time of immersion. Therefore, the observed change in stiffness in the bicomponent fabrics must be due to the degradation of the PGA component yarns upon immersion.

Table IX. Fabric Stiffness (lbs.) of the Eight Tubular Knitted Fabrics Upon Immersion.

Sample Code	Time (Days) 0	14	30	60	90	120
K1	0.31 (±0.1271)	0.33 (±0.1158)	0.24 (±0.0365)	0.28 (±0.0299)	0.23 (±0.0450)	0.30 (±0.1638)
K2	1.43 (±0.1223)	2.27 (±0.5764)	3.21 (±0.1238)	0.74 (±0.2423)	0.37 (±0.0556)	0.35 (±0.0618)
K3	4.68 (±1.3409)	6.47 (±1.3944)	7.24 (±1.7500)	1.41 (±0.3608)	1.94 (±0.4361)	1.51 (±0.1873)
K4	8.12 (±1.3761)	9.56 (±1.7702)	----	4.93 (±1.2200)	6.53 (±1.7872)	6.21 (±1.9093)
K5	0.11 (±0.0330)	0.11 (±0.0200)	0.18 (±0.0141)	0.44 (±0.0556)	0.09 (±0.0141)	0.13 (±0.0299)
K6	2.24 (±0.1565)	2.67 (±0.4171)	3.02 (±0.3624)	0.93 (±0.2142)	0.13 (±0.0271)	0.16 (±0.0320)
K7	4.70 (±1.3126)	6.50 (±1.3587)	5.92 (±0.5655)	0.18 (±0.9500)	0.11 (±0.0096)	0.21 (±0.0506)
K8	----	----	----	9.13 (±2.0717)	2.49 (±0.7417)	2.08 (±0.3590)

Bursting Strength

As shown in Figures 5 and 6, all bicomponent fabrics exhibited an initial significant loss of bursting strength and then regained parts of their losses at a later time. In other words, a pattern of minimum was observed and occurred from 14 to 30 days. The amount of decrease in bursting strength was the highest in K4 and K8 due to their highest PGA composition. Their corresponding percents of loss at 14 days were 50% and 66% of their original values, respectively, while the bicomponent fabrics having the least amount of PGA yarns (K2 and K6) lost only 30% and 40% of their corresponding initial strength, respectively. The recovery of bursting strength usually occurred at 60 days and continuously improved thereafter. At the end of 120 days, K2, K3 and K7 fabrics regained up to 90% of their original bursting strength, while K6 and K8 showed only about half of their original bursting strength, and hence remained the weakest. This observation indicates that the fabrics of the 250 denier series showed better recovery in strength than the ones in the 200 denier series, and therefore had better overall bursting strength at the end of immersion.

The initial loss of bursting strength was the result of the partial degradation of PGA yarns in the bicomponent fabrics which weakened the fabrics. As PGA yarns degraded, fabrics with larger interstices were produced. As a result fewer yarns were available per unit area to resist the puncture of the steel ball. The timing of this initial loss of

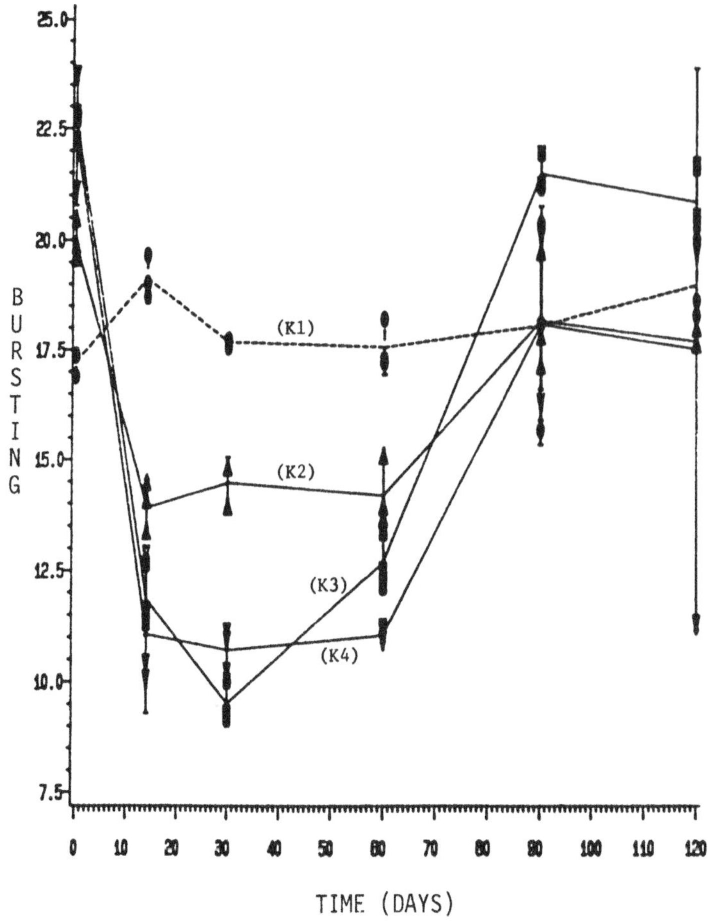

Figure 5. The changes of bursting strength of the bicomponent knitted fabrics upon immersion in phosphate buffer of pH = 7.4, 37°C. (●)(----) K1; (▲) K2; (■) K3; (▼) K4.

strength appeared to be consistent with the weight loss data and gross morphological observations.

The control fabrics (K1 and K5) behaved differently from the bicomponent fabrics. They showed an 11 to 18% increase in bursting strength at 14 days then decreased to the value close to their original bursting strength thereafter. Thus, the control fabrics at the end of immersion were as strong as they were initially.

The linear regression analysis of the bursting strength data indicated that there was no correlation between bursting strength and immersion period, suggesting that bursting strength of the fabrics was not significantly affected by the loss of PGA components as immersion proceeded. The R^2 values of all the fabrics in the linear regression plot ranged from 0.020 to 0.115. A R^2 value lower than 0.8 shows no correlation.

Suture Retention Strength

To examine further the integrity of the bicomponent fabrics after PGA

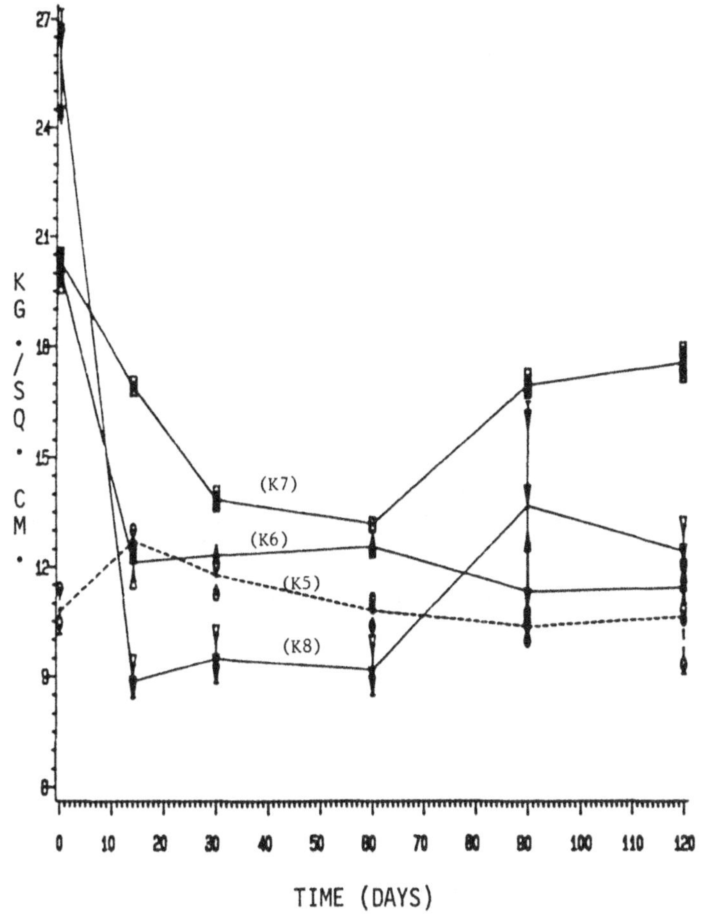

Figure 6. The changes of bursting strength of the bicomponent knitted fabrics upon immersion in phosphate buffer of pH = 7.4, 37°C. (○)(----) K5; (△) K6; (□) K7; (▽) K8.

degradation, suture retention strength was evaluated at both 2 and 5 mm distances from the edge of the fabrics.

Generally speaking, at the 5 mm distance, all fabric specimens showed an initial increase in suture retention strength at 14 to 30 days, as shown in Figures 7 and 8. Thereafter their strength decreased. At the end of immersion (120 days), all fabrics showed higher retention strength than their original values, and the percent of increment ranged from 142% (K1) to 8% (K7) depending on the fabric composition. The control fabrics had a much higher increase in retention strength than their corresponding component ones, particularly in the 250 denier series. Increasing PGA composition appeared to result in a smaller increase in suture retention strength. For example, the percentage of increment in suture retention strength for K2, K3 and K4 at 120 days were 51, 29 and 10% respectively. This relationship, however, was less obvious in the 200 denier series fabrics.

Upon immersion, the suture retention strength of the eight knitted fabrics at a 2 mm distance from the fabric edge was found to be less and more irregular in pattern than the 5 mm ones. This may be due to the

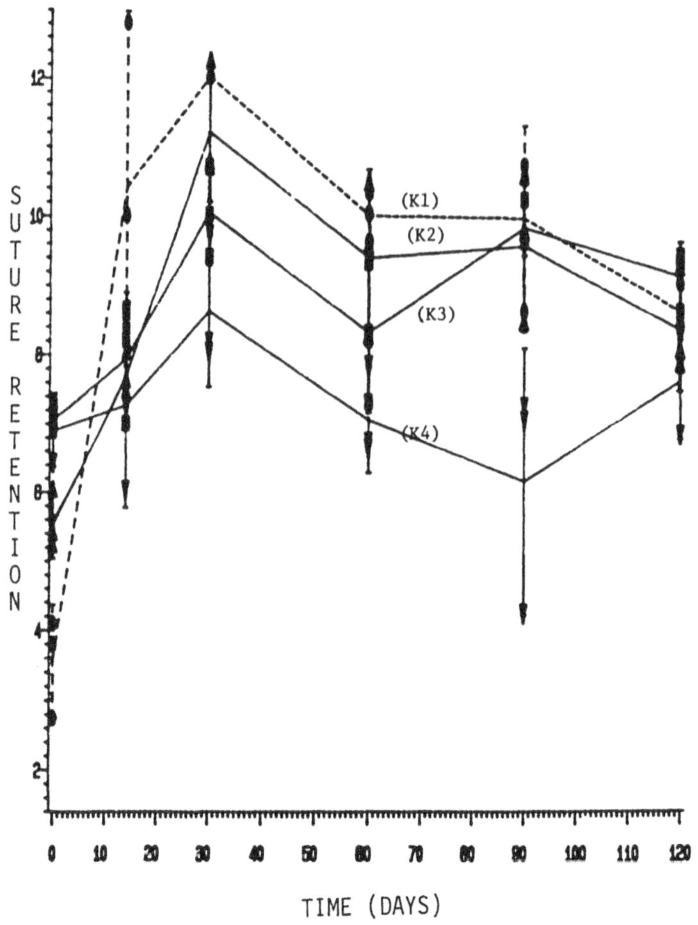

Figure 7. The changes of suture retention of the bicomponent knitted fabrics upon immersion in phosphate buffer of pH = 7.4, 37°C. The distance from the edge of the fabric was 5 mm. (●)(----) K5; (▲) K6; (■) K7; (▼) K8.

suture being too near the fabric edge causing the loops to unravel and fall off easily.

The linear regression analysis of the suture retention study indicated that there was no correlation between suture retention strength and immersion period. This suggests that suture retention strength was not significantly affected upon immersion. The R^2 values for 5 mm distance ranged from 0.006 to 0.256, while the same values for 2 mm distance ranged from 0.006 to 0.380. A R^2 value lower than 0.88 indicates no correlation between the two parameters in the plot.

Gross Morphological Observation Upon Immersion

The changes in gross morphology of the eight knitted fabrics after immersion were best observed by SEM micrographs. The thinning of the loops and their packing, resulting from the degradation of PGA yarns, were evident and are shown in Figures 9-11.

At 0 day, an increasing tightness of fabric structure and thickness

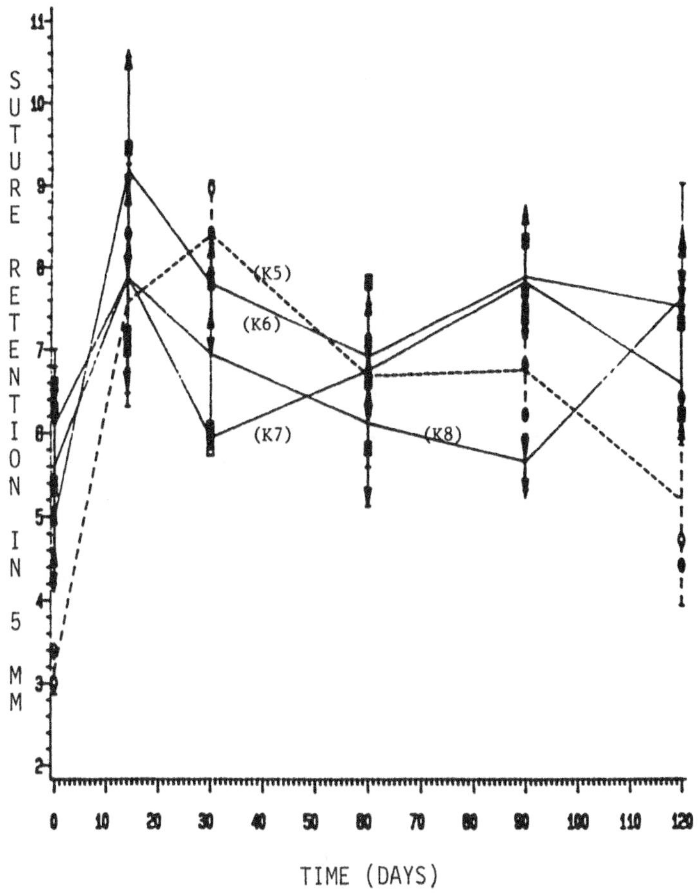

Figure 8. The changes of suture retention of the bicomponent knitted fabrics upon immersion in phosphate buffer of pH = 7.4, 37°C. The distance from the edge of the fabric was 5 mm. (○)(----) K5; (△) K6; (□) K7; (▽) K8.

of the knitted loops were found as more PGA yarns incorporated in the structure. As a result, the pores within the loops also decreased with increasing PGA yarns composition.

As immersion proceeded to 14 days, broken filaments in the knitted loops were evident in all bicomponent fabrics, indicating that a partial degradation of the PGA yarns occurred. This breakage of PGA yarns was more profound at 30 days and still evident in bicomponents fabrics having higher PGA compositions (e.g., K4, K8) at the end of the immersion period, 120 days. The bicomponent fabrics having the least amount of PGA yarns (e.g., K2 and K6) showed very little or no PGA yarn breakage after 60 days, indicating that nearly total PGA yarn degradation occurred in K2 and K6 at 60 days. As a result of this PGA yarn degradation, the knitted loops decreased in thickness and became thinner, and larger pore sizes were observed. It appeared, however, that the fabrics went through a dimensional change along with the thinning of the loops. This dimensional change, evident in the increasing loop packing per unit area, made the fabric structure tighter and less porous, and therefore partially neutralized the increment of porosity resulting from the PGA yarn degradation.

Figure 9. The changes of gross morphology of bicomponent knitted fabrics as a function of PGA/Dacron composition at 0 day. Upper left control K5; upper right K6; lower left K7; lower right K8.

DISCUSSION

The concept of using bicomponent fabrics for achieving low bleeding porosity at implantation and high healing porosity during the healing stage was first reported by Wesolowski in early 1960 (10). Basically, there were three different types of design in the Wesolowski study - collagen coated nylon or Dacron mesh, a taffeta tube with Dacron warp yarns and various composition ratios of Dacron to collagen filaments as filling yarns, and a taffeta tube with Dacron warp yarns and collagen core wrapped with Dacron yarns as filling yarns. However, because of uneven and untimely absorption of the collagen component, or the loss of structural integrity after collagen absorption, or the strong tissue reaction to collagen absorption, this first generation of bicomponent fabrics was not successful, even though some of the specimens had a 500x increase in water porosity.

To advance the composite vascular graft concept further, the incorporation of more biocompatible linear synthetic biodegradable aliphatic polyester fibers (e.g., polylactide, polyglycolide) as the component yarns in vascular graft fabrics was reported by Ruderman, et al. (11), Bowald, et al. (12-14), Greisler (15,16), and Gogolewski, et al. (17,18). Ruderman et al. used 40 denier PLA fibers in the filling direction of tightly woven fabrics with a final composition of 24% PLA and 76% Dacron

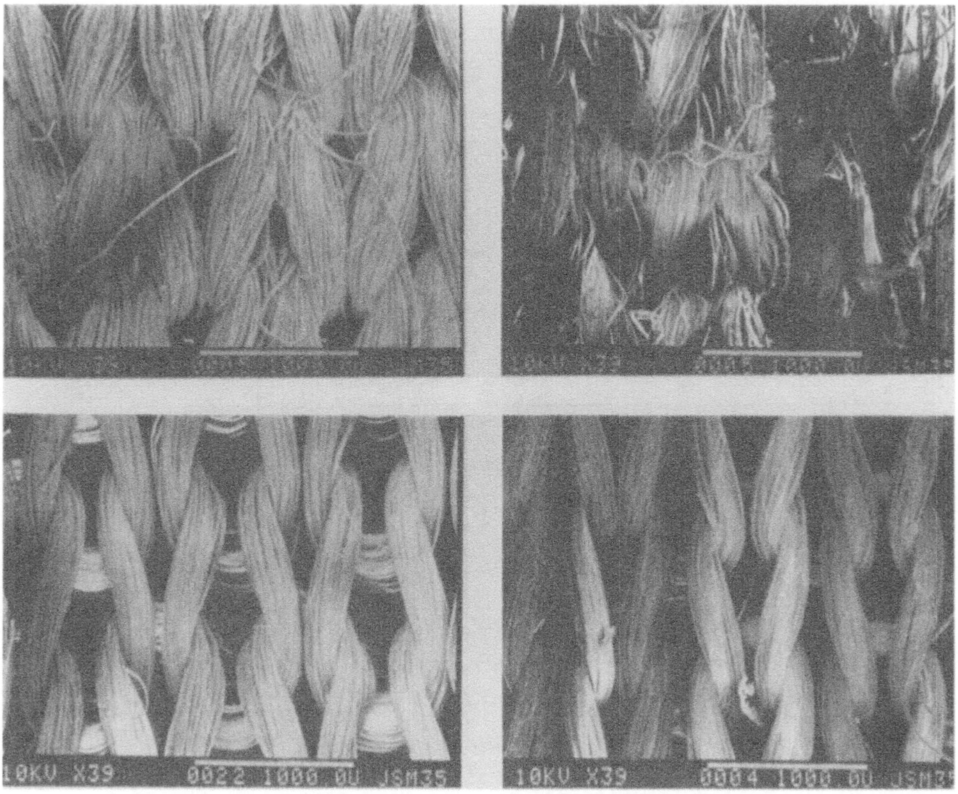

Figure 10. The changes of gross morphology of the bicomponent knitted fabrics (K7) upon immersion in phosphate buffer of pH = 7.4, 37°C. Upper left 0 days; upper right 60 days; lower left 90 days; lower right 120 days.

by weight (11). Its initial water porosity was 1066 ml/min/cm^2 at 120 mmHg pressure, and reached to 3297 ml/min/cm^2 after post-extraction of PLA component yarns. After approximately 100 days implantation in a canine abdominal aorta, all the PLA composite grafts were patent with no microscopic indication of calcification. The histological data revealed excellent ingrowth and adherence of external tissue with a very thin, well adherent pseudoendothelium.

Bowald, et al. used poly(glycolide-lactide) copolymer mesh fabrics or a combination of PGL and Dacron mesh fabrics either as a patch graft to repair of cardiac defects, or a tube to replace an aortic defect (12-14). In all of their cases (pigs), the biodegradation and the composite mesh fabrics allowed the arterial tissue regeneration with the formation of a smooth non-thrombogenic luminal surface following PGL absorption. This regeneration capability existed even with a graft as long as 15 cm. The newly formed tissue wall appeared to have adequate strength throughout the experimental period and was more resistant to lipid and calcium deposition than the fabrics made from Dacron or PTFE alone.

This arterial tissue regeneration capability over biodegradable scaffolds was further demonstrated in the Greisler, et al. studies in rabbit aortas (15,16). They implanted woven fabrics made from polyglycolide in rabbit aortas for 7-12 months. The results indicated that the biodegrad-

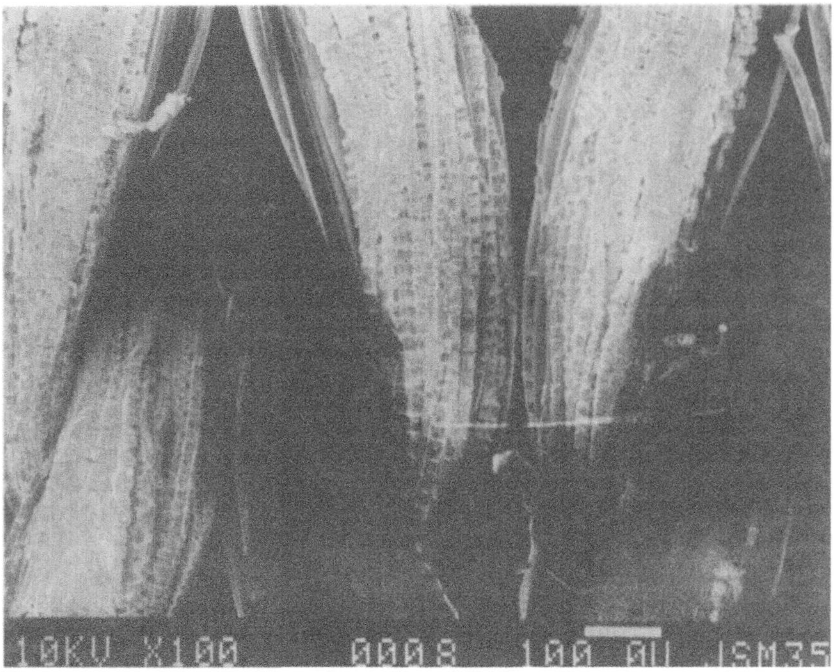

Figure 11. SEM micrograph of bicomponent knitted fabric K4 at 90 days of immersion in phosphate buffer of pH = 7.4, 37°C. The distinctive circumferential cracks and longitudinal microfibrils connecting two adjacent blocks were evident.

able PGA grafts were replaced by regenerating endothelialized blood vessels without thrombosis. The regenerated tissue consisted of an endothelial lined neointina, circumferentially oriented smooth muscle like myofibroblasts and dense fibroconnective tissue underneath the neointina, and an outer adventitial vascularized connective tissue layer. In a similar but more controlled and quantitative study, the same authors demonstrated that a neointina generated in the PGA graft was 3.2 times thicker (0.875 mm) than the grafts made from Dacron fibers alone (0.262 mm). All regenerated graft specimens were able to withstand three to five times systolic pressure after one week to three months, irrespective of the continuous degradation of PGA fibers during this period.

Recently, Gogolewski, et al. advanced the composite vascular grafts concept to a totally different aspect (17,18). Instead of varying the yarns in fabric structure in order to achieve biodegradability, Gogolewski, et al. produced a biodegradable microporous prosthetic vascular graft from a mixture of L-poly-lactide and polyethylene adipate and MDI based polyurethane polymer solutions. Formation of the neo-artery composed of intima, media and adventitia was found to be associated with the gradual biodegradation of the prosthesis. The neo-artery has the cellular structure similar to that of the original natural rat artery, particularly with regard to the presence of elastin fibers in the media, and hence is mechanically compatible with the adjacent vascular tissue.

All of the above studies indicate the common phenomenon that vascular grafts, which are either totally or partially biodegradable, are able to actively regenerate the arterial tissue. This arterial tissue regenerative capability appears to be strongly associated with the biodegradable polymers incorporated in the grafts. So far, linear aliphatic polyesters, such as PGA, PLA or their copolymers are the predominant biodegradable polymers in composite graft design. The ability for this class of polymers to regenerate tissues is not limited to soft tissues. Recently, Hollinger demonstrated that a biodegradable copolymer consisting of PLA and PGA at a composite ratio of 1:1 showed, after 7, 14, 21 and 28 days implantation in the tibias of rats, an accelerated rate of healing of the artificially created bony defects (19). It is believed the degradation products of these linear aliphatic polyesters, e.g., lactic and glycolic acids, must somehow play an unknown role in the stimulation of the regeneration of wounded tissues. Actually, Hunt, et al., has shown that lactate levels in the wound were extremely high, and high lactate would stimulate collagen synthesis in cell culture (20). This observation makes the use of polylactide and polyglycolide quite attractive in the wound repair of injured or diseased tissues of either a soft or hard nature.

In our study, we chose PGA as the biodegradable component in the vascular graft design not only for its demonstrated tissue regeneration capability but also for its relatively well studied degradation phenomena (21-25). An understanding of the degradation phenomena of the biodegradable component in the graft is important, because it is critical to match the rate of tissue ingrowth with the rate of degradation of the biodegradable component to maximize the use of the composite graft concept.

In this study, the overall changes in fabric properties of the eight knitted specimens were the result of the bicomponent composition of the fabrics and the duration of immersion. As immersion times increased, the PGA component was degraded, as evident from the SEM, weight loss, water porosity and changes in fabric diameter data. But, the alteration in physical and mechanical properties of the bicomponent jersey tubular knitted fabrics was not only caused by PGA degradation, but also the inherent relaxation shrinkage of the knitted fabrics. In other words, there are two opposite factors that greatly affected the changes in the properties of the bicomponent fabrics, relaxation shrinkage of knitted fabrics and absorption of the PGA component during immersion.

Relaxation shrinkage is the occurrence of dimensional changes in the fabric as it goes to its stable configuration. In a study by Hurt, it is assumed that if the stable configuration of the knitted loop is a right-angle diamond from a lattice model, any other shape will result in inherent shrinkage (26). The fabrics were under tension during the knitting process and once removed from the machine, the fabrics tended to relax. The single jersey knit fabrics made in this study are very unstable fabrics and relaxation shrinkage occurred more readily once they were removed from the knitting machine. Relaxation shrinkage tended to occur more freely when fabrics were immersed in water because water acted as lubricants and facilitated the movement of fibers and yarns in the structure.

As relaxation shrinkage occurred, the dimension as well as physical and mechanical characteristics of the knitted fabrics changed. Consequently, the number of courses increased while the number of wales decreased until the most stable fabric structure was obtained. These changes in width and length of the knitted fabrics would alter the shape of the knit loop and make the loops pack closely together, resulting in increasing thickness, packing factor, weight per unit area, stiffness and bursting strength. This increase, however, was at the expense of decreasing pore size, and hence porosity.

On the other hand, as the PGA component yarns in the bicomponent knitted fabrics started to degrade and were slowly absorbed as immersion progressed, the yarns in the fabrics became thinner causing their weight, thickness, packing factor, diameter, stiffness and bursting strength to decrease. The thinning of the yarns would also make the fabric interstices larger, which results in higher water porosity, and more relaxation shrinkage would occur because there were fewer frictional restraints applied to the yarns.

The overall effect depended on the relative magnitude of these two opposite factors. In this study, the overall effects are that as the bicomponent fabrics became more porous, their bursting and suture retention strength properties were reduced. Moreover, thickness, stiffness and packing factor were decreased. This was the result of the partial degradation of the PGA yarns which began at 14 days. However, as total absorption of PGA took place around 60-90 days, relaxation shrinkage of the fabrics readily occurred. This resulted in decreased porosity, a slightly decreased bursting and suture retention strength, and a slightly increased thickness, stiffness and packing factor. Although both mechanical and physical properties of the fabrics were altered by relaxation shrinkage and PGA absorption, the bicomponent fabrics eventually had a higher water porosity at 120 days than their initial values.

This overall increase in water porosity of the bicomponent fabrics makes feasible the concept of blending biodegradable and non-biodegradable yarns together in the fabric design, to achieve the two contradictory requirements (i.e., initial low bleeding porosity and late high healing porosity).

Although all of the bicomponent fabrics at 120 days exhibited water porosity lower than the optimum porosity of the graft, which was suggested to be in the range of 7,000 to 10,000 $cc/min/cm^2$ (27), their ability to increase water porosity with time makes them better future graft materials than the current single component commercial graft materials. As compared with the reported water porosity of the commercial vascular grafts (Table X), the ability to increase porosity of the bicomponent fabrics with time is an obvious advantage for facilitating better vascular connective tissue ingrowth without the expense of initial bleeding porosity. It was, however, evident from the study, as observed in K4, that there was no advantage to use more PGA yarns in the fabrics. The properties of fabrics were not improved when more than two PGA yarns are incorporated in the structure. Whether this _in vitro_ observation is true in the _in vivo_ situation is highly questionable because relaxation shrinkage would not occur _in vivo_ because the edges of the fabrics are securely fastened on the surrounding tissue. This secure fastening would not allow yarns to shift around as freely as in the _in vitro_ condition.

Furthermore, the bicomponent fabrics have another advantage over single component fabrics. The amount of materials remaining in the tissue will be less than the single component fabric due to the gradual absorption of the biodegradable component. It is well known that undesirable tissue reactions to implanted materials depend on the mass of the implant in addition to other factors. Hence, it would be ideal to design implants having biodegradable components to reduce the mass of an implant.

The SEM observation of the surface morphology of the bicomponent fabrics revealed a rather unique and interesting morphology of PGA yarns as they degrade. As observed in Figure 11, the degraded PGA fibers exhibited a so called shish kebab pattern after 90 days immersion in a buffer solution. In other words, the absorption of PGA fibers occurred selectively along the axis of the fibers and resulted in a series of blocks of

Table X. Comparison of Water Porosity Between Commercial and Newly Made Bicomponent Weft Knitted Surgical Fabrics at 120 mm Hg*.

Type	Initial Water Porosity (0 day, ml.min/cm^2)	Final Water Porosity (120 days, ml/min/cm^2)
DeBakey Vascular D	2,250	no change
DeBakey Standard	2,530	no change
DeBakey Ultra Lightweight	2,650	no change
Lopor	3,000	no change
Microknit	3,340	no change
Sauvage Guideline	3,420	no change
Milliknit	5,300	no change
K2	1,700	3,350 (97% increase)
K3	1,800	2,200 (22% increase)
K4	2,550	2,500 (2% decrease)
K6	1,950	4,200 (115% increase)
K7	1,800	3,900 (117% increase)
K8	1,500	2,450 (63% increase)

* The values of water porosity of commercial surgical fabrics are from Ref. 9.

PGA segments arranged relatively regularly along the fiber axis. The most unique aspect of this arrangement was the series of microfibrills which connected the two adjacent blocks of PGA segments together. Therefore, the whole fiber was not cut into fragments by hydrolytic degradation at the time of our SEM observation. This morphologic pattern of PGA fiber degradation is different from the previously reported observation (28). In that previous study, surface cracks were first observed on PGA fibers by day 40 immersion and the fibers were found to be chopped into small fragments. The majority of these fragments had relatively even and smooth cross sectional ends. There were no microfibrills protruding from the cross sectional ends of the cracks to connect the two adjacent portions of a PGA fiber together. A possible explanation to this discrepancy in morphology is that the PGA fibers in the knitted fabrics were subject to external stress during the fabrication of the tubular knit. This external stress may alter the degradation phenomena and subsequently be reflected in the observed PGA fiber morphology. In a recently published study, it was found that if any external stress or strain was applied to PGA fibers, it would accelerate the PGA fiber degradation (29). Whether this acceleration effect is related to the observed fiber morphology remains to be determined.

ACKNOWLEDGEMENT

The authors are grateful for the research grants provided by the

American Heart Association, New York Chapter, and the College of Human Ecology, Cornell University.

DEDICATION

This paper is specially dedicated to Shirley C. Chu who unexpectedly passed away recently. As the wife of Chih-Chang Chu, her devotion and dedication to her husband and family during the past two decades made Chih-Chang's research and teaching career possible, productive and successful. Her love of and care for people were beyond description and will be remembered forever by her family, relatives and friends.

REFERENCES

1. W. A. Dale, "Arterial Grafts: 1900-1978", In: "Graft Materials in Vascular Surgery", Herbert Dardik, Ed., Symposia Specialists Inc., Miami. FL, 1978.
2. R. S. Lord, "The Search for an Ideal Arterial Substitute", Australia-New Zealand Journal of Surgery, 44, 362-369 (1974).
3. N. Rosenberg, A. Simpson and R. Brown, :A Circumferentially Elastic Arterial Prosthesis: Three Year Study of a Dacron-Spandex Graft in the Dog", J. Surg. Res., 34, 7-16 (1983).
4. L. R. Sauvage, K. Berger and S. J. Wood,. et al. "Grafts for the 80's", Bob Hope International Heart Research Institute, Seattle, WA (1980).
5. ASTM D1777-64, "Measuring Thickness of Textile Materials", Annual Book of ASTM Standards, ASTM, Philadelphia, PA, Part 32, pp. 383 (1980).
6. ASTM D4032-81, "Stiffness of Fabric by the Circular Bend Procedure", Annual Book of ASTM Standards, ASTM, Philadelphia, PA, Part 32, pp 892 (1980).
7. S. A. Wesolowski, C. Fries, J. McMahon and A. Martinez, "Evaluation of a New Vascular Prosthesis with Optimal Specifications", Surgery, 59, 40-56 (1966).
8. ASTM D3787-80a, "Bursting Strength of Knitted Goods - Constant - Rate - of Traverse (CRT) Ball Burst Test", Annual Book of ASTM Standards, ASTM, Philadelphia, PA, Part 32, pp. 815 (1980).
9. W. King, P. Blias, R. Guidoin, E. Prowse, M. Marcois, C. Gosselin and H. P. Noel, "Polyethylene Terephthalate (Dacron) Vascular Prostheses - Materials and Fabric Construction Aspects", In: "Biocompatibility of Clinical Implant Materials", D. F. Williams, Ed., Vol. II., Chapter 8, CRC Press, Boca Raton, FL (1981).
10. S. A. Wesolowski, C. Fries, R. Domingo, W. Liebig and P. Sawyer, "The Compound Prosthetic Vascular Graft: A Pathologic Survey", Surgery, 53, 19 (1963).
11. R. J. Ruderman, A. F. Hegyeli, B. G. Hattler and F. Leonard, "A Partially Biodegradable Vascular Prostheses", Trans. Amer. Soc. Artif. Int. Org., 28, 30-33 (1972).
12. S. Bowald, C. Busch and I. Eriksson, "Arterial Regeneration Following Polyglactin 910 Suture Mesh Grafting", Surgery, 86(5), 722-729 (1979).
13. S. Bowald, C. Busch and I. Eriksson, "Absorbable Material in Vascular Prostheses: A New Device", Acta. Chir. Scand., 146, 391-395 (1980).
14. S. Bowald, C. Busch, I. Eriksson and T. Åberg, "Repair of Cardiac Defects with Absorbable Material", Scand. J. Thor. Cardiovas. Surg., 15, 91-94 (1981).
15. H. Greisler, "Arterial Regeneration Over Absorbable Prosthesis", Arch. Surg., 117, 1425-1430 (1982).
16. H. Greisler, D. U. Kim, J. B. Price and A. B. Voorhees, "Arterial

Regeneration Activity After Prosthetic Implantation", Arch. Surg., 120, 315-323 (1985).
17. S. Gogolewski and A. J. Pennings, "Groningen Biodegradable Vascular Prostheses, " In: "Biomaterials and Biomechanics 1983", Eds., P. Ducheyne, G. Vander Pene and A. E. Aubert, Elsevier Sci. Publ., B. V. Amsterdam (1984).
18. S. Gogolewski and A. J. Pennings, "Growth of a Neo-Artery Induced by a Biodegradable Polymeric Vascular Prosthesis ", Makromol. Chem., Rapid Commun., 4, 213-219 (1983).
19. J. O. Hollinger, "Preliminary Report on the Osteogenic Potential of a Biodegradable Copolymer of Polylactide (PLA) and Polyglycolic (PGA)", J. Biomed. Mater. Res., 17, 71-82 (1983).
20. T. K. Hunt, W. B. Conolly, S. B. Aronson and P. Goldstein, "Anaerobic Metabolism and Wound Healing: A Hypothesis for the Initiation and Cessation of Collagen Synthesis in Wounds", Am. J. Surg., 135, 328 (1978).
21. C. C. Chu, "The Degradation and Biocompatibility of Suture Materials", In: "CRC Critical Reviews in Biocompatibility", David F. Williams, Ed., Vol. 1, Issue 3, 261-322 (1985), CRC Press, Boca Raton, FL.
22. A. R. Katz and R. J. Turner, "Evaluation of Tensile and Absorption Properties of Polyglycolic Acid Sutures", Surg. Gynecol. Obstet., 131, 701 (1970).
23. A. M. Reed and D. K. Gilding, "Biodegradable Polymers for Use in Surgery Poly(glycolic)/Poly(lactic acid) Homo and Copolymers II. In Vitro Degradation", Polymer, 22(4), 494 (1981).
24. C. C. Chu, "Hydrolytic Degradation of Polyglycolic Acid: Tensile Strength and Crystallinity Study", J. Appl. Polym. Sci., 26, 1727-1734 (1981).
25. C. C. Chu and May Louie, "A Chemical Means to Examine the Degradation Phenomena of Polyglycolic Acid Fibers", J. Appl. Polym. Sci., 30, 3133-3141 (1985).
26. F. N. Hurt, "The Stabilization of Knitted Fabrics", "Textile Institute and Industry", 230-233 (1966).
27. S. A. Wesolowski, W. Lieberg, W. Golaski and C. Fries, "Knitted Arterial Prosthesis with Optimum Characteristics", Circulation 28, 825 (1963).
28. C. C. Chu and N. D. Campbell, "Scanning Electron Microscopic Study of the Hydrolytic Degradation of Polyglycolic Acid", J. Biomed. Mater. Res., 16(4), 417-430 (1980).
29. N. D. Miller and D. F. Williams, "The In Vivo and In Vitro Degradation of Poly(glycolic acid) Suture Materials As a Function of Applied Strain", Biomaterials, 5 (6), 365-368 (1984).

SODIUM NEUTRALIZATION IN A POLYETHYLENE IONOMER AND

BLOOD/MATERIAL INTERACTIONS IN AN *IN VIVO* CANINE MODEL

Jean A. Goggins[a,b], Anne Hiltner[b] and Richard D. Jones[c]

(a) Meadox Medicals, Inc.
Oakland, NJ 07436
(b) Departments of Biomedical Engineering and
Macromolecular Science
Case Western Reserve University
Cleveland, Ohio 44106
(c) Division of Surgical Research
St. Luke's Hospital
Cleveland, Ohio 44104

An ethylene/methacrylic acid copolymer was used to study the influence of sodium neutralization on blood/material interactions in a canine intra-arterial model. Material properties evaluated were sodium concentration, thermal characteristics, surface morphology, surface energy, surface clustering of the methacrylate component, hydrophilicity, sodium release, and acid group dissociation. The surface properties of the polyethylene control, 0% Na, and 50% Na samples tended to be similar; these properties were different from the 100% Na sample. The 100% Na ionomer was more hydrophilic, it had a higher polar component for its surface energy, and it was unique in exhibiting discrete anionic clusters 1 to 10 μm in diameter on its surface. These differences were manifested *in vivo* by platelet aggregation and thrombus formation on the polyethylene, 0% Na, and 50% Na implants, while the 100% Na implants'surface was predominately covered by singly adherent platelets. It is proposed that the lower levels of thrombus formation on the 100% Na ionomer are due to the cluster development in the neutralized methacrylic component. These cluster regions are polar and hydrophilic. In an aqueous environment, the polymer surface contains discrete, negatively charged regions containing mobile polymer chains. Implants that have these clustered regions on their surface exhibit platelet adhesion, but little thrombus formation.

INTRODUCTION

The surface of a polymeric material influences the response elicited when exposed to circulating blood. It is extremely important that the surface of a blood contacting material be thromboresistant to prevent vascular occulusion or generation of emboli. Most currently utilized materials depend on an ultimate pseudoneointimal lining, although some components, such as valve cages, remain clean and apparently passive in nature.

Previous research in this laboratory showed that when the carboxylic acid component of an α-amino acid copolymer, poly(γ-benzyl L-glutamate co-L-leucine) 50:50, was neutralized with sodium, thrombus formation was reduced (1). Other work indicated a relationship between surface charge and amount of thrombus formation for the α-amino acid copolymer, glutamic acid-leucine (2,3). It has been stated that surfaces that act as electron acceptors from blood at existing electrochemical potential will tend to be thrombogenic, while surfaces that act as electron donors will prevent thrombosis (4). Others report that some electronegative surfaces can be thrombogenic (5,6).

The purpose of the present research was to determine whether or not sodium neutralization would reduce thrombus formation in a copolymer other than an α-amino acid copolymer.

A polyethylene/methacrylic acid ionomer was selected as the test material, and a polyethylene sample provided by the National Heart, Lung, and Blood Institute was selected as the primary control. Material properties evaluated were sodium concentration, thermal characteristics, surface morphology, surface energy, surface clustering of the methacrylic component, hydrophilicity, sodium release, and acid group dissociation.

MATERIALS AND METHODS

Material Preparation

All copolymer and ionomer samples used in this study were from a single lot of Surlyn®, (S-1601, lot #875784L E. I. du Pont de Nemours, Inc., Wilmington, DE) kindly provided by Dr. T. R. Earnest, Jr. Nominal values for the properties of this ionomer resin may be found in a technical letter issued by the manufacturer (7). The polymer, as received, contained approximately 10% by weight (4.1 mole %) methacrylic acid and approximately 50% of the acid groups were neutralized to the sodium salt (8). Samples of this material are referred to as 50% Na. It was refluxed in equal volumes of either 1N HCL or 1N NaOH and tetrahydrofuran to produce the acid form of the copolymer (0% Na) and the completely neutralized form of the ionomer (100% Na), respectively (9). The contents of the refluxing flask were decanted into a Buchner funnel and the polymer was rinsed several times with methanol. After drawing off the final methanol rinse, the polymer pieces were set aside and allowed to dry.

A primary polyethylene control (Series #AB00228-00232, ABIOMED, Danvers, MA) as specified by the National Heart, Lung and Blood Institute (NHLBI) was used as received for comparison with the experimental polymer.

Polymer solutions were made by dissolving samples in a 60:40 mixture of perchloroethylene:n-butyl alcohol (Fisher Scientific) at temperatures greater than 70°C and less than 95°C (9). Polymer films 100 μm in thickness were cast onto glass plates from solutions of approximately 8% w/v concentration. The film casting temperature was maintained at 85 ± 5°C. All films were rinsed in methanol and placed in a vacuum oven at 55°C for 24 hours to ensure solvent removal. They were stored in a vacuum desiccator at room temperature until they were used. The air contacting side of the film was used as the blood contacting surface and for all surface characterization.

Material Characterization

After the reflux procedure, sodium content was measured by flame emission photometry (Chemistry department, St. Luke's Hospital, Cleveland, OH) and Fourier Transform Infrared Spectroscopy (FTIR), (FTS-14, Digilab, Cambridge, MA). Thermal properties were monitored by Differential Scanning Calorimetry (DSC), (DSC-2 Differential Scanning Calorimeter, Perkin Elmer, Norwalk, Conn.).

The characteristic topology of each sample was recorded by Scanning Electron Microscopy (SEM), (T. V. Mini-SEM^R, Mountain View, CA).

Contact angle measurements were made using the sessile drop technique on sterilized samples. Polymer critical surface tension was computed according to the method of Zisman (10). The polymer CSFEF (Composite Surface Free Energy Function) (11) and its polar and dispersive components were also evaluated according to the method described by Kaelble (12).

Surface anionic groups were visualized in the following manner: polymer samples were placed in distilled water, pH 6.7 at 37°C for 60 minutes. They were then transferred to a filtered, 0.1M silver nitrate solution for one minute, rinsed in H_2O, air dried, mounted on aluminum stubs and coated with palladium. A scanning microanalyzer was used to generate an image formed by back-scattered electrons (Analytical Microscope, JSM-840, JOEL, Peabody, MA). When stained regions were observed, they were analyzed for silver by Energy Dispersive X-Ray Analysis (EDXA), (Cambridge Stereoscan S4-20, Cambridge, England equipped with EDAX 9100 Quantitative System, Prairie View, IL).

Water absorption and sodium release by samples after incubation in distilled water (37°C, pH 6.8) were also monitored.

In Vivo Model

Adult, mongrel dogs weighing between 18 kg and 40 kg, with negative heartworm tests, were used in all experiments. Hematocrit and activated clotting times were measured in each animal and ranges permitted were 32-46% (13) for the former and 50-120 seconds for the latter. Animals were anesthetized with intravenous pentobarbital sodium (35 mg/kg), shaved, and prepared for surgery using aseptic techniques. They were allowed to breathe freely during surgery, no mechanical ventilation was provided. Cumming et al. have reported that mechanical ventilation inhibits platelet deposition on artificial materials.

Polymer films chosen for implant use were 100 μm in thickness. A rectangular segment 2 mm x 5 mm was cut from the sample and the glass-contacting side was attached to a double-armed 6-0 braided silk suture with a polyurethane adhesive. Sterilization was done by standard hospital procedure: ethylene oxide gas for 4.5 hours at 54.4^{o}C followed by vacuum degassing for 7.5 hours at 54.4^{o}C (St. Luke's Hospital, Cleveland, OH).

The implantation method used was originally developed by Van Kampen (15,16). Eight to ten centimeter segments of the carotid and femoral arteries were dissected free bilaterally. Implant segments were placed against the luminal surface of the vessel wall. Vessel dimensions ranged from 3 mm to 5 mm inner diameter. Each implant was wetted with sterile saline solution immediately before implantation.

Polymer implants were exposed to flowing blood in vivo for 60 minutes, after which they were harvested and prepared for evaluation by scanning electron microscopy. The method used by Helmus (2) was employed for scanning electron microscopic evaluation of the implant surface. This method provided data on the percentage implant surface area covered by each of the following categories: exposed polymer, singly adherent platelets, singly adherent white blood cells, dense fibrin, or thrombus (platelet and/or white cell aggregates, strands of fibrin and trapped red cells). All data were evaluated using Biomedical Data Analysis Programs, (Biomedical Data Processing (BMDP), Health Sciences Computing Facility, UCLA, Los Angeles, CA, Program revised 1979, 1981).

RESULTS

Material Characterization

Flame emission photometry indicated that the 0% Na copolymer cast from solution and the primary polyethylene control contained virtually no sodium. The 50% Na samples cast from solution and the 50% Na samples as received from the manufacturer both contained approximately 1% sodium by weight, with the samples cast from solution always containing slightly less (0.2%) than their original counterparts. An average value of 2.3% sodium was measured in the 100% Na samples cast from solution. The expected stoichiometric weight of sodium in the fully neutralized sample was approximately 2.13%, based on the mole and weight percentages (4.1 and 10 respectively) of the methacrylic component. Therefore, the original (as received) ionomer did have 50% of its methacrylic acid groups neutralized, but this amount decreased slightly after solution-casting. The 100% Na sample may have been slightly supersaturated.

Spectra obtained by FTIR of the polymer samples cast from solution clearly demonstrated the decrease in the hydrogen-bonded carbonyl peak at 1700 cm^{-1} and the development of a peak between 1560 cm^{-1} and 1580 cm^{-1} (asymmetric carboxylate stretching motion) (9) with increasing sodium content, Figure 1.

Melting temperature, crystallization temperature, melting enthalpy, and crystallization enthalpy values were all reduced by increasing sodium neutralization. The melting temperature changed only slightly, while the crystallization temperature indicated considerable supercooling with increasing sodium content. The supercooling that occurs during the crystallization temperature determination has been attributed to the increase in melt viscosity which occurs upon ionization (17). The degree of crystallinity of the copolymer and ionomer was less than the polyethylene homopolymer. The surface topology of all samples was similar at the 1 μm level.

The critical surface tension, the composite surface free energy function (CSFEF), its polar and dispersive components and the ratio of the polar component to the CSFEF are listed in Table I. The critical surface tension and the CSFEF decreased with increasing sodium concentration in the ionomer. The polar component of the surface energy of the polyethylene control, the 0% Na copolymer, and the 50% Na ionomer were similar; however, the polar component of the surface energy of the 100% Na ionomer was almost two times these other values. There were no significant differences between the surface energy properties of the polyethylene control and the 0% Na copolymer.

Backscattered electron images of silver-stained polymer surfaces indicated that the 100% Na samples had discrete anionic regions or clusters ranging in size from less than 1 μm to almost 10 μm in diameter scattered over the sample surface. Figure 2 shows backscattered images of

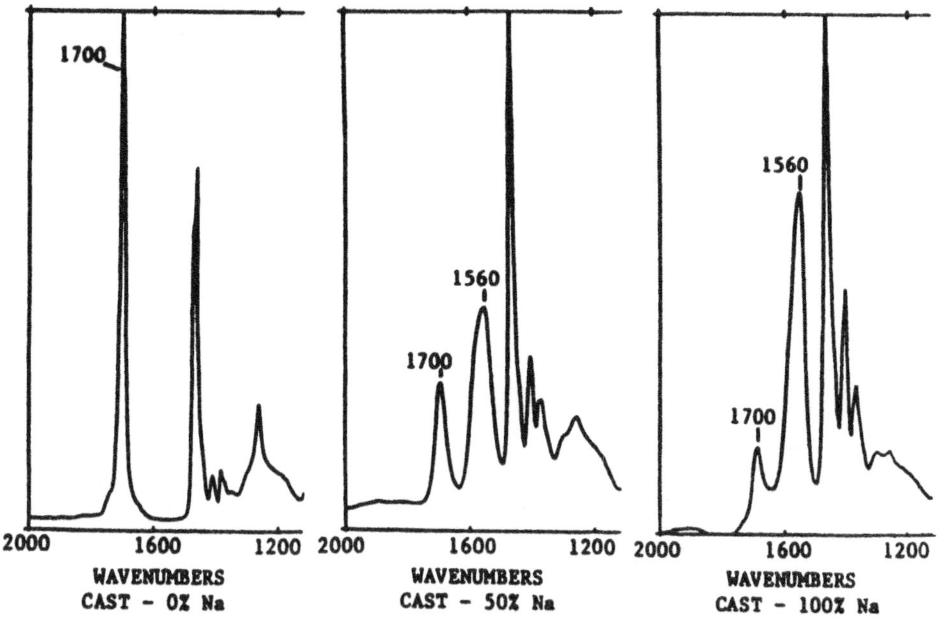

Figure 1. FTIR Spectra for thin films (25 μm thickness) cast from solution. As the amount of sodium increases, the hydrogen bonded carbonyl peak at 1700 wavenumbers diminishes and a new peak developes and increases at 1560 wavenumbers (asymmetric carboxylate stretching motion).

a typical sample of 50% Na (Figure 2a) and 100% Na, (Figure 2b.) EDXA of the cluster regions was positive for silver. Backscattering images of polyethylene and 0% Na were similar to the 50% Na samples; no silver stained clusters were visible.

Characterization of dynamic properties tended to separate the 100% Na samples from the other samples. The primary polyethylene control and 0% Na showed no significant increase in weight due to water uptake, while the 50% Na samples increased in weight by 5.7% at 24 hours and the 100% Na samples increased their weight in a range from 18 to 20% within 24 hours. This latter sample did not continue to absorb water with increasing time in solution, apparently having reached equilibrium by 24 hours, Figure 3. The lower crystallinity of the 100% Na sample, its higher sodium content and the clustering of the sodium-rich methacrylate regions all tended to increase its hydrophilicity (9,18,19).

Sodium release in water followed the water absorption pattern. Sodium was gradually lost from the 100% Na samples until, at 24 hours, an average 1.8% sodium remained (85% of that expected in a fully neutralized sample). No further sodium loss was detected for samples in water for as long as 7 days. The weight percentage of sodium in the 50% Na samples dropped from an initial average of 0.8% to 0.5% sodium at twenty-four hours, Figure 3. Polyethylene controls and 0% Na samples showed no change in sodium content regardless of the length of time they were incubated in distilled water.

Table I. Comparison of Surface Energy Properties.

Sample	Critical Surface Tension	Polar	Dispersive	CSFEF*	Polar / CSFEF
PE** NHLBI	34.1	2.9	39.0	41.9	0.07
PE Control	35.1 R = 0.97	5.7	36.2	41.8	0.14
0% Na	36.0 R = 0.96	6.9	35.2	42.1	0.16
50% Na	32.5 R = 0.97	7.1	29.2	36.3	0.20
100% Na	30.6 R = 0.96	13.2	23.1	36.3	0.36

* Composite Surface Free Energy Function = Polar Component + dispersive component.

** Values reported in NIH Contract Report (31).

All Values are in dynes/cm. R is the correlation coefficient for the regression analysis used to compute the critical surface tension.

Since the expected stoichiometric sodium concentration for the fully neutralized ionomer was 2.13%, the 100% Na ionomer maintained a high sodium concentration and it was significantly greater, in an aqueous environment, than the other samples.

In Vivo Model

Figure 4 indicates how sodium neutralization affected the percentage of implant surface that was covered by platelets, exposed polymer and thrombus. The primary polyethylene control, 0% Na and 50% Na implants elicited similar responses. Their surfaces were frequently covered with large amounts of thrombus, Figures 4 and 5B, D, F, and H. The upper portion of Figure 5B illustrates the type of thrombus most commonly encountered on 0% Na implants. White blood cells are visible near the thrombus and were always observed in areas where thrombus had pulled away from the implant surface during sample dehydration for scanning electron microscopy. Platelet deposits are located in the lower portion of Figure 5B and smaller groups of aggregated cells are present scattered over the remainder of the implant surface. Figures 5D, F, and H illustrate typical regions of platelet aggregation observed on the 0% Na implants. Statistically, the 100% Na implants were significantly different (p less than 0.05) from the other treatments for both the percentage of thrombus and singly adherent platelets on the implant surface, Figures 4 and 5A, C, E, and G. Combining platelet and thrombus percentages eliminated their individual differences, Figure 4. There was no significant difference for

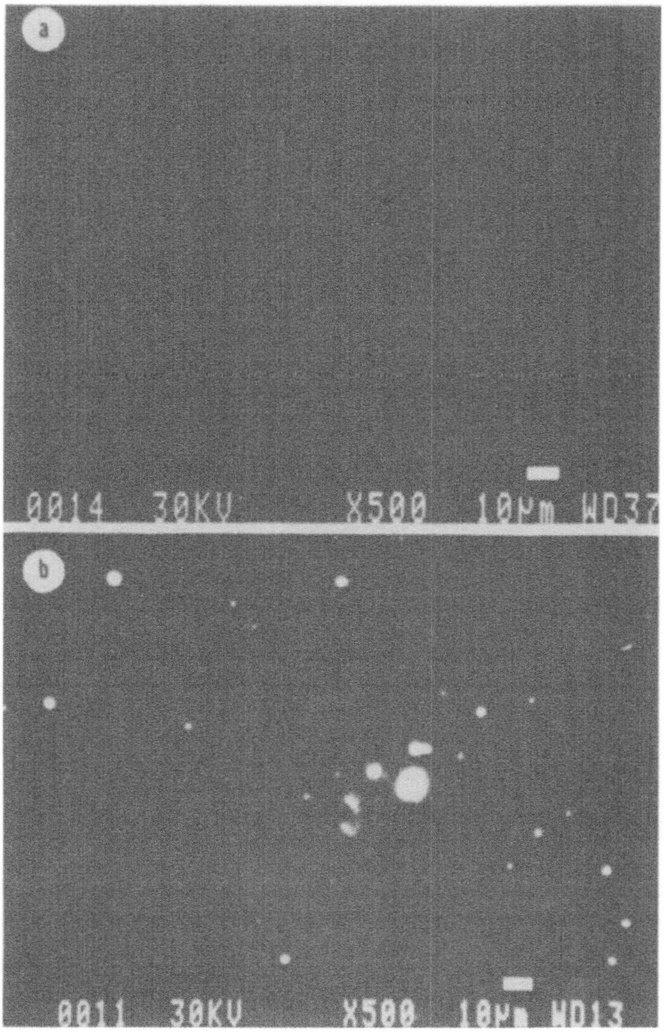

Figure 2. Representative backscattered electron images of a silver stained (a) 50% Na surface and (b) 100% Na surface.

exposed polymer surfaces among the implants, however when these values were combined with the platelet values, the level of difference between the 100% Na sample and the others was enhanced (p less than 0.025).

DISCUSSION

In this study, two different cellular responses were observed. In one, surfaces were clear (exposed polymer surface) or contained singly adherent platelets, Figures 5A, C, E, and G. In the other, aggregated cells (thrombus) (Figures 5B, D, F, and H were dominant. The <u>in vivo</u> blood response to polyethylene, the 0% Na copolymer and the 50% Na ionomer samples were similar. These implants tended to have large amounts of thrombus, accompanied by some white blood cells and fibrin, while the 100% Na samples often had their entire surface covered with singly adherent

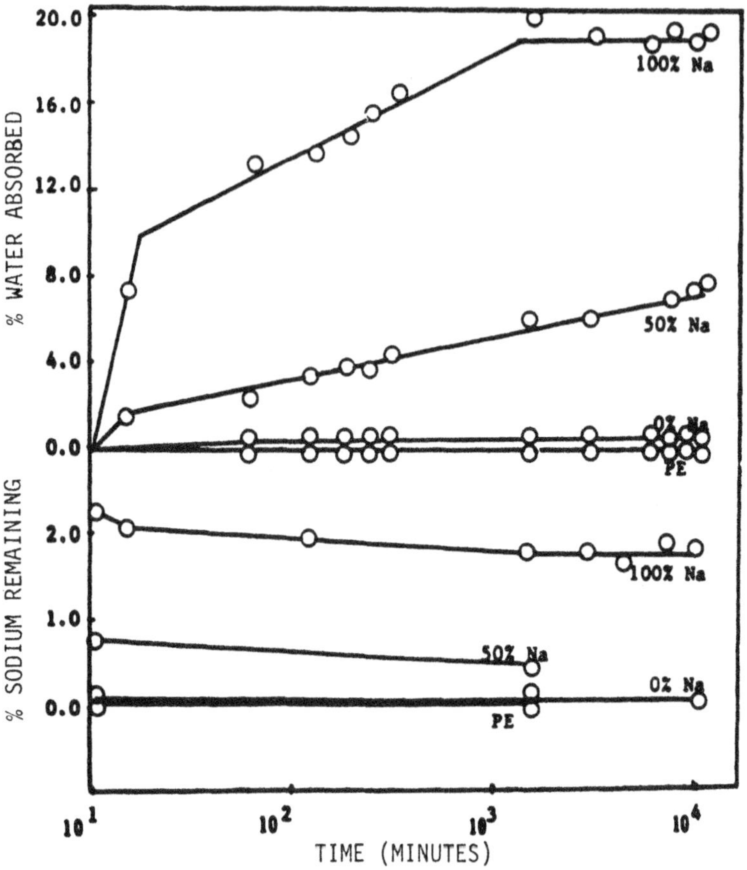

Figure 3. Percentage change in polymer weight due to water absorption at different times, upper section. Weight percentage of sodium remaining in polymer samples after extraction in water, lower section.

platelets. These findings are similar to those made by Van Kampen et al. who used poly(γ-benzyl-L-glutamate co L-leucine)-50:50, (Glu(OH):Leu-11). Glu(OH):Leu-11 has carboxylic acid groups that can be neutralized by sodium. Thrombus development on the acid form of Glu(OH):Leu-11 caused vessel occlusion within 20 minutes. Vessels containing implants of the sodium neutralized form of the copolymer remained patent for as long as seven days (1).

In evaluating the influence of sodium neutralization, the in vivo uniqueness of the 100% Na implants can also be found in the surface and dynamic properties of this ionomer. The ratio of the polar component of the surface energy to the total surface energy of polyethylene, 0% Na and 50% Na are all between 0.14 and 0.20, while the value for 100% Na is 0.36, Table I. Similarly, polyethylene, 0%Na and 50% Na absorb no more than 6% water, whereas 100% Na absorbs almost 20% water, Figure 3. Although the amount of sodium released by 50% Na and 100% Na samples is similar; 0.3% for the former and 0.5% for the latter, the amount remaining in the polymer is quite different; 0.5% and 1.8% respectively.

Our sodium release studies also indicated no measurable surface anionic groups and no hydrogen dissociation in the copolymer (0% Na) at physiological pH. In an earlier study (20), polyethylene controls and 0%

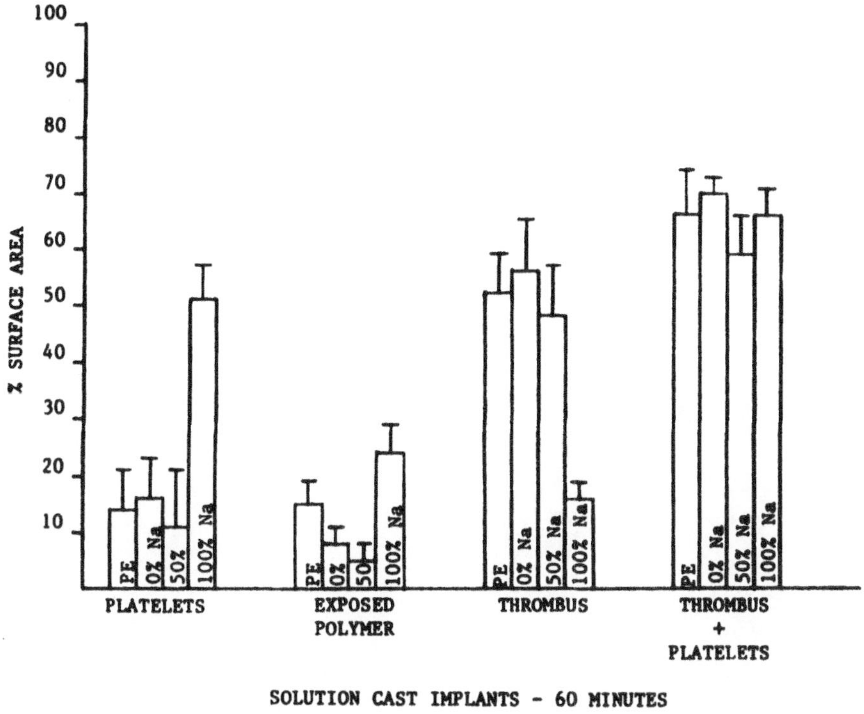

Figure 4. Percentage of implant surface covered by, from left to right; singly adherent platelets, exposed polymer, thrombus and the combination of platelets and thrombus - for solution-cast implants, following 60 minutes exposure to blood. Note that the platelet and thrombus response are inversely related.

Na samples showed no change in sodium content regardless of the length of time they were incubated in unbuffered saline, 0.1M phosphate buffered saline (ph 8.0), or 0.1N NaOH at pH 12. No water absorption was observed in these samples, with the exception of the 0% Na sample in sodium hydroxide, which increased its weight by 3% after 3 days, but did not absorb measurable amounts of sodium during this time. Since it might otherwise be assumed that some acid groups would dissociate and leave behind a negatively charged surface, this information further distinguishes the fully neutralized ionomer (100% Na) from the copolymer (0% Na). The similarity in surface energy properties between the polyethylene control and the 0% Na copolymer suggests the possible sequestering of the methacrylic component in the copolymer, leaving a surface rich in polyethylene.

Backscattered electron images of polymer samples after exposure to silver nitrate indicated that only the 100% Na ionomer developed ionized clusters of the methacrylate component, ranging in size from 1 to 10 μm in diameter. This ability to cluster has been reported previously. Longworth described a model for ionomers containing ionic clusters 100 A in diameter or greater in which the introduction of water tended to plastize the ionic segments. Transmission electron micrographs of 1 μm films of an ethylene:methacrylic acid copolymer cast onto RbOH solution contained stained clusters 1 μm in diameter and smaller (9). The discrepancy in cluster size in the current study may be due to two factors: first, our samples were 100 μm thick and in this bulk form more of the amorphous

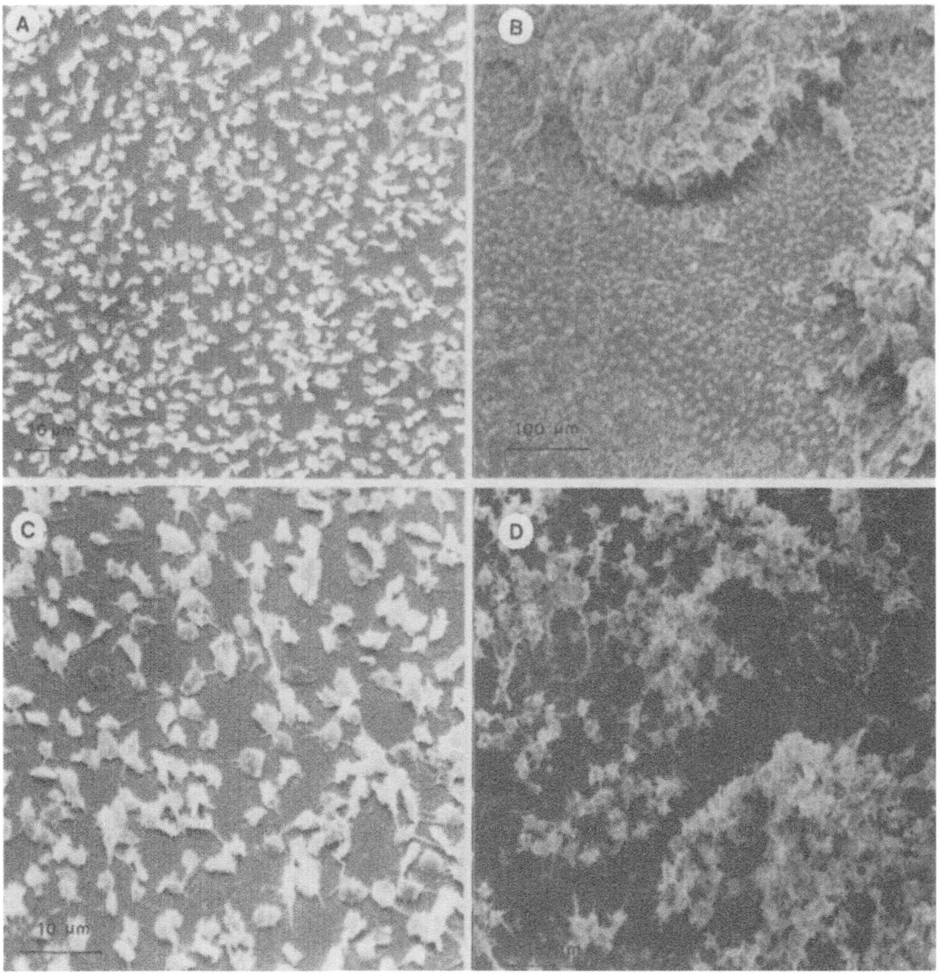

Figure 5. Scanning electron micrographs comparing 100% Na ionomer (A,C,E,G) [this page] and 0% Na copolymer (B,D,F,H) [facing page] implants after 60 minutes exposure to flowing blood. The paired micrographs progress from low to high magnification.

methacrylic component may be present at the sample surface, permitting larger cluster development; second, our samples were hydrated prior to staining allowing the methacrylic component to plasticize.

The fully neutralized ionomer takes up more water than other samples. This water is associated with the methacrylic component and causes the clusters to swell; it also lowers the glass transition temperature of the ionic segments (9,18). These factors should increase the mobility of the polymer chains in this region, permitting them to adjust to the _in vivo_ environment. The implanted ionomer equilibrated _in vivo_ because it was wetted with saline only immediately before implantaion.

Therefore, both characterization and implantation of these materials indicate that the 100% Na sample is different from the other samples. Fully neutralized ionomers are known to develop microphase separation, or

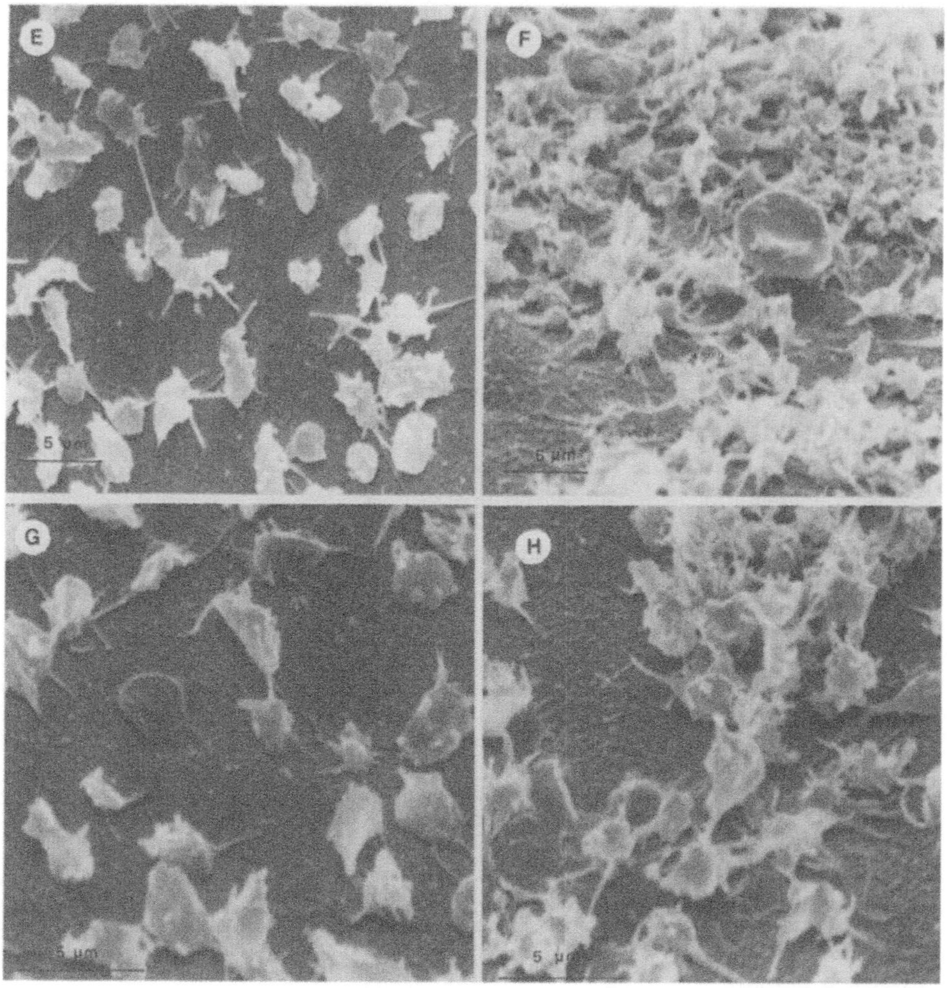

Figure 5. Continued.

cluster-like regions, which have a dramatic effect on polymer properties and account for the similarity of ionomers to block copolymers (21). We would like to suggest that the in vivo cellular response to this fully neutralized ionomer is attributed to the visible clustering of the methacrylic component. These clusters create a more hydrophilic, polar surface with anionic regions dispersed throughout the polyethylene component. Our in vivo sodium release results show very rapid ionization of the methacrylic groups which suggests that in vivo surface clusters become negatively charged as soon as they come into contact with the blood. Negatively charged surfaces have been credited with being both detrimental and beneficial. They enhance Factor XII activation in the presence of kallikrien and high molecular weight kininogen (22), but platelets and all blood cells and the vascular wall itself are negatively charged, so this condition is not sufficient in itself to cause thrombus formation in vivo. The vessel wall has a negative charge density one or two orders of magnitude greater than the cells found in the blood and non-adhesion of platelets to endothelium is attributed mainly to electrostatic repulsions between the negative charges present on both (23). Simionescu et al. reported that the vascular endothelium contains heterogeneously dispersed negatively charged sites (24).

Since it is platelets and thrombus that are affected by sodium neutralization in our experiments (Figure 4), it is probably platelet activation that initiates thrombus formation on the polyethylene, 0% Na and 50% Na implants. In normal hemostasis, platelet activation has been attributed to contact with collagen or thrombin (23). Platelets that adhere to the 100% Na implants do not appear activated and platelet adhesion itself does not necessarily lead to a thrombogenic response (25). Coleman et al. noted that clotting times were prolonged when singly adherent platelets were present on a surface (26).

There is general agreement that artificial materials in blood first interact with the plasma proteins and then with formed elements (9). The type of protein which is adsorbed (27,28) and/or any conformation changes induced by the material (29,30) may activate platelets and the coagulation response. It is possible that the 100% Na surface, like the heterogeneously dispersed negatively charged surface of the endothelium, does not directly activate platelets. Alternatively, the hydrated clusters may either adsorb non-thrombogenic proteins and/or not denature adherent proteins but, through limited molecular motion, adjust to prevent protein denaturation and subsequent platelet activation.

CONCLUSION

An ethylene/methacrylic acid copolymer was used to study the influence of sodium neutralization on blood/material interactions in a canine intra-arterial model. Material properties evaluated were sodium concentration, thermal characteristics, surface morphology, surface clustering of the methacrylate component, hydrophilicity, sodium release, and acid group dissociation.

The surface properties of the polyethylene control, 0% Na and 50% Na samples tended to be similar; these properties were different from the 100% Na sample. The 100% Na ionomer was more hydrophilic, it had a higher polar component for its surface energy, and it was unique in exhibiting discrete anionic clusters 1 to 10 um in diameter on its surface.

These differences were manifested <u>in vivo</u> by platelet aggregation and thrombus formation on the polyethylene, 0% Na, and 50% Na implants, while the 100% Na implants' surface was predominately covered by singly adherent platelets.

These observations are in agreement with the findings of Van Kampen et al. that sodium neutralization of polymeric acid groups reduces the thrombotic response of whole blood to these materials (1).

It is proposed that the lower levels of thrombus formation on the 100% Na ionomer are due to the cluster development in the neutralized methacrylic component. These cluster regions are polar and hydrophilic. In an aqueous environment, the polymer surface contains discrete, negatively charged regions containing mobile polymer chains. Implants that have these clustered regions on their surface exhibit platelet adhesion, but little thrombus formation.

ACKNOWLEDGEMENTS

The authors wish to acknowledge support for this research by the Saint Luke's Hospital Association of Cleveland, OH, and the Timkin Honors

Fellowship and Cross-Jones Medical Research Foundation Fellowship for J.A.G.

REFERENCES

1. C. L. Van Kampen, D. F. Gibbons and R. D. Jones, J. Biomed. Mater. Res., 13, 517-541 (1979).
2. M. N. Helmus, D. F. Gibbons and R. D. Jones, J. Biomed. Mater. Res., 18, 165-183 (1984).
3. M. N. Helmus, Ph.D. Dissertation, Case Western Reserve University, (1980).
4. P. N. Sawyer, Ann. N.Y. Acad. Sci., 416, 561-583 (1983).
5. R. I. Leininger, CRC Critical Reviews in Bioengineering, 333-381 (1972).
6. J. S. Greengard and J. H. Griffin, in: "Perspectives in Hemostasis", J. Fareed, et al, Ed., Pergamon Press, New York, 1981, pp. 21-30.
7. Technical letter #E-40378. E.I. DuPont de Nemours, Polymer Product Department, Ethylene Polymers Division, Wilmington, DE.
8. T. R. Earnest, Jr., Personal Communication.
9. R. Longworth, in: "Ionic Polymers", L. Holliday, Ed., Applied Science, London, 1975, pp. 72-172.
10. W. A. Zisman, in: "Contact Angle, Wettability and Adhesion", F. M. Fowkes, Ed., Amer. Chem. Soc., Wash., DC, 1964, pp. 1-51.
11. E. Nyilas, W. A. Morton, R. D. Cumming, D. M. Lederman, T.-H. Chiu and R. E. Baier, J. Biomed. Mater. Res. Sympos., 11, No. 8, 51-68 (1977).
12. D. H. Kaelble, J. Adhesion, 2, 66-81 (1970).
13. "Duke's Physiology of Domestic Animals", M. J. Swenson, Ed., Cornell Univ. Press, Ithaca, NY, 1970.
14. R. D. Cumming, D. Kong, M. Peterson and W. D. Watkins, Thromb. Haemostasis, 50(1), 276 (1983).
15. C. L. Van Kampen, Ph.D. Dissertation, Case Western Reserve University, (1977).
16. C. L. Van Kampen, D. F. Gibbons and R. D. Jones, Biomat. Med. Dev. Art. Org., 6, 37-56 (1978).
17. W. J. MacKnight, L. W. McKenna and B. E. Read, J. Appl. Phy., 38, 4208-4212 (1967).
18. A. Eisenberg and M. King in: "Ion-Containing Polymers", Acad. Press, NY, 1977, pp 15-64.
19. W. J. MacKnight and T. R. Earnest, Jr., J. Polym. Sci.: Macromolecular Reviews, 16, 41-122 (1981).
20. J. A. Goggins, Ph.D. Dissertation, Case Western Reserve University, (1985).
21. C. G. Bazuin and A. Eisenberg, J. Chem. Education, 58(11), 938-943 (1981).
22. B. Osterud, Scand. J. Haematol., 32(4):337-345 (1984).
23. G. V. R. Born, N. A. Begent and N. J. Cusack, in: "Human Blood Coagulation, Haemostasis and Thrombosis", R. Biggs, C. R. Rizza, Eds., Blackwell Scientific, Oxford, England, 1984, pp 319-348.
24. N. Simionescu, M. Simionescu and G. E. Palade, J. Cell Biol., 90, 605-613 (1981).
25. R. E. Baier, Ann. N. Y. Acad. Sci., 416, 34-57 (1983).
26. D. L. Coleman, D. E. Gregonis and J. D. Andrade, J. Biomed. Mater. Res., 16, 381-398 (1982).
27. D. J. Lyman, K. Knutson, B. McNeill and K. Shibatani, Trans. Amer. Soc. Artif. Int. Organs, 21, 49-54 (1975).
28. S. W. Kim, R. G. Lee, H. Oster, D. Coleman, J. D. Andrade, D. J. Lentz and D. Olsen, Trans. Amer. Soc. Artif. Intern. Organs, 20, 449-455 (1974).

29. S. A. Barenberg and K. A. Mauritz, in: "Biomaterials: Interfacial Phenomena and Applications", S. L. Cooper, N. A. Peppas, Eds., Amer. Chem. Soc., Wash., DC, 1982, pp. 195-208.
30. S. A. Barenberg, Ann. N.Y. Acad. Sci., 416, (1983).
31. E. Wong, E. Kupske and S. Vaughan, NIH Contract Report #NO1-HV-92932-3, May, 1982.

SURFACE CHARACTERISTICS OF POLYETHYLENE OXIDE-POLYSTYRENE MULTIBLOCK COPOLYMERS

David Grainger, Teruo Okano and Sung Wan Kim

University of Utah
Department of Pharmaceutics
Salt Lake City, Utah 84112

Multiblock copolymers of hydrophilic polyethylene oxide and hydrophobic polystyrene were synthesized in order to study the influence of various surface properties on the adsorption of albumin from solution. These model surfaces display microphase separated structure and have the ability to be altered on a molecular scale to change surface characteristics. Compositional changes were shown to influence hydrophilicity/hydrophobicity, surface mobility, domain morphology, albumin adsorption values, binding constants, and adsorbed albumin surface stability. Changes in surface and interfacial energies as measured by Wilhelmy plate contact angle analysis could be correlated to hydrophilic/hydrophobic compositional ratios. In addition, adsorption isotherms, binding constants, and protein surface exchange rates could be similarly correlated to the same compositional trends. Increasing the hydrophilic ratios of the copolymer caused increased surface mobility (increased hysteresis), decreased receding contact angle, decreasing albumin adsorption, decreased binding constants, and increased adsorption protein stability over time. These results are attributed to unique surface behavior induced by the formation of microdomain surface morphology.

INTRODUCTION

Thrombosis and immunogenic responses have long been recognized as chronic problems compromising the success of blood-contacting applications for many implantable synthetic biomaterials. Many attempts have been made recently to understand and control specific interactions with implanted foreign materials. Of especially intensive interest has been the study of blood protein adsorption at material interfaces. Theories correlating protein adsorption and subsequent biological interactions to surface charge (1), protein and surface dynamics (2), surface composition (3,4), and surface energy (5,6) have been advanced by many researchers. It is hoped that these theories will provide insight allowing designs of new nonthrombogenic and biocompatible materials for implant use. Normal vascular endothelium, the ideal nonthrombogenic surface, is considered to be structurally separated into hydrophilic and hydrophobic microdomains

(7). Many copolymers have been synthesized that contain mutually incompatible segments and show similar microdomain structure. Such synthetic polymer surfaces that exhibit hetrogeneous microphase separated surface structure are reported to play a significant role in material-blood compatibility. It is hypothesized that improved biocompatibility is a result of the direct influence of heterogeneous microdomain hydrophilic/hydrophobic balance on adsorption of proteins from plasma (3,4).

It is well known that foreign surfaces contacted with blood are immediately covered with an array of adsorbed plasma proteins. It is also known that certain primary plasma proteins have preferential affinity for either hydrophilic or hydrophobic surfaces, suggesting that surface energies and surface heterogeneity influence adsorption (8). Preferential adsorption of certain blood proteins to hydrophilic or hydrophobic loci on the implanted substrate is thought to create an organized adsorbed protein layer (8-10). Subsequent events in the coagulation cascades such as platelet adhesion, aggregation, activation, and degranulation have been shown to be controlled or hindered, at least in part, by this adsorbed protein patchwork (9,10). It is hypothesized that the organization of adsorbed proteins by the heterogeneous morphology inhibits platelet membrane activation and thereby suppresses binding and degranulation reactions that initiate thrombus (11). Selective adsorption of blood proteins onto surfaces having microdomain structure therefore may help elucidate important protein-polymer interactions that determine the thrombogenicity of implanted materials. Since protein adsorption and the ultimate biocompatibility of implanted materials are dependent upon the surface properties of these materials, the characterization of such surfaces is considered very important for development of new blood-contacting implants.

This paper presents the initial results of synthesis of a hydrophilic/hydrophobic microphase separated multiblock copolymer used as a model surface to investigate polymer-blood interactions. To study the effects of microdomain surface structure, surface energy and mobility, and compositional hydrophilic/hydrophobic effects on biological interactions, a series of microphase separated block copolymers were synthesized and characterized. The value of these copolymer systems lies in their ability to be tailored on a molecular scale to change hydrophilic/hydrophobic balance, bulk and surface compositions, wettability (surface mobility), and microdomain structured surface morphology. It is hoped that by examining how altering these properties changes certain biological interactions that the mechanisms of cellular and protein response to blood-contacting surfaces may be more clearly elucidated.

EXPERIMENTAL

Purification of Materials

Chloroform and dioxane were distilled by conventional methods. Styrene was distilled under reduced pressure and the fraction at bp 65°C/33mm Hg was collected. Dihydroxy-terminated polyethylene oxide (PEO) samples of various molecular weights considered potentially valuable for biomaterials applications (12) were melted and dried under vacuum at 65°C.

Synthesis of Polyethylene Oxide - Polystyrene Multiblock Copolymer

(a) Synthesis of Isocyanate-Telechelic Oligostyrene

p,p'-Diisocyanate Diphenyl Disulfide (IPDS) was prepared as described by Okano and coworkers (13). The desired amounts of styrene monomer and IPDS were weighed into quartz ampoules. The ampouled solutions were then immersed in liquid nitrogen and evacuated to 10^{-2} mm Hg. The ampoules were isolated from the system and allowed to warm sufficiently to melt the contents. The solutions were subsequently refrozen in liquid nitrogen and the degassing operation repeated a total of three times. Then the ampoules were sealed under vacuum at 10^{-2} mm Hg. After immersing and stirring in a water bath at 30°C, the ampoules were irradiated with ultraviolet light (Rayonet 100, Southern New England Ultraviolet) for various times at a distance of 20 cm. The polymerized contents were diluted with benzene and poured into 95% hexanes to precipitate the isocyanate telechelic oligostyrene. The precipitates were rinsed and dried in a vacuum desiccator.

(b) Characterization of Oligomers

A series of molecular weights of dihydroxy-terminated PEO were obtained from Sigma Chemical Company. The intrinsic viscosity of each molecular weight was measured with an Ubbelohde viscometer in double-distilled water at 35° ± 0.2°C and molecular weights were calculated by the Mark-Houwink equation where $[\eta] = 1.66 \times 10^{-4}M^{0.82}$(14). The intrinsic viscosity of oligostyrene was determined with the Ubbelohde viscometer in benzene at 25° ± 0.2°C. Its molecular weight was calculated from another Mark-Houwink equation where $[\eta] = 3.64 \times 10^{-4}M^{0.64}$ (15).

(c) Multiblock Copolymerization

Under conditions of [OH]/[NCO] = 1.0, isocyanate-telechelic oligostyrene and dihydroxy-terminated PEO were allowed to react under nitrogen purge in a solution polymerization of refluxed chloroform stirred at 50°C for five days as shown in Scheme 1. The products were recovered by a double precipitation procedure involving initial precipitation in 9:1 hexane: ether, filtering, drying, dissolution in dioxane, precipitation in double-distilled H_2O and then filtering and drying. The procedure removed unreacted chains of both species as monitored by IR spectroscopy. Final copolymer product yields ranged from 25-65% depending on composition.

Homopolymer polystyrene to be used as reference standard was synthesized in a bulk polymerization at 60°C using AIBN as the initiator.

(d) Analysis of copolymer

The structure of the polymers was confirmed using infrared spectroscopy (Beckman 620MX), proton NMR and ^{13}C NMR (JEOL FX270), differential scanning calorimetry (DSC) (Perkin-Elmer 2) and elemental analysis.

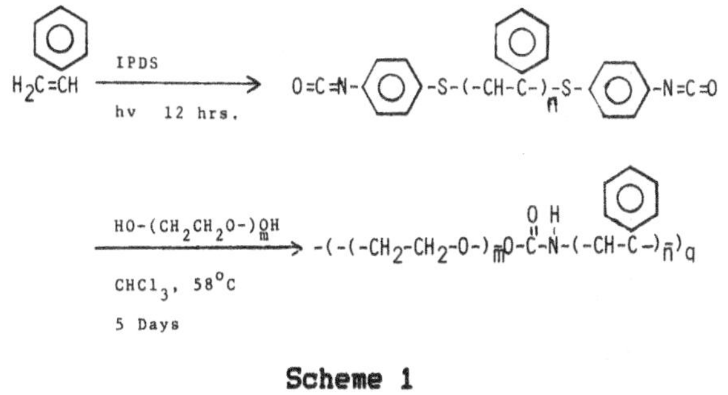

Scheme 1

Gel Permeation Chromatography (GPC)

A Waters Inc. (Farmingham, MA, USA) liquid chromatograph consisting of six columns packed with crosslinked porous polystyrene (Styragel) was used. The upper size exclusion limits of the porous packing were 10^5, 10^4, 10^3, 500 and 100 A. Elution conditions were as follows: eluant = THF (BHT) inhibited); flow rate = 1 ml/min; temperature = 30°C; polymer solutions = 0.7% in THF.

All sample data was computer analyzed with Nelson Analytical Software, version 7.01. The entire analysis was generously performed by Mr. R. Steffens (Ciba-Geigy Basic Research, Ardsley, NY).

Contact Angle Measurement

Chloroform solutions containing 1.0% solution of the polymers were prepared and filtered through 0.22 μm Teflon filters. Films were cast in triplicate from these solutions onto microscope slides cleaned in hot chromic acid, followed by an ethanol wash, a distilled water wash, then a Freon reflux. The dip-cast slides were dried overnight in a clean room at 20°C. Dynamic advancing and receding contact angles, as well as hysteresis data was collected on a Wilhelmy Plate contact angle apparatus (Wet-Tech, Biomaterials International) in double-distilled water by the method of Smith and coworkers (16) for copolymers of various compositions and the pure polystyrene standard. In order to examine hydration effects on the polymers, contact angle analysis was performed for a series of hydration times in double-distilled water from three to 48 hours.

Electron Microscopic Observation

Polymer solutions of 0.1% in chloroform were flowed over carbon-coated copper grids mounted on microscopic slides. Thin films for transmission electron microscope observation were prepared by evaporating chloroform at room temperature, vacuum drying, and then fixation in osmium tetroxide vapor (1.0 wt% aqueous solution) for eight hours. The microstructures were observed with an electron microscope (JEOL JEM 200cx).

Protein Adsorption

Human serum albumin (Sigma Chemical, Fraction V) was used as supplied and adsorption was studied utilizing a column method similar to that of Walton and coworkers (17). Glass beads (0.10 mm diameter) were cleaned using the same procedure as for microscope slides above and subsequently coated two separate times with copolymer using a filtered 0.2% solution in chloroform. After drying at room temperature and then under vacuum, the beads were equilibrated in buffer and packed into jacketed glass columns (0.6 x 15 cm) generating a total surface area of 2560 cm^2. Constant flow rate and temperature were maintained using an adjustable gravity field and temperature-controlled water jackets. Albumin solution of various concentrations in 0.15 M PBS, pH 7.4 were flowed into the column and eluted protein fractions were collected on the effluent end and monitored using ultraviolet light absorbance (Perkin-Elmer Lamda 7) at 280 rm. Relative adsorption isotherms were constructed for each applied albumin concentration. Dynamic "Langmuir-type" isotherms were constructed by plotting quantitative adsorption at each applied concentration versus protein concentration. Isotherms were constructed for uncoated glass beads and polystyrene coated beads for comparison and control.

Protein Exchange

Human serum albumin (HSA) was radiolabeled using reductive alkylation with ^{14}C-formaldehyde (Dupont - New England Nuclear) and sodium cyanoborohydride by the techniques of Dottavio-Martin and Ravel (18). Ultraviolet spectra scans demonstrated that the labeled albumin maintained the same structure as unlabeled albumin. Experiments designed to permit the adsorption, desorption and exchange of labeled and unlabeled albumin were conducted using identical polymer-coated bead columns from earlier experiments. The procedure involved initial adsorption of protein by flowing a mixed solution of labeled and unlabeled albumin (1 mg/ml) into the bead column until adsorption saturation or equilibrium was attained. A mixture of unlabeled albumin of identical concentration was subsequently eluted through the column and allowed to exchange with the previously adsorbed labeled mixture. Elution fractions were collected and analyzed by both UV adsorption and scintillation counting (Beckman LSC 7500).

The effect of timed adsorption incubation periods on subsequent exchange profiles of labeled HSA was investigated using identical materials and methods except that labeled solutions of albumin were allowed to "sit" under "no-flow" conditions in the column for various lengths of time before being allowed to exchange by in-flow of unlabeled albumin. This incubation technique would demonstrate changes in binding affinities or states of denaturation of adsorbed proteins over time. Labeled HSA was flowed into the coated-bead columns until a saturated adsorption value on the surface was detected. Flow was stopped and the column containing the buffered protein solutions was allowed to incubate at constant temperature for 15 minutes to one hour. Unlabeled HSA solutions were then eluted through the column and fractions were collected. Exchange and desorption values were calculated from monitoring UV absorbance and radioactivity of eluted fractions as mentioned previously.

RESULTS AND DISCUSSION

Infrared spectra confirm the formation of the desired copolymer. Figure 1 shows overlays of the IR spectra for both the telechelic oligosytrene and PEO-PSt multiblock copolymer. Carboxyl bands at 1740 cm^{-1} confirm formation of isocyanated oligostyrene (curve b). As expected, the strong isocyanate peak disappears in the copolymer spectra (curve a), replaced by covalent urethane linkage formation between chains. This is evidenced by formation of N-H stretch (3300 cm^{-1}), C-N stretch (1060 cm^{-1}), and C-O stretch (1100 cm^{-1}) from PEO chains. Proton NMR confirmed the presence of phenyl groups (6.5-7 ppm), methylene-ether PEO backbone (3.4-3.8 ppm), and PSt backbone (1.4-2 ppm). ^{13}C NMR correlated well with proton NMR and additionally, shows a carbonyl peak downshifted to 170 ppm as evidence of the covalent urethane linkage. Figure 2 shows overlayed proton NMR data for PEO, PSt, and the copolymer.

GPC curves are displayed in Figure 3. Curve A shows shoulders on each side of a narrow peak, indicating a wide dispersion of molecular weights and wide degrees of conversion. Curves B and C show higher molecular weight and no shouldering but the peaks are not narrow and display tails off the lower molecular weight sides. Copolymer C was the product of improved polymerization and purification procedures and resulted in higher conversions and higher molecular weight. Improvements presently implemented are hoped to improve purifications, polymerizations and dispersions even further.

Table I contains Wilhelmy Plate contact angle data from nonhydrated samples. This is a dynamic surface analysis method producing advancing

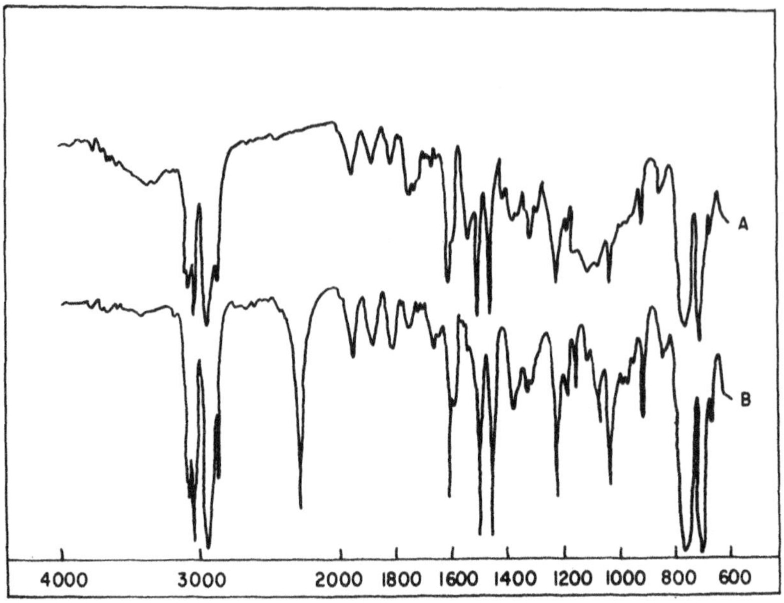

Figure 1. Infrared spectra for PEO-PSt multiblock copolymer (A) and telechelic oligostyrene (B).

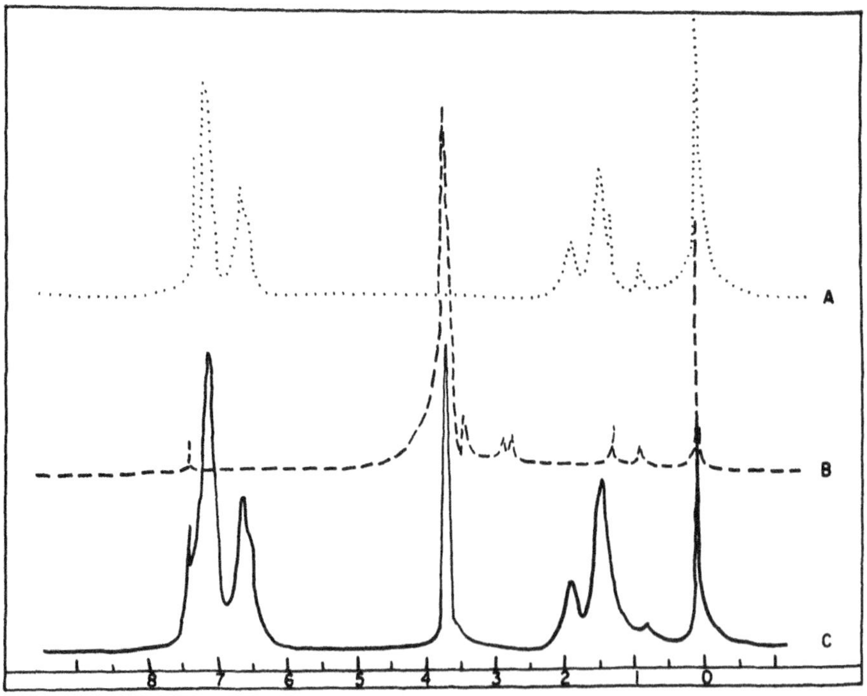

Figure 2. Proton NMR spectra for polystyrene (A), polyethylene oxide (B) and PEO-PSt multiblock copolymer (C).

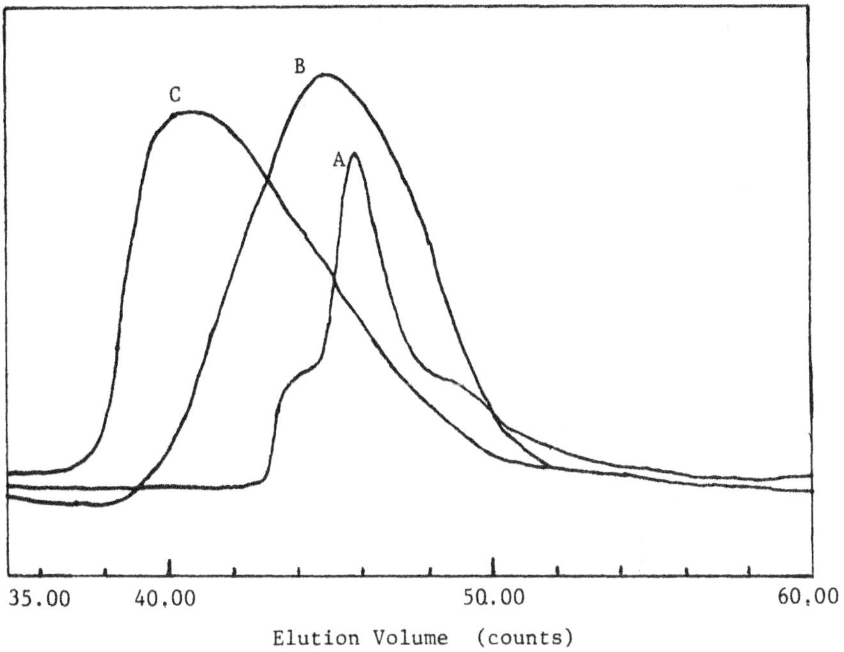

Curve	PEO M_w	PSt M_w	Copolymer M_w	Block Ratio (PEO:PSt)	Yield
A	8130	4300	12300	1:1	52.5%
B	1000	8600	19000	2:2	59.0%
C	6450	8600	45000	3:3	43.0%

Figure 3. GPC results for three copolymer compositions.

and receding contact angles which can provide considerable insight into the nature of the copolymer surfaces. Contact angles are predominately a function of functional groups and atoms in the exposed outermost 10 A of solid surface.

Contact angle hysteresis traces from the Wilhelmy plate technique, as in Figure 4, are a direct result of the mobility of surface polymer molecules and side chains when dynamic interaction with the liquid medium (water) is the only disturbance present (19). Minimization of the interfacial energy is the driving force for the changes in the orientation of surface chains between air and water interfaces as the sample is alternately exposed to each medium. Important assumptions involved in this technique are understood and have been addressed by Andrade, Smith and Gregonis (20). It has proven difficult, however, to assess the validity or approximations of these assumptions to certain experimental conditions. Furthermore, interpreting contact angle phenemona in terms of apparently applicable assumptions can be arbitrary and rather ambiguous. Generally, contact angle data provides a generous amount of valuable information about material surfaces and interfacial behavior and is considered a valuable characterization technique.

In heterogenous materials, such as these copolymers, the advancing contact angle represents the low energy (non-wetting) domains while the receding angles represent high energy (wetting) phase, producing a hysteresis loop (20). As can be seen in Table I, as the polystyrene hydrophobic fraction of the bulk composition decreases, advancing angles

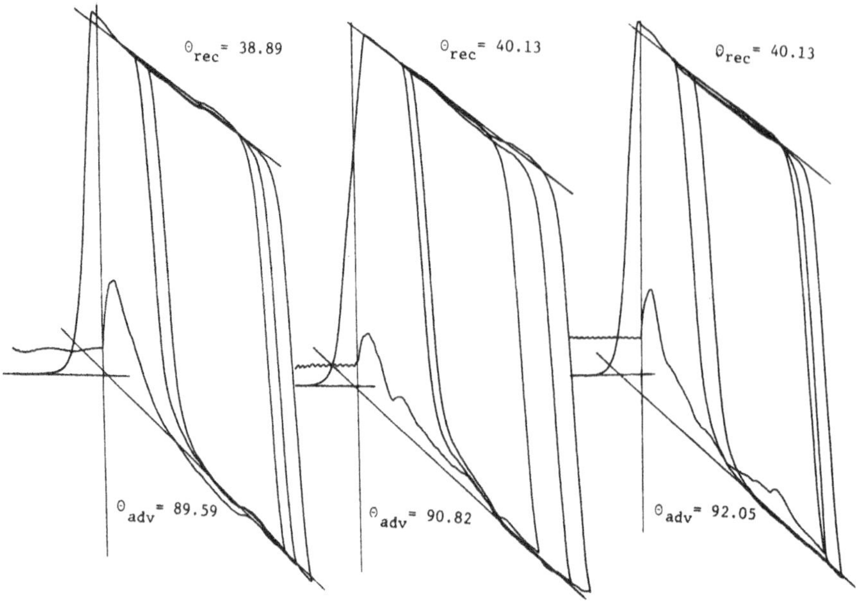

Figure 4. Hysteresis loops from Wilhelmy Plate contact angle analysis.

change only slightly while receding angles decrease dramatically. Hysteresis increases substantially with increasing PEO content. A number of factors could contribute to these hysteresis effects. Under optical microscopy, polymer films for contact angle experiments looked smooth and generally acceptable. Therefore, surface roughness is not a factor. Heterogeneous microdomain structure is most likely a contributor to hysteresis. Although Neumann and Good (21) conclude that heterogeneities smaller than 0.1 μm contribute little to hysteresis, TEM micrographs (discussed later) show microdomain structures to be of various sizes. In addition, hydrophobic domains are evidenced to increase advancing angles while hydrophilic domains decrease receding angles (22). Thus, as PEO domains increase in size and number with decreasing PSt content, this

Table I. Wilhelmy-Plate Contact Angle Data in Wate (no hydration) (± SD, n = 6).

Wt% PSt	Advancing Θ	Receding Θ	Hysteresis
100%	103.52 ± 4.23	69.02 ± 4.54	34.50
89%	106.92 ± 3.93	64.89 ± 3.00	42.03
59%	105.90 ± 0.88	64.60 ± 1.20	41.30
53%	103.78 ± 0.87	60.91 ± 1.62	42.87
35%	101.68 ± 1.28	29.30 ± 1.68	72.38
32%	112.70 ± 0.44	28.76 ± 0.49	83.94

phenomenon is observed in our system. indicating that microdomain structure may significantly affect surface energies by changes in either morphology and/or chain segment mobilities.

Water is a difficult medium for contact angle because of its small molecular volume allowing penetration and swelling of surfaces (20). It is, of course, most appropriate to use in analysis of biomedical materials such as this copolymer since it is the key component of all biological systems. Penetration of water molecules into the surface would be enhanced by increasing hydrophilic PEO content. Such swelling causes hysteresis by decreasing the surface energy in contact with water and also increasing the free volume of polymer chains to allow relaxation or restructuring that minimizes surface energy. Because these copolymers have no pendant side chains capable of β-type relaxation, contact angle hysteresis may be attributed solely to backbone reorientation. Hydrophilic chains of PEO thermodynamically favor water over air interfaces and if allowed to relax, will bury into the bulk polymer when exposed to air and exposed preferentially to water when immersed in this medium. Hydrophobic PSt will behave in an opposite manner. Thus, Wilhelmy plate methods induce constant surface reorientation as the polymer surface is exposed alternately to air and water interfaces, causing the contact angle hysteresis phenomenon. PSt is known to reorient in contact with water (23) and PEO is known to swell in water such that hysteresis may be expected to vary with composition. Structural interactions may be altered due to changes in composition that allow increased chain mobilization, reorientation and possible morphological changes with increasing PEO content.

Hydrated contact angle measurements using double-distilled water with 0.2% sodium azide (antimicrobial agent) demonstrate considerable change of surface character with hydration. Figure 5 shows changes in hysteresis with hydration time. All polymer compositions show an increase in hysteresis to approximately the 12 hour mark. From this point, the copolymers of low PEO content decrease hysteresis to the 48 hour point. Copolymers

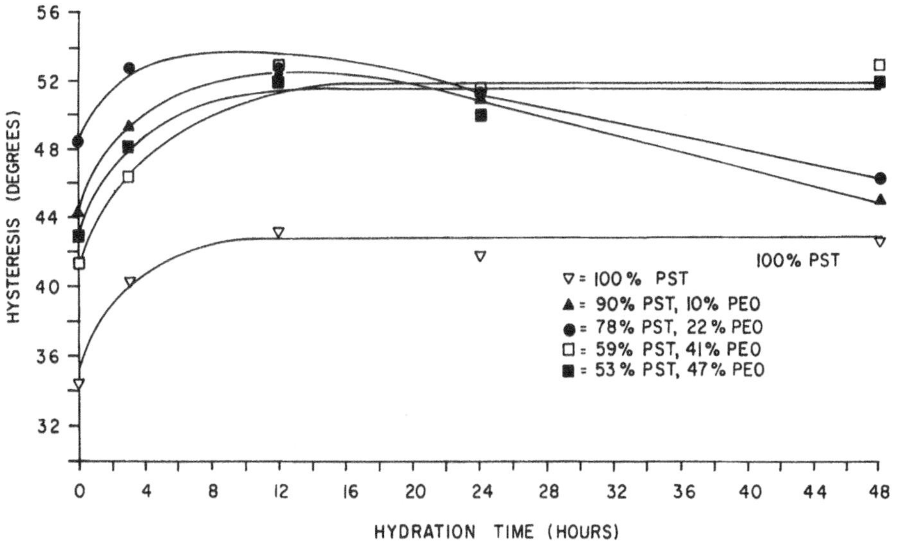

Figure 5. Contact angle hysteresis as a function of hydration time.

of high PEO content maintain plateau levels of high hysteresis value while PSt homopolymer plateaus at a much lower hysteresis value. While advancing angles showed few hydration trends correlated to composition, receding angles, shown in Figure 6, demonstrated consistant decreases with hydration time. These changes correspond to the same time points exhibiting inflections in the hysteresis curves (12 hour point). As mentioned previously, receding angles are indicative of the PEO domains. Thus, these phenomena may be directly attributed to plasticizing effects of water penetrating the polymer surface through hydrophilic PEO domains, allowing relaxation and higher degrees of mobilization for copolymer chains located near the surface. Major restructuring of microdomains may also occur during hydration soaks, minimizing interfacial energy and permitting hydrophobic PSt domains to bury into the bulk of the film, away from penetrating water molecules. PEO domains may swell to compose a larger percentage of polymer surface exposed to water. In support of this, and in agreement with the hydrophilic-hydrophobic ratio hypothesis (24), Ratner, Weathersby and Hoffman (25) report in HEMA-EMA copolymers that amounts of hydrophilic HEMA decreased significantly in the surface upon dehydration while amounts of hydrophobic EMA remained unchanged.

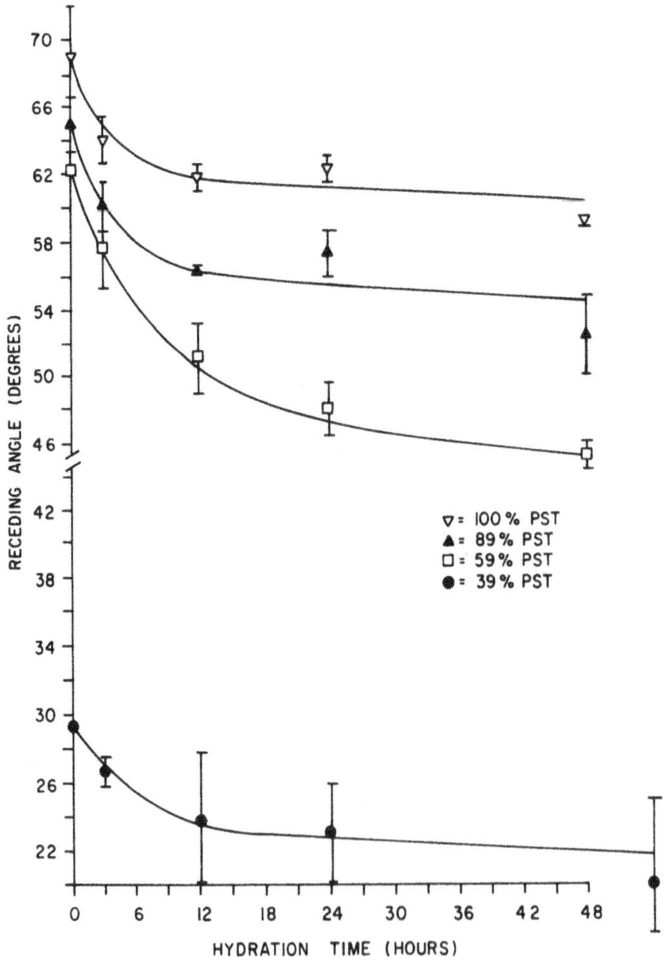

Figure 6. Receding angle as a function of hydration time.

Hydration phenomena may also be significant with respect to the optimum polar/apolar ratio hypothesis (26,27), which suggests that hemocompatibility is a direct function of relative polar and apolar contributions to surface free energy. Coleman, Gregonis and Andrade (28) extended this idea to correlate polar and dispersion contributions to interfacial free energy as affected by water content in biomedical hydrogel systems. Empirically, they observed that water content profoundly alters interfacial energies in aqueous/hydrogel systems and specifically, that water content could control interfacial free energy and zeta potentials in addition to polar and apolar components of surface free energy. With regard to hemocompatibility, they discovered a convincing correlation demonstrating improved blood compatibility as the dispersion component begins to dominate surface energetics. In addition, it has been reported (27,28) that, for biomaterials in general, a high dispersion component, low polarity substrate will promote formation of a strongly adhered, stable protein layer while surfaces of low dispersion, high polarity, have characteristically weak plasma protein affinity and unstable protein adhesion.

Although we are aware of numerous attempts (5,26-28) to correlate surface energy as measured on a macroscale to antithrombogenic character, work by our coworkers (8-10) on amphiphilic copolymers has not been able to correlate surface free energy nor any of its components to hemocompatibility. Physiological biocompatibility cannot be completely assessed simply in terms or polar and dispersive components of surface energy. These contributions, however, are probably involved in bioadhesion and material-blood interactions and it is for this reason that we are attempting to utilize these parameters. From results of our block copolymer systems thus far, we propose that measured polar/apolar ratios or hydrophilic/hydrophobic ratios on the level of assembly of chain segments or polymer molecules is required to correlate surface character to biological response. Macrotechniques such as contact angle and critical surface tension measurements used to probe surface and interfacial energies can be used to show trends and are difficult to apply on a molecular scale. In copolymer systems, this results in "mean value" measurements that average the contributing energies of incompatible microdomain structures. More importantly, proteins and platelets sense substrate microenvironments often on the order of nanometers that may not always be related to measurements of macroscopic techniques. The environments sensed by blood constituents may be considered to be completely different than those characterized by surface probing techniques, especially in heterogeneous block copolymer systems assembled on the order of molecules.

Water adsorption and absorption into our multiblock copolymer system therefore most probably does alter interfacial energies, as demonstrated by the contact angle data with respect to hydrophilic/hydrophobic and polar/apolar surface ratios. Over the hydration time course, dispersion and polar components of the copolymer change, influenced by actual surface restructuring and chain relaxations, causing the surface to assume new hydrophilic/hydrophobic character. However, we are not able to correlate those changes directly to antithrombogenic character.

The changes in contact angle data, and hence surface energies, with compositions are predictable since hydrophilic functions are influenced not only by composition but also by the microstructure of the copolymer. It has been suggested that the hydrophilic/hydrophobic character of such amphiphilic copolymers is governed by the aggregation state of each segment (13), in this case the size and geometry of the PEO and PSt domains. Results of transmission electron microscopy studies on various copolymers, shown in Figure 7, demonstrate size and morphology of microdomain to be a function of composition. Although DSC showed no evidence

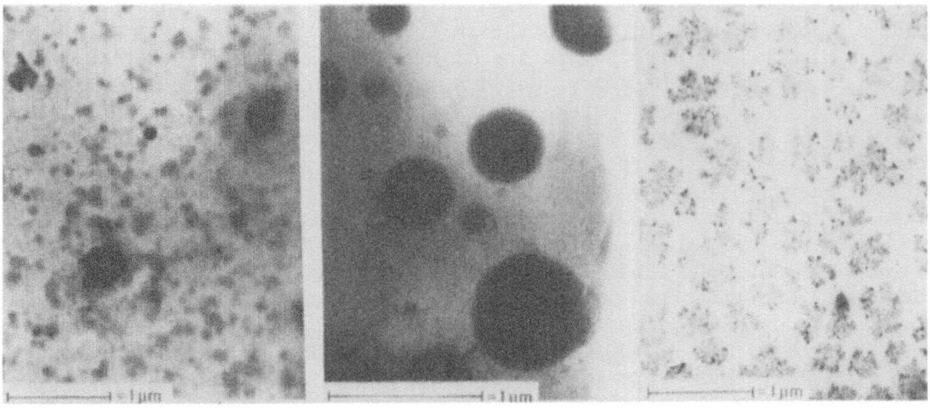

Figure 7. Transmission electron micrographs of copolymers stained with OsO_4:
A) PSt M_w = 4300, PEO M_w = 8130
B) PSt M_w = 8600, PEO M_w = 6450
C) PSt M_w = 9160, PEO M_w = 6450

of crystalline morphology, TEM photographs clearly show spherulitic formations of PEO stained with OsO_4 for some copolymer compositions (micrograph C). It is proposed that there exists some critical PEO chain length, possibly associated with a related critical PSt chain length that permits chain folding and mobility such to permit crystallization. Also, Merrill and Salzman (12) report changes in PEO solid states from typically crystalline to amorphous when exposed to water. Crystallization states seen in TEM work may therefore be an artifact of sample preparation techniques that keep the sample anhydrous. Most other compositions exhibit "island-sea" morphology as isolated, round islands of PEO in a sea of PSt. Alternate lamellae morphology has been reported (9,10) to offer optimum antithrombogenic character as measured by platelet adhesion and activation but these morphologies have yet to be encountered in our system.

Many factors can affect block copolymer morphologies. Casting conditions such as temperature, solvent evaporation rate, and film thickness are well-known influences upon microdomain morphology. Solvents, whose thermodynamic compatibility with polymers is measured by derived solubility parameters, greatly affect developing morphologies as the solvent is evaporated. Interblock chain-chain compatibilities, measured as interaction parameters, also affect microdomain structure. Surface heterogeneities produced by microdomains have been implicated in surface energetics as discussed in relation to contact angles. Microphase-separated structure has also been evidenced directly to influence protein adsorption (3,4,8) and is implicated in explaining our albumin adsorption results. It is anticipated that varying these morphological influences will allow formation of many types of microstructures to allow thorough investigation of microdomain influence on material-blood responses.

The column method is a solution depletion technique that permits examination of protein adsorption onto any designed substrate surface coated on beads that provide a very large surface area. Dynamic albumin adsorption isotherms for bare glass, polystyrene, and three PEO-PSt multiblock copolymers are displayed in Figure 8. Langmuirian-type behavior is demonstrated. Quantitation of the amount of protein adsorbed on

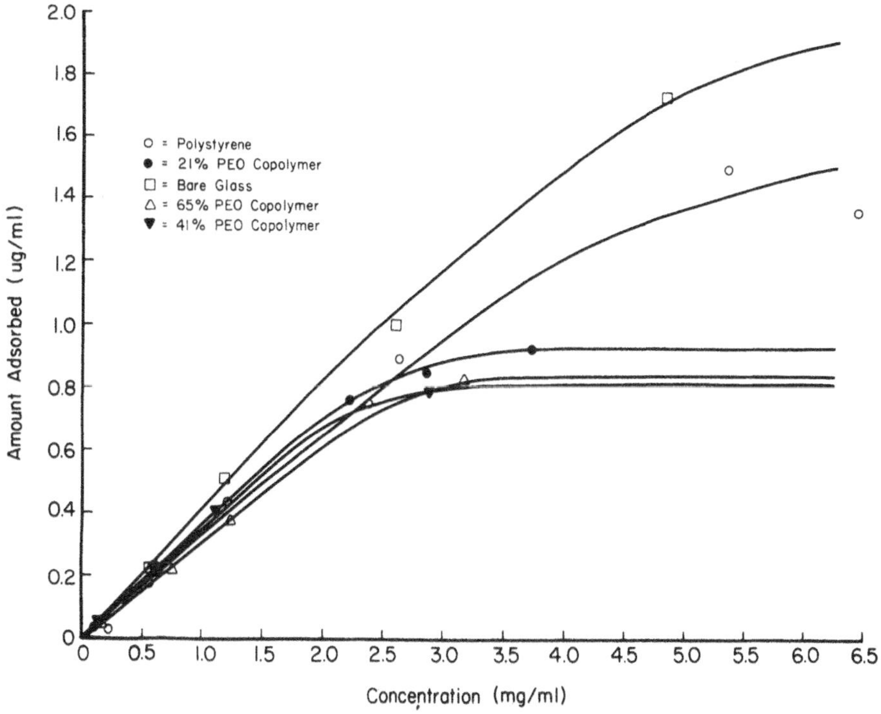

Figure 8. Languirian-type dynamic isotherm for albumin adsorption.

each surface indicates that the copolymers adsorb less albumin than either glass or polystyrene. Adsorption amounts per area at equilibrium for the copolymers correlate well with typical values found by other workers (17,29) (3-8 $\mu g/cm^2$). Values for polystyrene and glass are slightly higher which suggests multilayer formation (9-12 $\mu g/cm^2$). Linear type Langmuir plots constructed to analyze binding constants for the surfaces demonstrate good experimental data fitting. Table II presents calculated binding constants and correlation coefficients derived from linear Langmuir plots of adsorption data. The Langmuir equation is derived from the Klotz form:

$$C/S = C/n + 1/Kn$$

where C = protein solution concentration, S = amount of protein adsorbed at concentration C, n = limiting plateau value of S, and K = equilibruium binding constant. A plot of C/S vs. C should be a straight line of slope 1/n, ordinate intercept of 1/Kn and abscissa intercept of -1/K (30).

The copolymer surfaces have lower albumin binding constants than polystyrene and show increasing binding constants with increasing PSt content. This correlates well with observed low binding behavior of proteins with PEO surfaces (12) and proposed hydrophobic nature of albumin binding (31). The mechanism governing these adsorption difference has yet to be elucidated. Figure 9 displays changes in both binding constant, K, and hydrated contact angle hysteresis with copolymer PEO content. The hyperbolic nature of these curves, although opposite in curvature, is symmetrical and suggests some type of inverse correlation between surface mobility and binding constants.

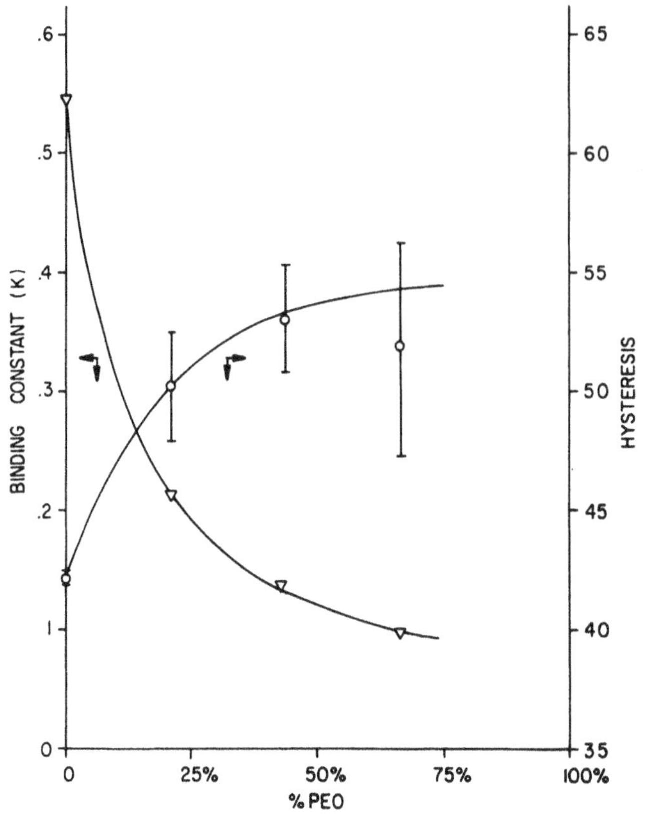

Figure 9. Correlation of contact angle hysteresis and albumin binding constants as a function of copolymer composition

Table II. Albumin Binding Constants on Column Substrates.

Substrates	Binding Constant (K)	Correlation Coefficient (r)
100% PSt	.54	.967
79% PSt, 21% PEO	.21	.942
59% PSt, 41% PEO	.13	.950
35% PSt, 65% PEO	.09	.929
Glass	.06	.980

It is suggested that PEO domains decrease the overall binding constant of the copolymer surface and that microdomain structure promotes a type of organization of adsorbed protein where albumin adsorbs to hydrophobic PSt sites preferentially over PEO. Recent work (12,32,33) has shown that PEO-rich surfaces greatly decrease protein adsorption. Mobile PEO grafted chains (33), as well as bulk PEO segments in homopolymer (32) and copolymer (12) form, have shown minimal protein adsorption. It has been proposed that this may be due to extreme chain mobility (33) and lack of molecular binding sites on PEO (12) to allow protein binding. In agreement with our results, Furosava (34) has shown in systems of PEO-PSt block copolymers that plasma protein adsorption is significantly reduced with increasing PEO content.

Adsorption values between the multiblock copolymers and PSt and glass can also be explained in terms of the observations of Norde and Lyklema (31,35) who established that the interactions between surfaces and protein increases with increasing surface hydrophobicity, and adsorption increases with increasing hydrophobicity of protein involved.

As shown in Figure 8, polystyrene, known to be non-wetting and hydrophobic by our contact angle analysis, shows the highest polymeric adsorption values. Albumin is also highly hydrophobic so that this result may be due to entropic hydrophobic interactions tending to minimize surface energies. Glass is generally regarded as hydrophilic and often charged. The main adsorptive force on glass beads, therefore may be some ionic interaction with albumin. Zeta potential measurements (streaming potential) would be helpful to assess surface charge involved in this assertion. The series of copolymers shows lower adsorption than either glass or polystyrene and follows the trend that decreasing surface hydrophobicity (increasing PEO content) decreases protein adsorption. This also corresponds to the polar/apolar ratio hypothesis. PSt has a high dispersive contribution and negligible polar contribution to surface interfacial energy (36). While polar and apolar contributions are not known for PEO because of its wetting and solubility problems encountered in critical analysis, it can be assumed that the ether backbone of PEO would have considerably more polar character than PSt. We, therefore, can correlate HSA adsorption values and binding constants to these polar-apolar ratios on a molecular scale. In agreement with others (27,28) we found higher binding affinities and higher adsorption values with increasing PSt fraction or increasing dispersive contribution to interfacial energy. Okano's observation (4) that albumin preferentially adsorbs to hydrophobic domains in hydrophilic/hydrophobic microphase separated block copolymer systems appears not to be supported unless different domain sizes, numbers, and morphologies observed in each multiblock composition are assumed to adsorb protein differently.

In light of supporting research and the unique character of PEO both alone and in combination with incompatible segments in microphilic copolymers it is not surprising that protein adsorption varies with surface character and composition.

A three-parameter graph detailing relative exchange rates of labeled albumin under exchange with unlabeled albumin during the first five minutes of exchange is shown in Figure 10. This plot displays percent of labeled protein exchanged from the surface after adsorption and during the exchange process as a function of incubation time. The general trend for all surfaces is a relative decrease in amount of albumin exchanged as incubation time increases. At zero incubation time (immediate exchange after adsorption), all surfaces show nearly the same adsorption/exchange values. Large decreases in adsorbed protein in this first five minutes of

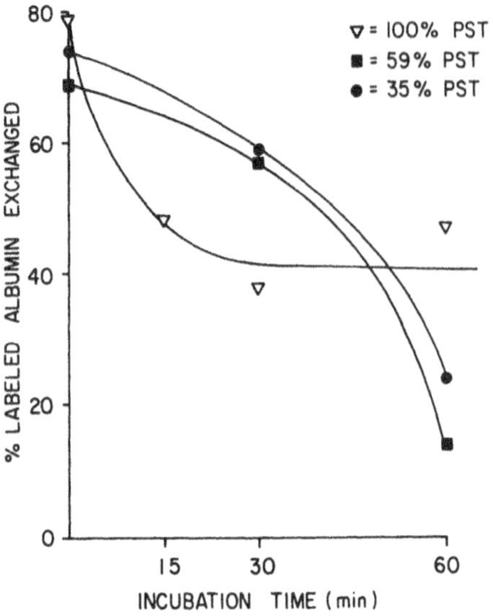

Figure 10. Albumin exchange from polymeric substrates as a function of incubation time.

exchange indicate that protein is reversibly bound to all surfaces in a similar manner.

The PSt exchange profile shows a different behavior than that of the copolymers. This curve shows an inverted hyperbolic shape with increasing incubation time. At 15 minutes or shorter incubation times, the majority of albumin (50-80%) is exchanged within the first five minutes, indicating a large degree of loose or reversible adsorption on the surface. At greater incubation times, however, the exchange value is slowed then stabilized, suggesting formation of some sort of stable adsorbed layer on the surface, due possibly to a limited, adhesive denaturing of albumin onto the surface. Copolymer curves, on the other hand, show a different curve form. A flatter initial slope suggests that a consisently stable, limited protein layer is formed immediately, even though the majority of what is initially adsorbed is exchanged (70%). This indicates a higher degree of reversible adsorption than PSt at lower incubation times. Exchange values remain high and do not change drastically from zero to 30 minutes incubation time in contrast to the PSt homopolymer case. After 30 minutes incubation, exchange values drop at an increasing rate to levels much below the homopolymer case and do not appear to stabilize. We propose that the copolymer surface allows increasing stabilization of albumin adsorption during these times, permitting less exchange and increased adhesion, promoting organized adsorbed structures, and possibly causing protein denaturation on the surface. It is suggested that these changes in apparent albumin binding/exchange values may be due to protein reorganization on the surface. The reorganization is induced by gradual bound protein denaturation "kicking off" more reversibly or loosely bound proteins to occupy those substrate sites in an irreversible manner. Larger exchange values for the copolymers at incubation times up to 50 minutes appear to demonstrate the influence of polar PEO microdomain structures on decreasing albumin stability and adhesion. At greater incubation times, albumin denaturation could nullify this effect by promoting irreversible adsorption, overriding the PEO influence. It is suggested that the microdomain structure of PEO, or more generally this heterogeneous microdomain structured surface, promotes reversible protein adsorption longer than PSt homopolymers due to its affect on binding constants and disparate protein adsorption to energetically-different domain. This copolymer effect is diminished with increasing incubation time, allowing eventual formation of a tightly-bound adsorbed layer.

Work by Soderquist and Walton (37) has done much to explore the kinetics and conformation of desorbing proteins. Given the similarity between their system and ours, it is helpful to attempt to correlate their results to our data. We note the differences between their desorption process in a pure sense with our exchange system. The exchange condition implies that desorption of labeled proteins is a competing process with unlabeled proteins attempting to adsorb. Where a complete or partial desorption in non-populated buffer might allow readsorption without competition, exchange denotes an almost compulsory, kinetic argument as to which protein, labeled or unlabeled, will ultimately remain adsorbed. Andrade (31) does justice to this concept very effectively.

The desorption process depends upon incubation or contact time of protein and substrate (37). In agreement with this, we have noticed that longer incubation times slow the desorption/exchange process. In addition, desorbed protein will predictably have some degree of denaturation that severely compromises its apparent ability to readsorb. The longer a protein resides on a biomaterial, the more substantial are structural changes induced in the protein by the adsorption process (37). It is thought that these structural transitions occur entropically in order to optimize interaction with the surface. Despite these adhesive transitions, proteins will slowly and irreversibly desorb from the surface, in our case probably replaced by protein in solution. Overall, the rate and ultimate values of desorption or exchange becomes a function of the incubation time unless, as has been shown for PEO, that adhesion is not favored.

CONCLUSIONS

Multiblock copolymers, synthesized as model amphiphilic surfaces on a molecular scale, show unique surface characteristics and morphologies. Alteration of copolymer composition and hydrophilic/hydrophobic ratios by changing constituent segment lengths allows "tailoring" of surface properties. Contact angle analysis demonstrates changes in surface and interfacial energies with both composition and hydration. Hysteresis increases and receding angles decrease with increasing PEO content and increasing hydration time. These changes can be correlated with protein adsorption values and adsorption stability. Protein adsorption and calculated binding constants decrease with increasing PEO content. In addition, changes in binding constants with PEO content correlate inversely with changes in hysteresis. Specifically, copolymers retain more of their original adsorbed albumin layer with increasing incubation time. Homopolymer PSt shows little difference in desorption values after initial incubation periods. Strictly on the basis of surface properties, we can make no conclusions about ultimate hemocompatibility. It is proposed that the unique properties of PEO in the copolymer as well as its incorporation into microdomain structures cause these heterogeneous copolymers to have very different behavior than glass or PSt homopolymer surfaces. Furthermore, biological response may be able to be controlled or monitored as a function of heterogeneous surface morphology and energetics as shown in protein adsorption and exchange properties for copolymer compositions. Because the copolymer can be easily altered to provide different morphologies, hydrophilic/hydrophobic and polar/apolar ratios, these hypothesized influences on hemocompatibility may be studied in a unified and consistent manner. Molecular scale construction provides the advantage of creating hetrogeneous surface properties correlated to the microenvironment of physiological systems.

ACKNOWLEDGEMENTS

The authors wish to thank Professor Jan Feijen for valuable discussion and ideas. This work has been supported by NIH Grant HL 17623.

REFERENCES

1. W. Norde and J. Lyklema, J. Colloid Int. Sci., 66, 257 (1978).
2. J. D. Andrade, V. Hlady and R. Van Wagenen, Pure and Applied Chem., in press (1984).
3. Y. Imai, Kobunshi (High Polymers, Japan), 21, 569 (1972).
4. T. Okano, S. Nishiyama, I. Shinohara, T. Akaika and Y. Sakurai, Polymer J., 10, 223 (1978).
5. D. J. Lyman, W. Muir and I. J. Lee, Trans. Amer. Soc. Artif. Int. Organs, 11, 301 (1965).
6. D. L. Coleman, D. E. Gregonis and J. D. Andrade, J. Biomed. Mat. Res., 16, 381 (1982).
7. P. N. Sawyer, C. Burrowes, J. Ogoniak, A. O. Smith and S. A. Wasolowski, Trans. Amer. Soc. Artif. Int. Organs, 10, 316 (1964).
8. T. Okano, S. Nishiyama, I. Shinohara, T. Akaika, Y. Sakurai, K. Kataoka, T. Tsuruta, K. Kuwona and S. Miyata, Proc. ISAO, 253 (1979).
9. T. Okano, S. Nishiyama, I. Shinohara, T. Akaike, Y. Sakuma, K. Kataoka and T. Tsuruta, J. Biomed. Mat. Res., 15, 393 (1981).
10. T. Okano, S. Nishiyama, T. Akaike, Y. Sakurai, K. Kataoka, T. Tsuruta, K. Kuwana and S. Miyata, Proc. ISAO, 253 (1979).
11. T. Okano, M. Shimada, T. Aoyagi, I. Shinohara, K. Kataoka, K. Abe and Y. Sakurai, J. Biomed. Mat. Res., in press (1985).
12. E. W. Merrill and E. W. Salzman, ASAIO J., 6, 60 (1983).
13. T. Okano, M. Katayama and I. Shinohara, J. Applied Polymer Sci., 22, 369 (1978).
14. D. Thomas and A. Charlesby, J. Polymer Sci., 17, 195 (1960).
15. E. Tsuchida, I. Shinohara and S. Kambara, Kgoyao Kagakukaishi, 66, 329 (1963).
16. L. Smith, C. Doyle, D. Gregonis and J. D. Andrade, J. Applied Polymer Sci., 26, 1269 (1982).
17. A. Walton and B. Koltisko, in: "Biomaterials: Interfacial Phenomena and Application", S. L. Cooper, Ed., ACS, Washington, D.C., 1982, p. 245.
18. D. Dottavio-Martin and J. Ravel, Anal. Biochem., 87, 562 (1978).
19. F. J. Holly and M. F. Refojo, J. Biomed. Mat Res., 9, 315 (1975).
20. J. D. Andrade, L. Smith and D. E. Gregonis: "Surface and Interfacial Aspects of Biomedical Polymers, Vol. I: Surface Chemistry and Physics", J. D. Andrade, Ed., Plenum Press, New York, NY, 1985, p. 273.
21. A. Neumann and R. Good, J. Colloid Int. Sci., 38, 341 (1972).
22. R. Johnson and R. Dettre, Surface Colloid Sci., 2, 85 (1969).
23. R. Good and E. Kotsidas, J. Colloid Int. Sci., 66, 360 (1978).
24. B. D. Ratner, A. S. Hoffman, S. R. Hanson, L. A. Harker and J. D. Whiffen, J. Polymer Sci., Poly. Symp., 66, 363 (1979).
25. B. D. Ratner, P. K. Weathersby and A. S. Hoffman, J. Applied Polymer Sci., 22, 643 (1978).
26. E. Nyiles, W. A. Morton, D. M. Lederman, T. H. Chiu and R. D. Cumming, Trans. Amer. Soc. Artif. Organs, 21, 55 (1975).
27. D. H. Kaelble and J. Moacanin, Polymer, 18, 475 (1977).
28. D. L. Coleman, D. E. Gregonis and J. D. Andrade, J. Biomed. Mat Res., 16, 381 (1982).

29. R. G. Lee and S. W. Kim, J. Biomed. Mat Res., 8, 251 (1974).
30. P. Molyneaux, "Water Soluble Synthetic Polymers: Properties and Behavior", CRC Press, Boca Raton, Florida, 1984, p. 120, vol.2.
31. J. D. Andrade in: "Interfacial Aspects to Biomedical Polymers, vol. 2: Protein Adsorption", J. D. Andrade, Ed., Plenum Press, New York, NY, 1985 (in press).
32. D. E. Gregonis, D. E. Buerger, R. A. Van Wagenen, S. K. Hunter and J. D. Andrade, Biomaterials 1984, Trans. Soc. Biomaterials, 7, 766 (1984).
33. Y. Mori and S. Nagaoka, Trans. Amer. Soc. Artif. Organs, 28, 459 (1982).
34. K. Furasawa, Y. Shimura, K. Otobe, K. Atsumi and K. Tasuda, Ronbushi Kobushu, 4, 309 (1977).
35. W. Norde in: "Adhesion and Adsorption of Polymers", vol. 2, L-HY Lee, Ed., Plenum Press, New York, NY, 1980, pp. 801-826.
36. D. K. Owens and R. C. Wendt, J. Applied Polymer Sci., 13, 1741 (1969).
37. M. E. Soderquist and A. G. Walton, J. Colloid Int. Sci., 75, 386 (1980).

FIBRINOGEN/PLATELET INTERACTION WITH PGI_2 ANALOG-HEPARIN IMMOBILIZED ALBUMINATED POLYURETHANE

C. P. Sharma and George Joseph

Division of Biosurface Technology
BMT Wing, Sree Chitra Tirunal Institute for Medical Sciences and Technology
Trivandrum 695012, India

Fibrinogen adsorption and platelet adhesion studies were done on modified polyetherurethane urea vascular grafts fabricated by precipitation method on clean glass rods. Surface modification was done by immobilizing heparin and PGI_2 analog on albuminated polymer substrate. The results indicated that although these biologically modified surfaces adsorbed more fibrinogen, the platelet adherence was less. Therefore the antiplatelet activity of the immobilized biomolecules should also be given attention while interpreting the fibrinogen-platelet interrelationship towards thrombogenicity.

INTRODUCTION

One of the initial events taking place when blood comes in contact with an artificial surface is adsorption of plasma proteins (1,2) followed by the adhesion of platelets leading to thrombus formation (3,4). It has been suggested that platelets adhere where they find adsorbed fibrinogen (5). However, the biological activity of the substrate may also play an important role. To investigate this we have immobilized Heparin and PGI_2 analog on albuminated polyurethane. The fibrinogen adsorption and platelet adhesion on these surfaces are studied. Our observations suggest that while interpreting the phenomenon of platelet adhesion and fibrinogen adsorption, the antiplatelet activity of the biological substrate should also be given proper consideration.

EXPERIMENTAL

Materials and Methods

Polyetherurethane urea (PEUU) synthesized in our laboratory by solution polymerization (6); 10,10-difluoro-13-dehydroprostacyclin (DF_2-PGI_2) obtained from the Department of Chemistry, University of Chicago, U.S.A.: Albumin (Human, Fraction V. essentially fatty acid free, Sigma Co.); Fibrinogen (Human fraction I, over 95% clottable, Sigma Co.); γ-Globulin (Human Cohn Fraction II, Sigma Co.); Heparin (5000 I.U. per ml., Biological E. Ltd., India); Iodinated (^{125}I) Human Fibrinogen (Amersham, England). All other chemicals used were of AR grade.

Preparation of Surfaces

Vascular grafts of ~ 5 mm diameter were fabricated from PEUU dissolved in dimethyl formamide as described elsewhere (6,7). They were cut into pieces of length 3 cm, washed throughly in distilled water and then rinsed in ethanol. The grafts were dried at 60°C and the following surfaces were prepared.

1. Surface I

Unmodified PEUU.

2. Surface II

PEUU grafts were dipped in phosphate buffer pH 7.4 for 30 minutes to remove air/water interface and then exposed to 100 mg% albumin solution for 3 hours at room temp. (~30°C). They were taken out rinsed in buffer, vacuum dried for 12 hours and glow discharged treated (GDT) for 15 min. The grafts were exposed to albumin again for 3 hours and then crosslinked with glutaraldehyde (0.5% v/v, 1 hour exposure). These samples were exposed overnight at 4°C to heparin (100 mg %) and then rinsed in buffer.

3. Surface III

In the above modification instead of second exposure to albumin, the samples were exposed to heparin for six hours at 4°C and then exposed to heparin again as described above.

4. Surface IV

Instead of the final exposure to heparin in the case of surface III, the glutaraldehyde crosslinked samples were exposed to DF_2-PGI_2 (50 μg %), dissolved in phosphate buffer pH 7.4) for six hours at 4°C.

Glow Discharge Treatment

Edwards vacuum coating unit E306 A was employed for the glow discharge treatment. The plasma glow was generated with nitrogen gas at a pressure of 10^{-1} m bar. The grafts were held in a vertical position on a holder placed below the electrodes and the treatment was given for 15 minutes.

Platelet Adhesion Studies

Fresh citrated bovine blood was prepared into a platelet rich plasma (PRP) as discussed elsewhere (8). The surfaces were exposed to the platelet suspension for 15 minutes and rinsed with buffer under controlled flow rate. The platelets were fixed with 2.5% glutaraldehyde and stained with Coomassie Blue G. The platelet density was estimated using an optical microscope.

Labelled Studies

For studying the adsorption kinetics of fibrinogen, the surfaces were exposed to protein mixture containing 25mg% albumin, 15 mg% γ -globulin and 7.5 mg% fibrinogen, with a known amount of labelled fibrinogen in phosphate buffer pH 7.4. The exposure was done at 37°C and the experiments were run over a period of 24 hours. The samples were rinsed throughly three times in buffer taken in separate beakers and then in running buffer. The amount of fibrinogen adsorbed was estimated with the help of a γ-counter using the following relation.

$$\overline{\Gamma} = \frac{C_p \quad R_f}{A \quad R_s}$$

where $\overline{\Gamma}$ = Surface concentration ($\mu g\ cm^{-2}$)

C_p = Bulk concentration ($\mu g\ ml^{-1}$)

R_f = Count rate of surface

A = Total area of surface (cm^2)

R_s = Count rate per ml of protein solution

The desorption kinetics were studied over a period of 24 hours at 37°C.

RESULTS AND DISCUSSION

Table I indicates the fibrinogen adsorption and desorption on the modified surfaces. The fibrinogen adsorption on the modified surfaces (II, III and IV) is high compared to the bare polymer. The platelet adhesion on the modified surfaces is however low as seen from Table I. On the DF_2-PGI_2 modified surface, the number of platelets adhered is surprisingly more than what we had observed earlier (9). This may be because of the possible inactivation of the PGI_2 molecule during experiment and also due to the presence of heparin. In past we have also observed in the case of PGE_1 modified Biomer, increased adsorption of fibrinogen and reduced platelet adherence compared to the bare Biomer (10,11). It has been

Table I. Fibrinogen Adsorption and Platelet Adhesion.

Surfaces	Fibrinogen Adsorption $\mu g\ cm^{-2}$		Fibrinogen Desorption $\mu g\ cm^{-2}$ (after 3 hr adsorp.)		Number of Platelets/mm^2 ± S.D.
	15 min	3 hrs	15 min	3 hrs	
I. Bare PEUU	0.59	1.27	0.92	0.76	9.50 ± 1.60
II. I + Alb + Alb + Hep	1.23	2.19	1.87	1.76	3.75 ± 1.30
III. I + Alb + Hep + Hep.	1.29	2.35	1.98	1.77	4.00 ± 1.70
IV. I + Alb + Hep + DF_2-PGI_2	1.12	2.07	1.79	1.58	3.50 ± 1.30

suggested that platelets adhere where they find adsorbed fibrinogen (5). From our studies we feel that in the case of biologically modified surfaces, the antiplatelet activity of the biomolecules also deserve consideration while interpreting the fibrinogen-platelet interrelationship towards thrombogenicity.

ACKNOWLEDGEMENTS

We wish to express our thanks to Dr. Josef Fried, Department of Chemistry, University of Chicago for providing us with a gift of 10,10-Difluoro-13-dehydroprostacyclin. We appreciate the helpful discussions we had with Mr. Thomas Chandy. This work was supported by the Department of Atomic Energy, India.

REFERNCES

1. L. Vroman, A. Adams, M. Klings, G. Fischer, P. Munos and R. Solensky, Ann. N.Y. Acad. Sci., 283, 65-76 (1977).
2. D. J. Lyman, L. C. Metcalf, D. Albo, Jr., K. F. Richards and J. Lamb, Trans. Am. Soc. Artif. Intern. Organs., 20B, 474 (1974).
3. E. S. Lee and S. W. Kim, Trans. Am. Soc. Artif. Intern. Organs., 25, 124-131 (1979).
4. R. E. Baier, Ann. N.Y. Acad. Sci., 283, 17-36 (1977).
5. M. B. Zucker and L. Vroman, Proc. Soc. Exp. Biol. Med., 131, 318-320 (1969).
6. C. P. Sharma, Current Science, 52, 1057-1058 (1983).
7. C. P. Sharma and G. Kurien, J. Coll. Inter. Sci., 97 (1), 38-40 (1984).
8. C. P. Sharma and T. Chandy, J. Coll. Inter. Sci., 89, 479 (1982).
9. George Joseph and C. P. Sharma, Communicated to the Journal of Biomedical Mat. Research.
10. C. P. Sharma and N. V. Nirmala, J. Coll. Inter. Sci., 101 (2), 591 (1984).
11. C. P. Sharma and George Joseph "Protein adsorption on PGE_1 immobilized albuminated Biomer", accepted for presentation at the 5th International conference on surface and Colloid Science, Symposium on Protein and Polyelectrolyte Adsorption, Potsdam, New York, June 24-28, 1985.

ON THE BLOOD COMPATIBILITY OF A HEPARIN-POLYVINYL ALCOHOL HYDROGEL

Cynthia H. Cholakis, Wan F. Ip, W. Zingg* and Michael V. Sefton

Department of Chemical Engineering and Applied Chemistry
University of Toronto, Toronto, Ontario, M5S 1A4
*Division of Surgical Research Hospital for Sick Children
Toronto, Ontario, M5G 1X8

Heparin immobilized onto polyvinyl alcohol (PVA) by action of glutaraldehyde has been shown to be biologically active in a novel parallel flow arteriovenous shunt in dogs. The heparin-PVA hydrogel was applied as a coating to chromic acid etched polyethylene tubing. A pair of Y-connectors was used to divert < 2% of the flow in the AV shunt through the heparinized tube which remained patent for longer than two hours while a control tube without heparin was occluded within 30 minutes at a main shunt flow rate of 150 ml/min. This enhanced patency was achieved without significant release from the surface, consistent with earlier experiments attributing the thromboresistance of the heparin-PVA hydrogel to the formation of a surface bound inactive thrombin-antithrombin III complex. <u>In vitro</u> test systems used to assess the platelet reactivity of immobilized heparin have not revealed any effect in contrast to <u>ex vivo</u> studies where platelet adhesion and aggregation was significantly reduced by immobilized heparin. These demonstrate the utility of heparin PVA hydrogel for preparing materials with potential long term thromboresistance and for testing validity of the hypothesied limitations to the long term use of immobilized heparin.

INTRODUCTION

Heparinized materials are receiving a great deal of attention because of their potential for satisfying at least some of the need for nonthrombogenicity for blood contacting applications. The primary focus of our research, apart from the development and testing of a blood compatible polymer (heparin-PVA), has been to question the hypothesized limitations to the long term use of heparinized materials. The necessity of a heparin microenvironment, the potential passivation of immobilized heparin, the consumption of clotting factors and increased platelet consumption are serious theoretical limitations to the development of heparinized surfaces.

Ironically heparinized materials or those with a steady leakage of heparin are suitable for and have been used with success in various short term applications. This has led to the erroneous conclusion that heparin loss and a corresponding heparin microenvironment at the blood material interface, were indispensable for rendering the material effective for the prevention of thrombogenesis. While early attempts to immobilize heparin covalently were unsuccessful it was by no means clear that this was a necessary and universal conclusion of all covalent immobilization methods.

Heparin's ability to interrupt the intrinsic coagulation cascade may prevent the normal failure mode of devices exposed to low blood flow rates (i.e., red thrombus formation) but may expose an underlying platelet aggregation or white thrombus formation problem. Since the interaction of heparin with platelets is unclear (1) it has remained to be shown that immobilized heparin does not cause greater thrombosis under conditions where platelet deposition is more important than fibrin formation.

Our lab, has attempted to clarify the validity of these concerns using a heparin-poly(vinyl alcohol) hydrogel in which heparin is immobilized to a glutaraldehyde crosslinked PVA, through its terminal serine.

MATERIALS AND METHODS

Heparin-PVA

Heparin-polyvinyl alcohol (PVA) hydrogel was prepared as before (2) with 1% sodium heparin (porcine mucosal, 176 USP U/mg, Canada Packers Ltd., Toronto). Control gel was prepared without heparin.

Polyethylene tubing (1.14 mm ID x 1.57 mm OD, Intramedic PE160, Clay Adams Parsippany, NJ) was prepared for coating by glow discharge cleaning (Harrick Scientific, Ossining, NY) and surface activated in hot chromic acid (3) to obtain good adhesion between the hydrogel and the polyethylene (PE) substrate. Tubing was coated with a thin layer of heparin-PVA hydrogel by filling the tubing by hand and then displacing the excess with filtered, compressed air. Average dry coating thicknesses were in the range of 5-10 µm (mean 8.1 µm).

Parallel Flow Arterio-Venous Shunt (4)

To examine the *ex vivo* thromboresistance of the heparinized tubing, a SILASTIC chronic arteriovenous shunt (3.18 mm ID x 6.35 mm OD) was surgically inserted between the iliac artery and contralateral vein of a beagle dog with coumadin therapy (0.25-0.5 mg/kg orally) to aid in maintaining shunt patency during the postoperative period. Experiments were begun after anticoagulant therapy had been stopped and the hemostatic parameters had returned to normal. For an experiment, a smaller diameter (1.14 mm ID) heparinized or control coated PE tube (20 cm long) was inserted in parallel with the exteriorized portion of the SILASTIC shunt through a pair of Y-connectors (Figure 1) made from SILASTIC tubing. The length of the SILASTIC portion between the Y-connectors was also 20 cm. Flow rate in the shunt was monitored with a Doppler flow meter (Blood velocimeter BV380, Sonicaid, Fredericksburg, VA). After an occluding thrombus had developed or after a fixed period of time the experiment was terminated, the small diameter tube was removed and the chronic shunt reconnected until the next experiment. In this way the heparinized tubes were tested at low blood flow rates (3 ml/min) and physiologically relevant shear rates (350 s^{-1}) without surgery and so without its associated artifacts but in a test system that could be reused many times as long as the chronic SILASTIC shunt remained patent (typically more than

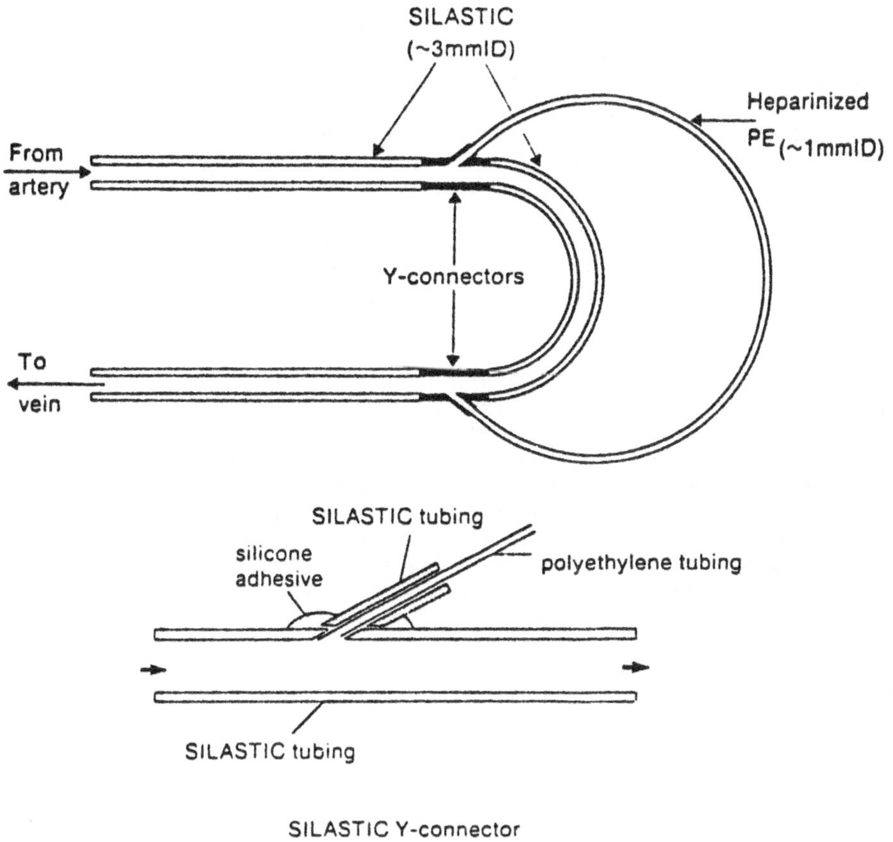

Figure 1. Schematic illustration of: (a) Parallel flow shunt. Less than 2% of the flow in the chronic shunt is diverted through the Y-connectors from the larger diameter tube into the 1 mm diameter heparinized or control coated polyethylene test section. (b) Detail of the SILASTIC Y-connector. Holes are punched in the ends of an 20 cm length of SILASTIC tubing into which the polyethylene tubing is secured with adhesive.

6 months). To assess platelet reactivity under high flow conditions HEP-PVA and PVA coated PE tubings (30 cm, 2.69 mm ID) were connected in series to the arterial and venous cannulae. A not insignificant advantage is that the same animal was used many times, reducing both the inter-animal variability in the results and the costs of experimentation.

Platelet Adhesion and Release

An *in vitro* dynamic test system developed by I. A. Feuerstein (5) was used to compare platelet adhesion and release between heparin-PVA and PVA surfaces. In this test system a suspension of human washed platelets was used at 37°C and prepared according to the technique of Mustard et al. (6). Cell suspensions (hematocrit = 40%) were drawn through heparin-PVA and PVA coated tubes (1.14 mm, 10 cm long) at a flow rate of 1 ml/min for 6 minutes. The number of ^{51}Cr-platelets associated with the surface and the degree of release of ^{3}H-serotonin from the adherent platelets was determined by counting segments of the rinsed tubes (5).

RESULTS

Patency of Heparinized Tubing

Heparin was released from the coated tubes at 2.5×10^{-4} $\mu g/cm^2/min$ (between 300 and 600 hours of washing). After this washing period the amount of heparin bound to the tubing was 14 $\mu g/cm^2$, as determined indirectly by mass balance.

The patency of heparin-PVA coated polyethylene tubing (1.14 mm ID) is compared with that of control PVA coated tubes (without heparin) in Table I. The flow rate in the PE tubing was approximately 1.7% of that in the chronic shunt assuming Poiseuille's Law. The patency time is the directly observed time that the blood continued to flow in the coated PE tubing. Results are reported as the average patency time and the fraction remaining patent at 2 hours. No PVA tubes remained patent at 2 hours.

The observed difference in patency between heparinized and nonheparinized tubing was greater at higher flow rates. The effect of flow rate was presumed to be the consequence of thrombotic processes occuring in the Y-connector. At sufficiently high flow rates the developing thrombus in the connector may be washed downstream before it has a chance to occlude the branch line. On the other hand at lower flow rates the heparin may provide inadequate protection to the PE tubing if platelets have been activated by prior exposure to the Y-connector. Nevertheless the parallel flow shunt system is still able to distinguish heparinized from nonheparinized tubing making it a useful system for our purposes.

Scanning electron microscopy was used to examine the tubes after removal from the parallel flow shunt. In the middle of the tubes there were only a few isolated platelets adhering to the heparinized tubing with little or no fibrin strands apparent. However, there was a more extensive and more uniform platelet deposit entrapped in fibrin on the control tubes.

Platelet Interaction

One aspect of the interaction of heparin with platelets is shown in the results in Table II. There appears to be little significant difference in the number of platelets adhering to heparin-PVA or PVA coated tubes in the Feuerstein test system. There was no significant difference in the degree of release of ^{14}C-serotonin by the adhering platelets,

Table I. Parallel Flow Shunt Patency.

Main Shunt Flow rate (mL/min)	PVA (a) (min ± SD)	HEP-PVA (min ± SD)	HEP-PVA (Fraction Patent at 2 hours)
130	20 ± 7 (n=2)	73 ± 40 (n=6)	0.17
150	30 ± 22 (n=6)	85 ± 30 (n=12	0.25
160	29 ± 7 (n=7)	116 ± 44 (n=4)	0.25 (>180 min)
170	60 (n=2)	170 ± 17 (n=3)	1.0 (>150 min)

(a) No PVA tubes remained patent at 2 hours.

Table II. In Vitro Platelet Adhesion and Release.

	Adhesion (Platelets/10^3 m^2)	Release by Adherent platelets (%)
PVA	14.7 ± 9.3	74 ± 10
hep-PVA	13.3 ± 4.6	79 ± 10

although the release values were relatively high considering the few platelets adhering to the gel.

Preliminary experiments have demonstrated no drop in canine systemic platelet count over 4 days of flow at 150 mL/min in HEP-PVA coated tubings (30 cm, 2.69 mm ID) in one animal. No evidence of platelet adhesion was noted under SEM. This was in sharp contrast to PVA coated tube where there was extensive white thrombus formation observed after only a few hours of flow.

DISCUSSION

The prolonged patency observed for heparin-PVA coated polyethylene tubes relative to the nonheparinized controls (Table I) and the reduction in fibrin deposits in the SEM are consistent with the previously reported *in vitro* evidence of the biological effectiveness of the immobilized heparin (2,4,7). The biological activity of heparin-PVA is rationalized in part by noting its point of coupling to PVA. Since heparin is bound primarily through the terminal serine at the end of the protein/carbohydrate linkage region it is presumed that it can assume its native conformation. This thromboresistence was achieved without significant release of heparin to create a microenvironment. The release of heparin at the very low measured rate is too small to create a heparin microenvironment at the blood-material interface.

In Feurestein's test system using washed platelet suspensions, there appeared to be little significant difference between the number of platelets adhering to heparin-PVA or PVA control hydrogels (Table II). Although few platelets adhered to the hydrogel, more than 70% of the platelets adherent to the hydrogel had released their ^{3}H-serotonin content. It is conceivable that more platelets had adhered but had embolized or that significant aggregation would have occured had apyrase not been added to the suspension. Bead column retention measurements have confirmed the absence of a significant enhancing effect of heparin on platelet aggregation.

It is our impression that the underlying substrate (here PVA) may be more important in defining the response of platelets *in vitro*. A highly reactive surface, for example, would result in significant platelet adhesion regardless of the presence or absence of immobilized heparin. Using the chronic A-V shunt, thrombocytopenia did not develop during a 4 day exposure to 25 cm^2 of heparin-PVA at a flow rate of 150 mL/min. The lack of platelet adhesion and aggregation on the heparinized surface was presumed due to the inactivation of thrombin, a strong platelet stimulant, at the surface. The PVA coated PE surfaces revealed a high degree of platelet reactivity under SEM after only a few hours of flow, because of the failure to inactivate the surface generated thrombin adequately. Strategies are currently being devised to immobilize antiplatelet agents (e.g., PGE) or to modify the PVA surface otherwise to enhance the platelet compatibility of both it and heparin PVA.

CONCLUSION

The results reported here, in conjunction with earlier results, indicate that immobilized heparin need not necessarily be lost from a surface in order to impart thromboresistance to the surface. For heparin-PVA, and perhaps for other covalent reactions that do not inactivate the heparin, the irreversibly bound heparin can accelerate the formation of a surface-bound inactive thrombin-antithrombin III complex. The long term use of heparinized materials requires stable immobilization techniques that do not compromise the biological function of the heparin. This investigation and others indicate that these requirements are not mutually exclusive, and that bound heparin can potentially retain its biological activity over the long term. Further studies underway in our laboratory point to the absence of adverse interactions with thrombin or antithrombin III that might compromise this long term bioactivity.

In vitro tests systems used to assess the platelet reactivity of immobilized heparin have not revealed any effect of the heparin. The *in vitro* results support the hypothesis that surface bound heparin has not significant effect on platelet aggregation and that the interaction of heparin-PVA with platelets is dominated by the reactivity of the PVA. The *ex vivo* studies, suggest that surface bound heparin's ability to inactivate thrombin and thereby lessen platelet activation via thrombin may play an important role in modulating the *ex vivo* response to platelets.

ACKNOWLEDGEMENTS

The authors acknowledge the financial support of NHLBI/DTB under grant HL24020 and the Ontario Heart Foundation, the benefit of collaboration with Professor I. A. Feuerstein and the technical assistance of Dr. G. Kent and staff of the Surgical Research Facility, Hospital for Sick Children.

REFERENCES

1. M. B. Zucker, Fed. Proc., 365, 47 (1977).
2. M. F. A. Goosen and M. V. Sefton, J. Biomed. Mater. Res., 17, 359 (1983).
3. R. Evangelista and M. V. Sefton, Proceedings IUPAC MACRO 82, 357 (1983)
4. W. F. Ip, W. Zingg and M. V. S. Sefton, Biomed. Mater. Res., in press.
5. G. A. Adams and I. A. Feuerstein, Am. J. Physio., 240 (Heart Circ. Physiol. 9), H95-H108 (1981).
6. R. L. Kinlough-Rathbone, J. F. Mustard, M. A. Packham, D. W. Perry, H. J. Reimers and J. P. Cazenave, Thromb. Haemostas., 37, 291 (1977).
7. M. F. A. Goosen and M. V. Sefton, J. Biomed. Mater. Res., 13, 347 (1979).
8. M. F. A. Goosen, M. V. Sefton and M. W. C. Hatton, Thromb. Res., 20, 543 (1980).

NEW HEPARINIZABLE MATERIALS: SURFACE GRAFTING OF POLY (AMIDO AMINE) CHAINS ON POLYURETHANE

Rolando Barbucci, Manuela Benvenuti, Paolo Ferruti and Mila Nocentini

C.R.I.S.M.A. University of Siena
Le Scotte - Nuovo Policlinico
53100 Siena, Italy

Poly(Amido-amine)s chains are capable of forming stable complexes with heparin thus their grafting on the surface of devices renders the surface non-thrombogenic. A calorimetric study performed on the heparin-poly(amido-amine)s system indicate that mainly electrostatic interactions between heparin and poly(amido-amine)s occur in aqueous solution at pH=7.4. Furthermore the FTIR spectrum of the solid heparin-poly(amido-amine) complex obtained from aqueous solution shows that it is not a simple mixture of the two polymers. Grafting of poly(amido-amine) N_2LL on polyurethane surface has been studied along with all the chemical steps. In addition, immobilized heparin was monitored by FTIR technique. Biological tests shows that the anticoagulant activity of heparin remains even if it is linked to poly(amido-amine) grafted on a PU surface. The sequence of SEM photos shows that after N_2LL treatment the PU surfaces are covered by platelets and fibrinogen. No deposit is observed, however, on heparinized surfaces.

INTRODUCTION

It is well known that mechanical limitations to the use of biomedical devices have been got over, the single most important obstacle preventing even more diffuse application of such devices remains biomaterial thrombogenicity. Two approaches have been taken into consideration to eliminate this problem: the surface modification approach (1) or the pharmaceutical approach (2). Long term antithrombogenicity can be reached by immobilization of heparin on surfaces (3). On the other hand, a rapid release of some amounts of heparin during the initial time is welcome to create immediately a heparin environment around the device to prevent thrombus formation. Our approach consists in grafting on the surface of the material, heparin complex polymers like the poly(amido-amine)s (4), by utilizing the reactive groups present in its chemical formula, which is shown below in Diagram 1. In this respect synthetic methods for grafting poly(amido-amine)s on the surface of glass, PVC, Silastic, Dacron, have been developed by us (5). As an advantage of this procedure we can mention the fact that it maintains the size, shape and mechanical per-

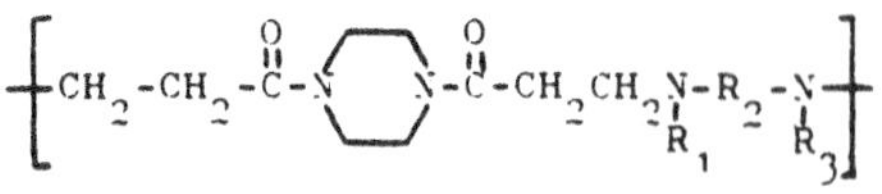

Diagram 1. General formula of poly(amido-amine)s.

formances of the starting articles, thus avoiding any problem connected with the manufacture of entirely new products.

RESULTS AND DISCUSSION

Structure and Properties of Poly(amido-amine)s

Linear poly(amido-amine)s are obtained by polyaddition of primary monoamines or bis-(secondary amine)s to bis-acrylamides (6). The reaction takes place readily in water and alcohols at room temperature and without added catalysts. These synthetic reactions are outlined below in Diagram 2. Poly(amido-amine)s carrying additional functions as side substituents can be easily obtained starting from the appropriate monomer too. In fact hydroxyl groups, tertiary amino groups, allyl groups, etc., if present in the monomers, do not interfere with the polymerization process (5).

Many poly(amido-amine)s are highly crystalline (6), and some crystallinity is even observed, in some cases, where a certain degree of irregularity has been introduced (6). As a rule, most poly(amido-amine)s are water soluble or at least highly hydrophylic. They are also soluble in many organic solvents. The intrinsic visocosities range from 0.15 to 0.9 dl/g and molecular weights (number average) from 5×10^3 to 5×10^4. In poly(amido-amine)s N_2, N_2L, N_2LL, Figure 1, the basicity constants and the protonation enthalpies of the aminic nitrogens of each repeating unit don't depend on the overall degree of ionization (4). The number of thermodynamic changes of protonation is equal to the number of nitrogen atoms present in the monomeric unit and they are similar to those of the corresponding non-macromolecular models and other mono, di, and triamines (7). In particular the lengthening of the aliphatic chain between the two tertiary nitrogen atoms determines an increase of the basicity constants, and the successive protonation constants decrease in the usual manner: $\log K_1 > \log K_2$. The enthalpy values follow a similar trend (Table I).

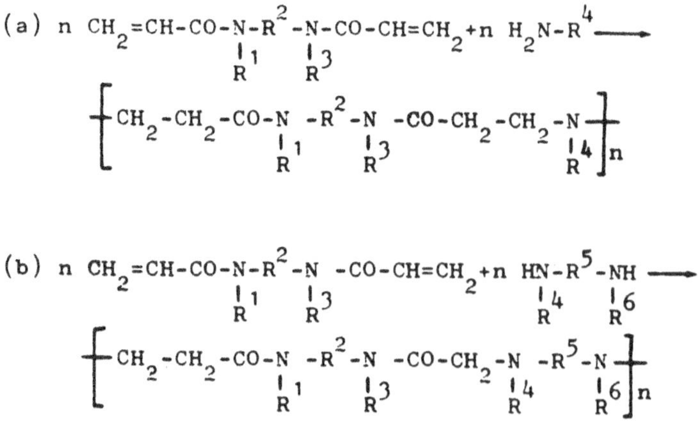

Diagram 2. Outline of synthetic reactions.

N_2

N_2L

N_2LL

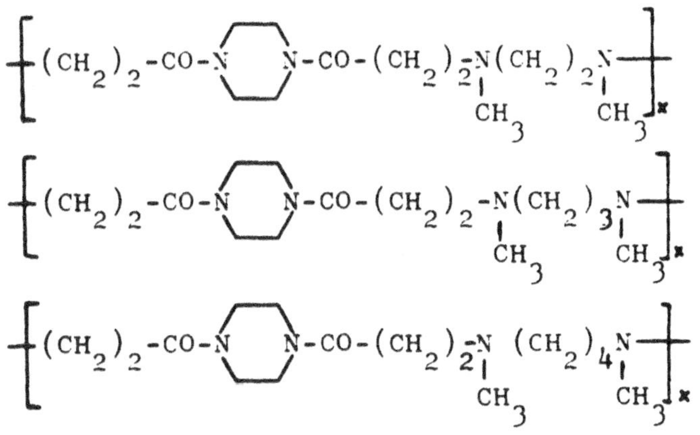

Figure 1. Poly(amido-amine) structures

Viscosimetric titrations prove that the conformational freedom is reduced for protonation (8) (Figure 2). All the polymers in the neutral form are most likely in a random coil structure due to the great number of accessible conformations of about the same energy.

The first protonation leads to the formation of a strong hydrogen bond between "onium" ions and carbonyl groups belonging to the same monomeric unit (Figure 3).

When the first protonation of all the monomeric units is complete, the above effect strongly reduces the conformational freedom of the whole polymer which tends to assume a rigid structure and this explains the jump of $\eta sp/c$ at $\alpha = 0.5$. The second protonation leads to an electrostatic repulsion between the positively charged onium ions belonging to the same monomeric unit and this effect further compels the polymer in a rigid more extended conformation. Consequently a second jump is observed

Table I. Thermodynamic Functions of Protonation at 25°C in 0.1 M NaCl

Compound	Reaction	log K(a)	$-\Delta G^o$(a,b)	$-\Delta H^o$(a,b)	ΔS^o(a,c)
N_2	$L + H^+ \rightleftharpoons LH^+$	8.09 (2)	11.03 (3)	7.35 (12)	12.3 (3)
	$LH^+ + H^+ \rightleftharpoons LH_2^{2+}$	4.54 (5)	6.19 (7)	6.09 (11)	0.3 (6)
N_2L	$L + H^+ \rightleftharpoons LH^+$	8.172 (6)	11.14 (1)	8.54 (23)	8.7 (7)
	$LH^+ + H^+ \rightleftharpoons LH_2^{2+}$	6.691 (9)	9.12 (1)	8.24 (28)	3.0 (9)
N_2LL	$L + H^+ \rightleftharpoons LH^+$	8.715 (8)	11.88 (1)	8.95 (14)	9.8 (4)
	$LH^+ + H^+ \rightleftharpoons LH_2^{2+}$	7.689 (9)	10.49 (1)	8.65 (18)	6.2 (6)

(a) Values in parentheses are standard deviations computed as the sum of the standard deviations for the different runs.
(b) Kcal/mole
(c) Kcal/mole

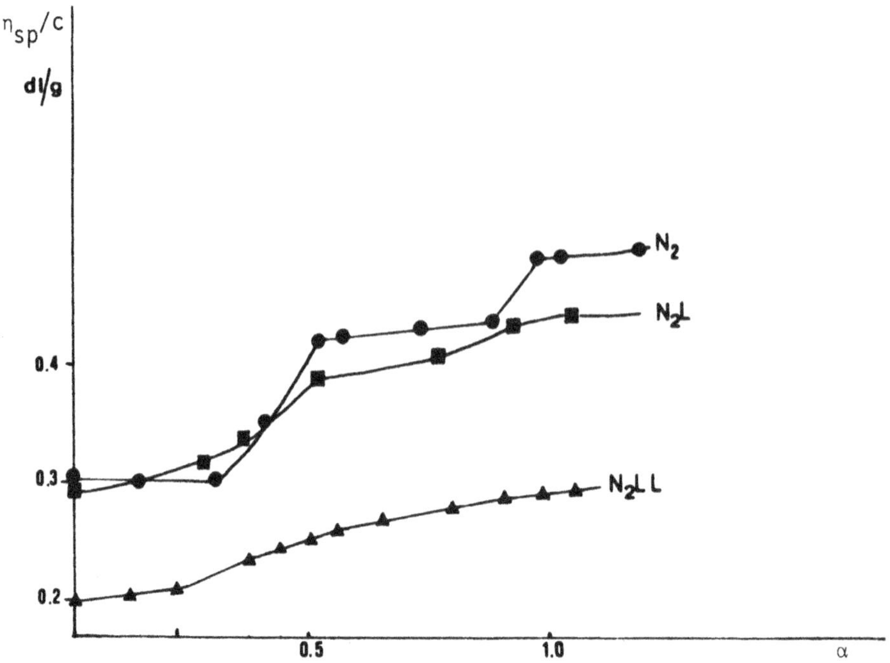

Figure 2. Viscosimetric titrations of polymers in 0.1 M NaCl.

in the η_{sp}/c vs α-curve. The reduction of conformational freedom obviously decreases with the lengthening of the aliphatic chain in each monomer. This explains the slight reduction of the two jumps of $_{sp}/c$ by increasing the number of methylenic groups between the amino nitrogens.

Interaction Heparin-Poly(amido-amine)

At pH 7.4 the repeating units of the poly(amido-amine)s may be present in three forms: the unprotonated one (L) the monoprotonated one (LH^+) and the diprotonated one (LH_2^{2+}). The relative amounts of these species have been calculated from the basicity constants for each polymer, and are reported in Table II. It may be calculated that the poly(amido-amine) deriving from N,N'-dimethyl-1,4-diaminobutane is fully protonated at physiological pH. The protonation of both basic nitrogens provides the longitudinal hydrogen bonds needed to stabilize a helical conformation (Figure 4). It is important to see that all the atoms lie outside the ideal cylinder on which the polymer is wound without steric hindrances, so the aminic nitrogens of the repeating units are in the

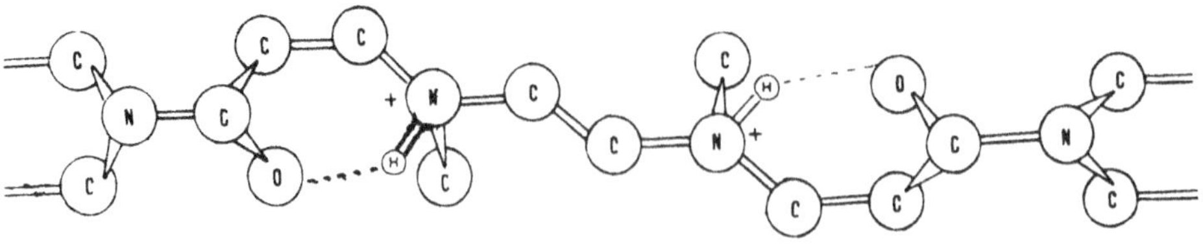

Figure 3. Possible rod-like conformation for the polymer N_2 after the second protonation.

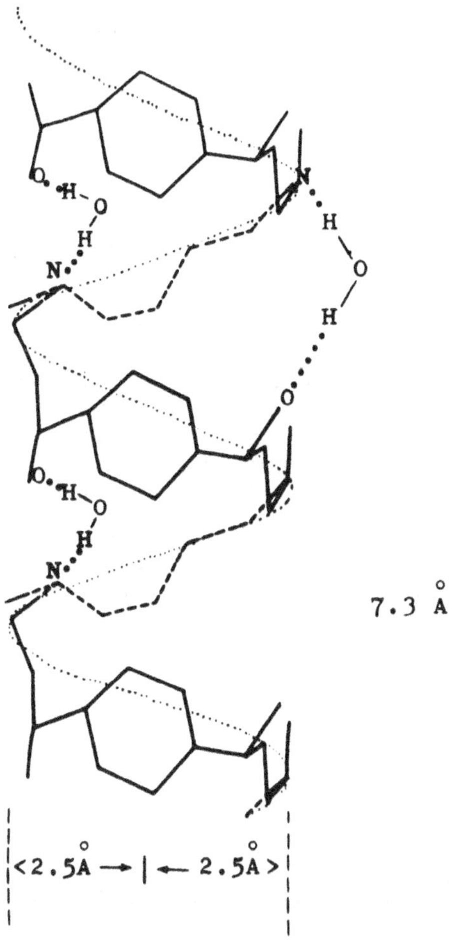

Figure 4. Hypothesized helical conformation of the poly(amido-amine) N_2LL in the diprotonated form.

most favorable position for their interaction with heparin.

A calorimetric study on the complex formation between heparin and poly(amido-amine) N_2 has been performed in water (9). The results are reported in Figure 5. It may be observed that the reaction is exothermic and its value decreases by increasing the amount of poly(amido-amine). In other words it follows the same trend observed during the neutralization with NaOH of some ionizing groups present in insulin (10). If one assumes that binding of metal ions by SO_3^-, COO^-, groups along the polysaccharide chains remains an intrinsically endothermic process (11), it may be deduced that the interaction between heparin and PAA promotes, via a screening of the fixed charges, an additional exothermic process. The heat effect also increases by decreasing the ionic strength of the medium and also by decreasing pH. All these results indicate the electrostatic interaction mainly takes place between heparin and poly(amido-amine). In fact the intensity of electrostatic interaction is lowered at high ionic strength because of the shielding effect, on the other hand lowering the pH increases the charge density on poly(amido-amine) which has no effect in the range considered on the charge density of heparin.

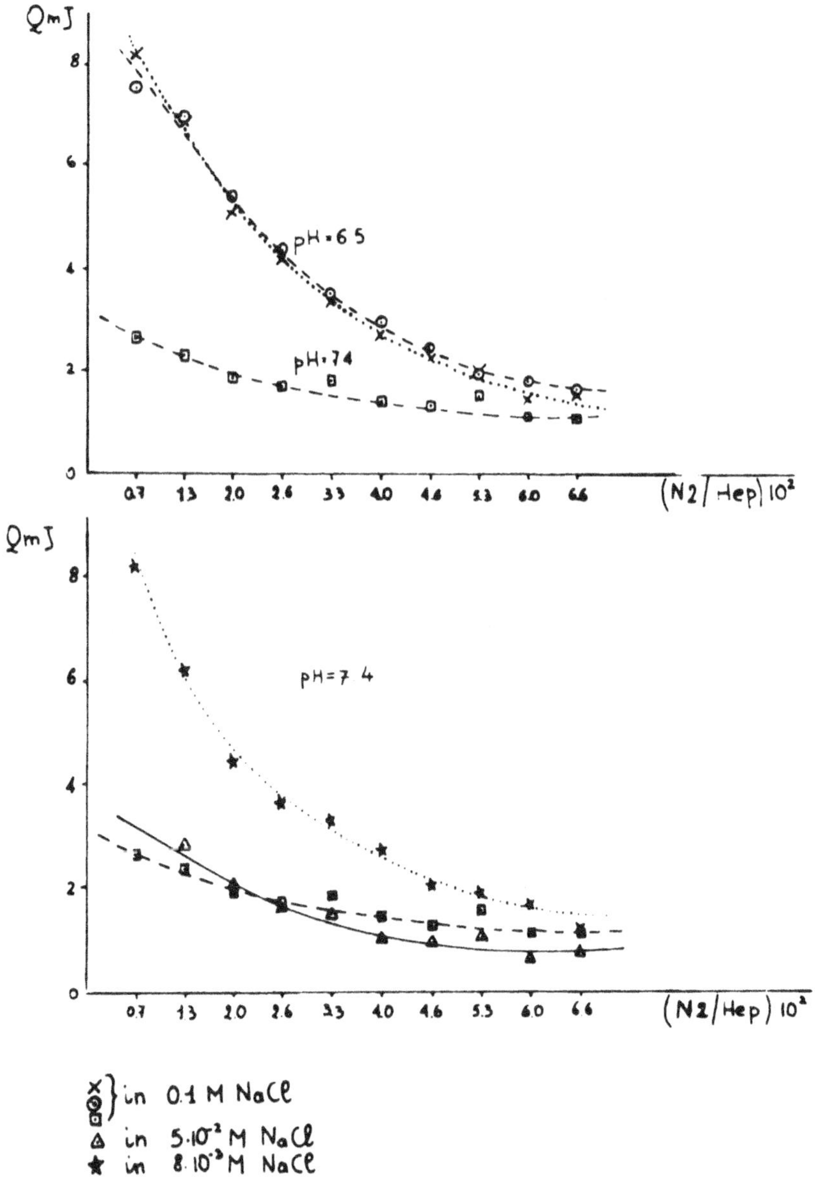

Figure 5. Heat determined for N_2-heparin complex formation at different N_2/heparin ratios.

Besides, a complex could be isolated, obtained by mixing heparin and poly(amido-amine) at pH 6; the elemental analysis correspond to a 1:1 complex. The FTIR spectrum of the solid complex has been reported in Figure 6. We can clearly observe the presence of the most intense bands of both components heparin and N_2; besides, a FTIR difference spectrum (Figure 7) was obtained by substrating the spectrum of the heparin from that of the complex. By comparing the difference spectrum with that of solid N_2 some inequalities are observed, which can be probably attributed to the interactions among different groups of the two polymers.

Grafting of Poly(amido-amine) N_2LL on Polyurethane Surfaces

The grafting reaction proceeds by the addition of diisocyanates to PU with the formation of an ureic bond (12), as shown in Diagram 3.

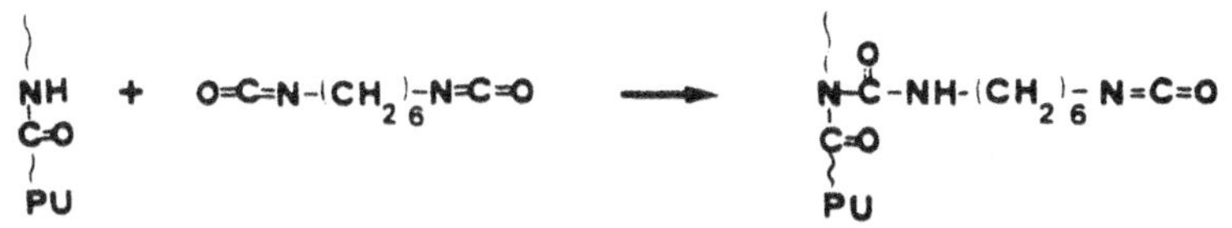

Diagram 3. Diisocyanate addition to PU.

The further reaction of a secondary amino-end-capped poly(amido-amine) N_2LL with the terminal isocyanate group leads to the grafting of this polymer as shown below in Diagram 4.

All the chemical steps of the grafting reaction of synthetic PU were monitored by FTIR technique (12). Figure 8 shows the transmittance spectrum of native PU and the spectrum of the same sample after diisocyanate treatment. The presence of a new band at 2270 cm^{-1}, assignable to the asymmetric stretching of the NCO group is evident. This band shows that grafting of the diisocyanate occurs. Substrating the spectrum of native PU from that of diisocyanate treated PU, it is possible to point out two bands assignable respectively to the amide I (1630 cm^{-1}) and to the amide II (1580 cm^{-1}) Figure 9. These above assignments clearly demonstrate that the chemical reaction of PU with diisocyanates occurs through the formation of ureic bonds, as described in the scheme.

The spectrum of the PU sample after the final treatment with N_2LL is shown in Figure 10 together with the spectrum of N_2LL solid. Comparing this spectrum with that of PU after diisocyanate treatment, we note a clear drop in intensity of the 2270 cm^{-1} band. The presence of N_2LL

Table II. Relative amounts of each species calculated from basicity constants.

Polymer	log K_1 (a) $L + H^+ \rightleftharpoons LH^+$	log K_2 (a) $LH^+ + H^+ \rightleftharpoons LH_2^{2+}$	distribution % of species repeating units at pH 7.4		
			uncharged (L)	monocharged (LH^+)	dicharged (LH_2^{2+})
N_2	8.09 (5)	4.54 (5)	20.1	79.8	0.1
N_2L	8.172 (6)	6.691 (9)	11.6	76.9	11.6
N_2LL	8.715 (8)	7.689 (9)	2.0	36.0	62.0

(a) Values in parentheses are standard deviations.

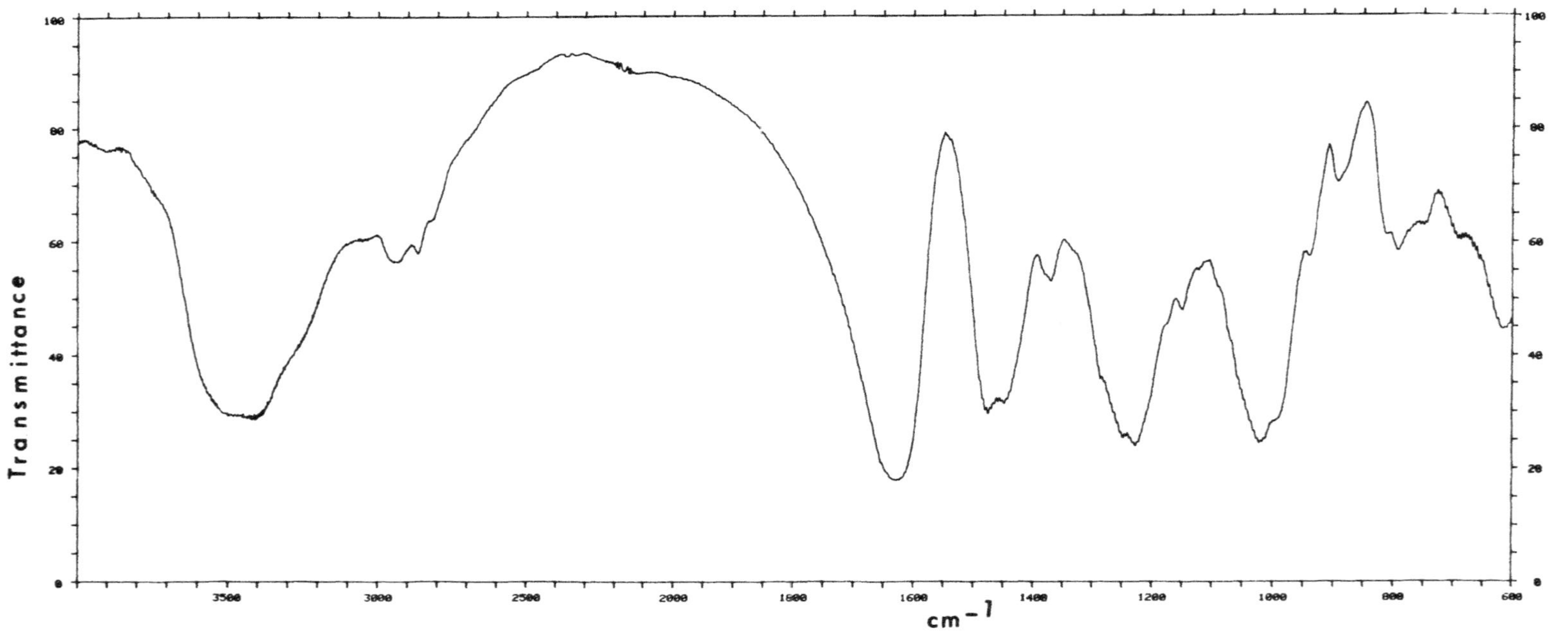

Figure 6. FTIR spectrum of the solid complex between heparin and N_2 polymer.

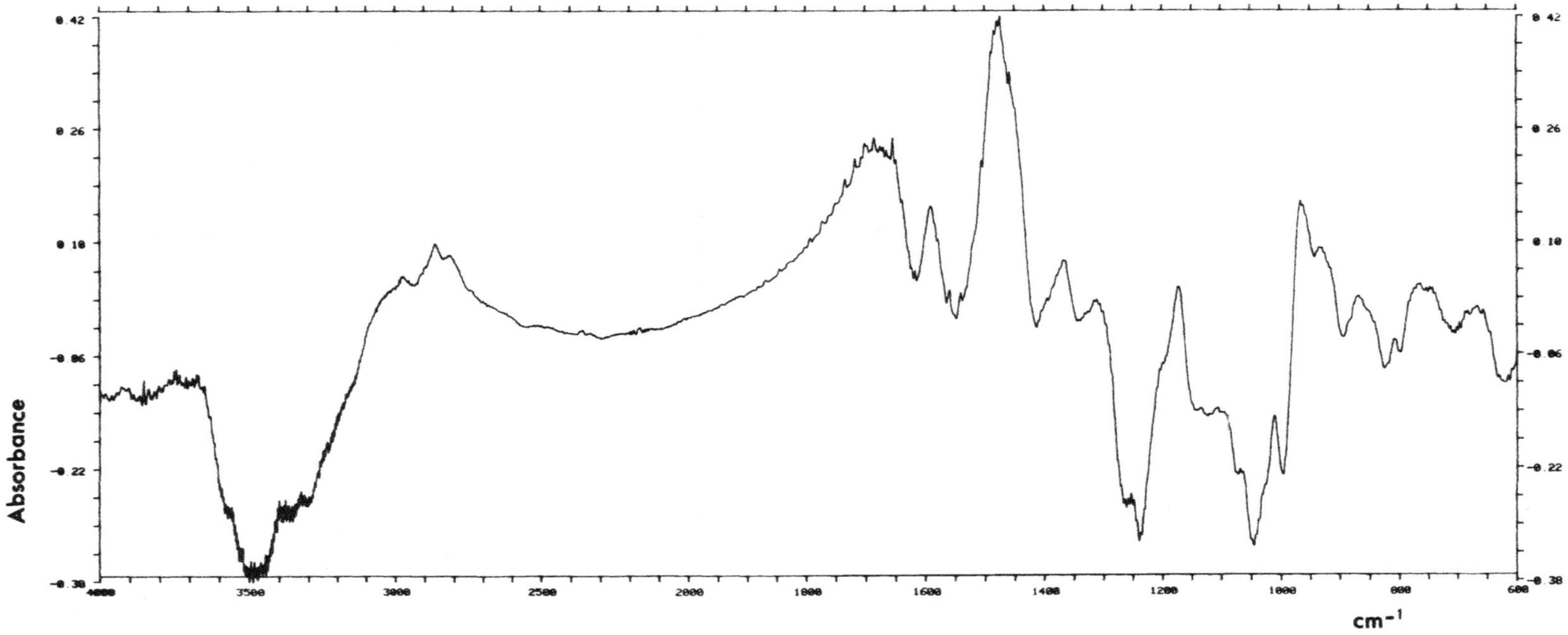

Figure 7. Difference spectrum obtained by substracting the heparin spectrum from that of the complex.

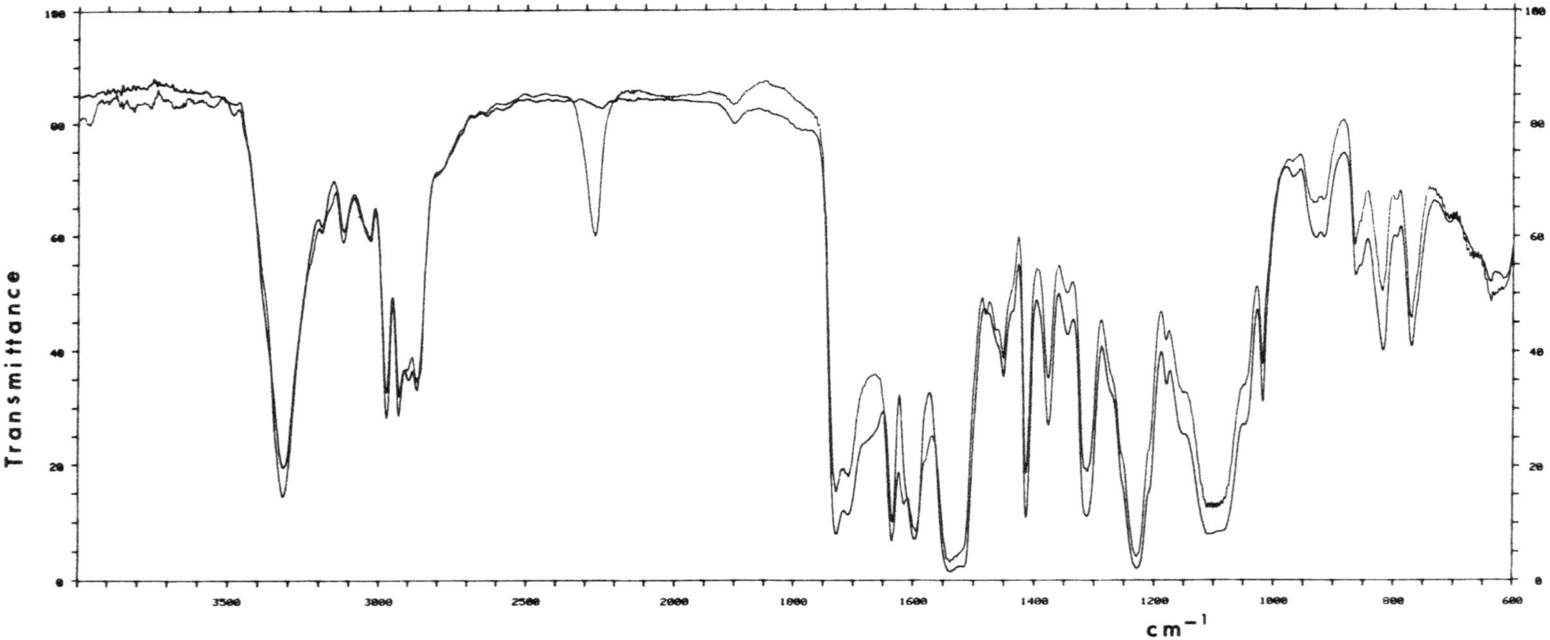

Figure 8. FTIR spectrum of native PU (----) and diisocyanate treated PU (- - - -).

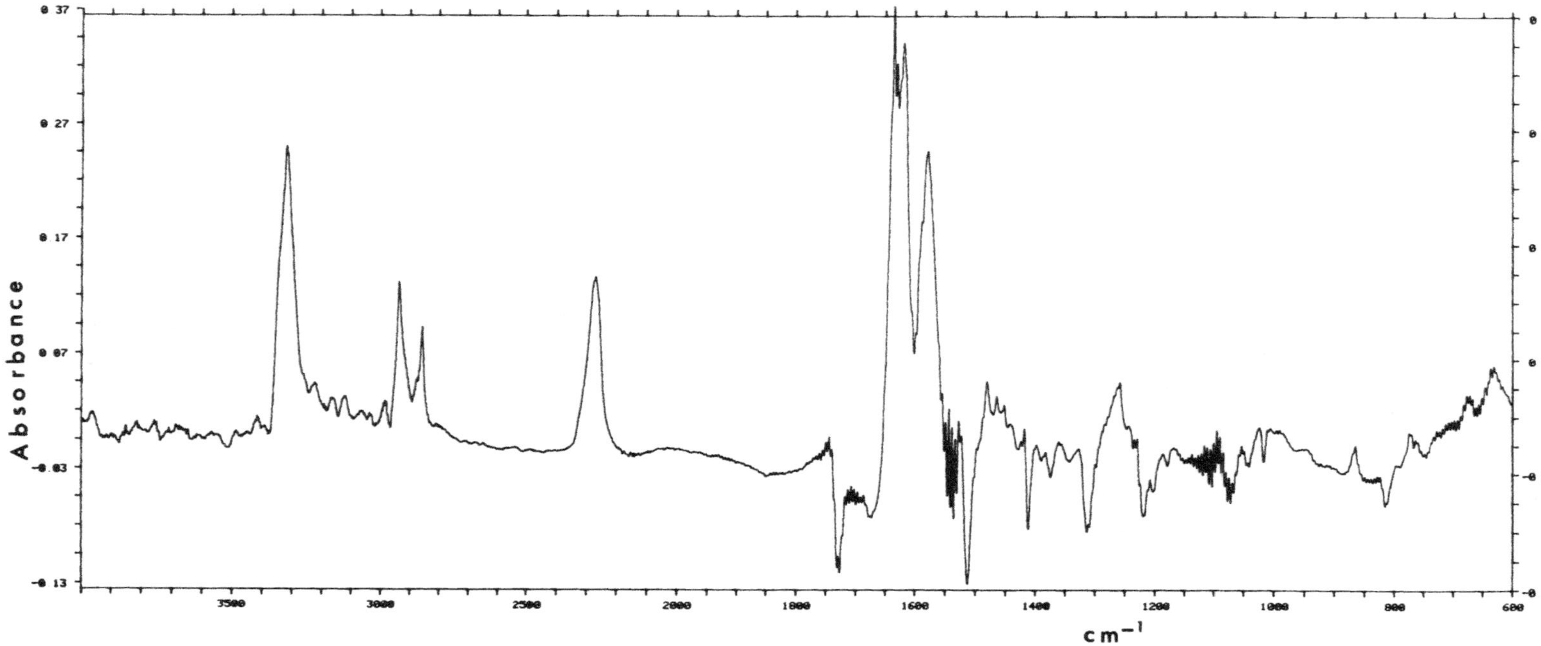

Figure 9. Difference spectrum of diisocyanate treated PU minus native PU.

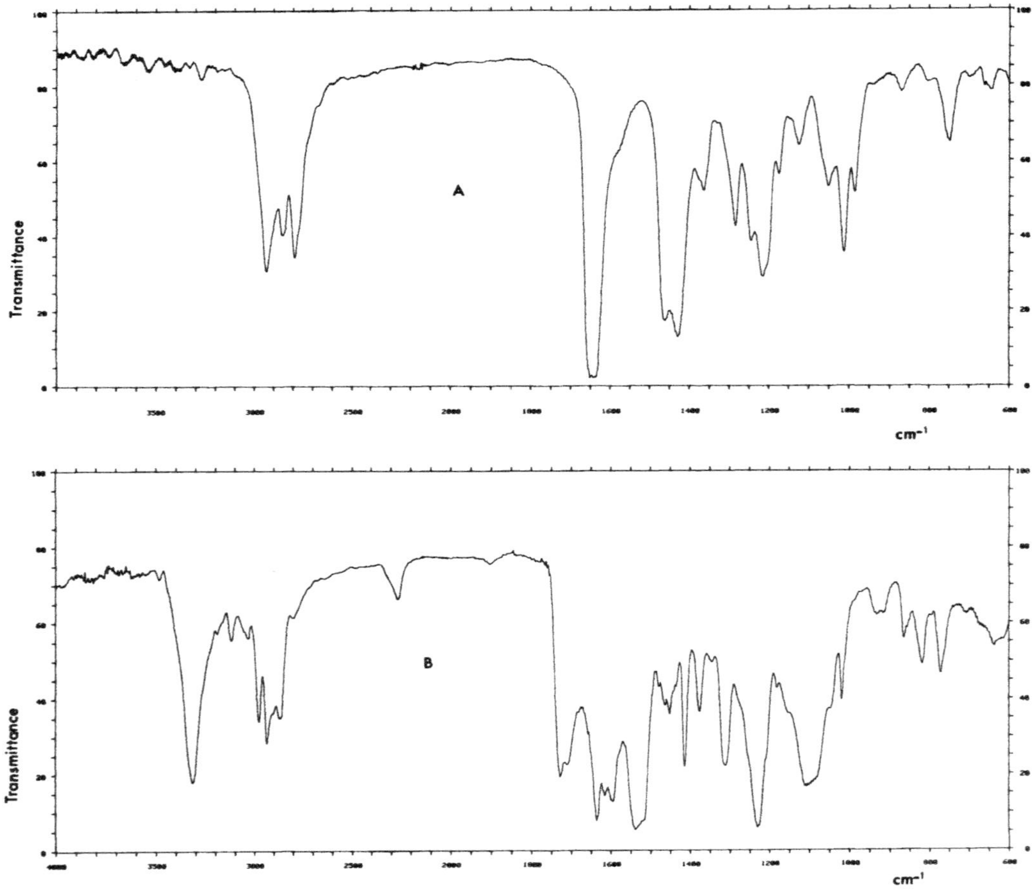

Figure 10. A: FTIR spectrum of N_2LL.
B: FTIR spectrum of N_2LL treated PU.

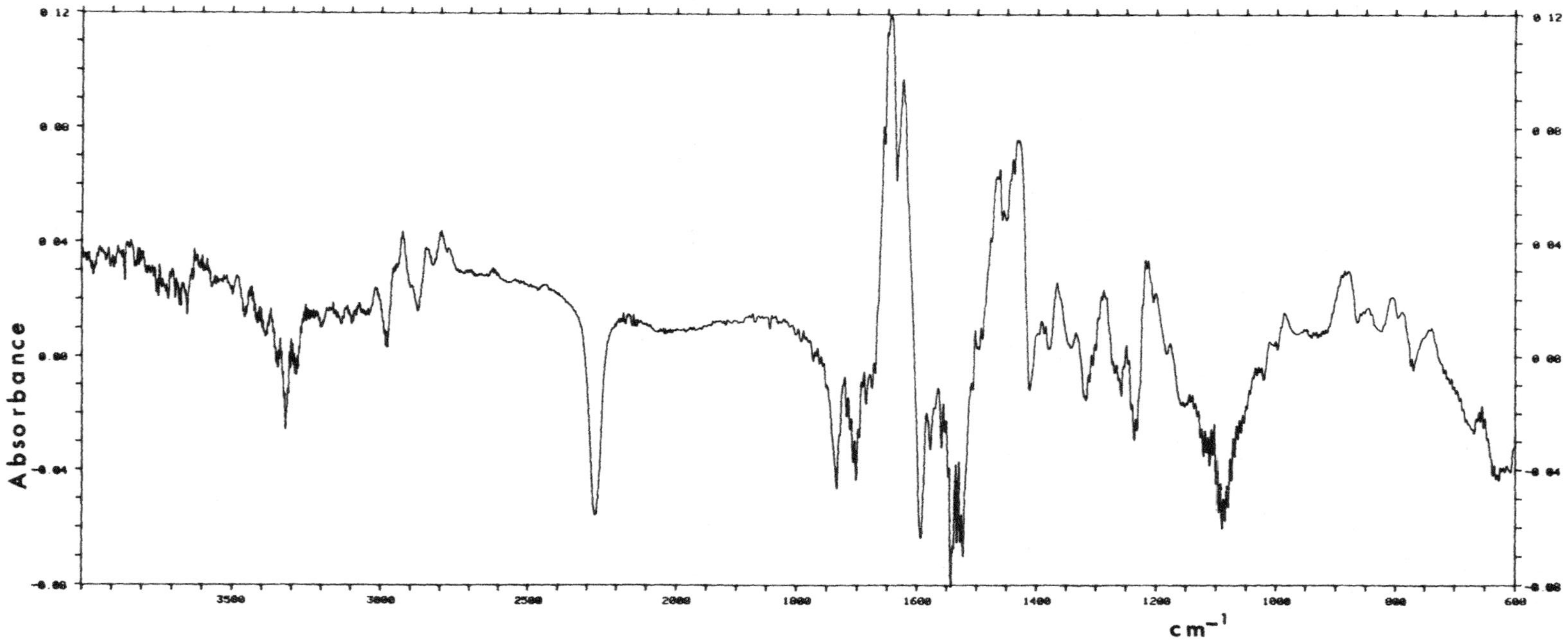

Figure 11. Difference spectrum of N_2LL treated PU minus diisocyanate treated PU.

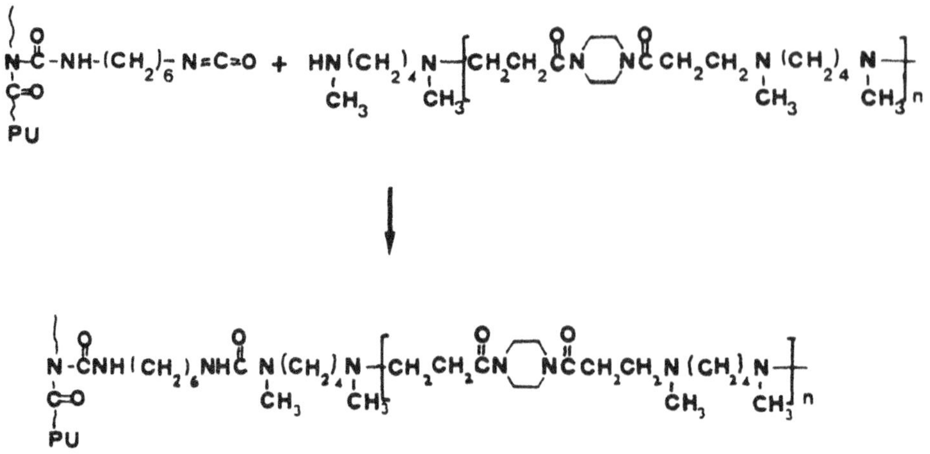

Diagram 4. Grafting reaction on modified PU.

chains on the surface of PU can be demonstrated by substracting the diisocyanate-grafted PU spectrum from that of N_2LL grafted PU. The difference spectrum, Figure 11, consists of three bands of medium intensity in the region 2800-3000 cm^{-1} and four intense bands at 1640, 1465, 1435 and 1220 cm^{-1}. These bands correspond to the most intense ones in the N_2LL spectrum.

Heparin Immobilization on Poly(amido-amine) Grafted PU Surface

Heparinization of the PU samples was performed by dipping the N_2LL grafted PU into a solution at pH 6. This treatment determines the complete protonation of all the basic nitrogens of N_2LL and renders them capable to form the electrostatic bond with the negatively charged heparin. After 12 hours of immersion the films were put in contact with a 0.5% heparin solution in a water/ethanol/acetic acid mixture.

Two different adsorptions of heparin occur on the surface (13), in the first, adsorbed heparin is bonded to N_2LL grafted on PU surface via an electrostatic bond and can be removed only by very drastic conditions such as 0.1 M NaOH, and even in this case desorption takes place gradually and requires a considerable time to be completed. In the second, heparin is only physically adsorbed on the first layer of heparin and can be eluted with saline, Figure 12. Only the first kind of heparin can be

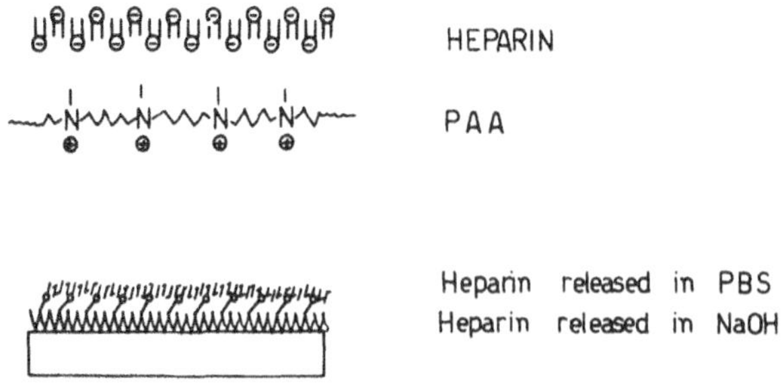

Figure 12. Heparin adsorbed on poly(amido-amine) grafted PU

Table III. Quanity of Complexed Heparin on PU Samples.

Samples		IU Hep	IU Hep/ cm^2 PU	mg Hep/ cm^2 PU	Polymer N_2LL Repeating Unit (µmoles/cm^2 PU)	mg Hep/ µmoles N_2LL	Average
	I	1.7	0.6	0.0042	0.07	0.06	
PU(a)	II	2.5	1.1	0.0069	0.13	0.05	0.06
	III	3.9	1.2	0.0078	0.12	0.06	

(a) PU: Pellethane 2363-80A (Upjohn Polymers, Netherlands)

exactly determined. Table III shows the results of biological tests on three heparinized PU samples after treatment with 0.1 M NaOH.

The constant heparin/poly(amido-amine) ratio observed for these PU samples leads us to suppose that the complex formation reaction proceeds with a definite stoichiometry, as observed in the solid heparin poly-(amido-amine) complex too. To demonstrate the presence of heparin on the PU surface after the third step, we have used ATR-FTIR technique. By subtracting the N_2LL treated PU spectrum from that of heparinized N_2LL treated PU film, the spectrum in Figure 13 was obtained. Only one band in the range from 1050 to 1000 cm^{-1} clearly appears, indicative of the presence of heparin.

Anticoagulant Activity of Heparin Linked to Poly(amido-amine) Grafted PU

The biological tests performed on the heparin released from PU surface, either by saline or by 0.1 M NaOH solution, are indicative of the fact that the anticoagulant activity of heparin is maintained unaltered once free from the surface. To determine the activity of heparin when it is still linked to the N_2LL grafted surface, the films of heparinized PU were dipped in platelet poor plasma (PPP). The APTT and TT, when measured in presence of the films, remain constant for several hours, thus indicating that no heparin release is occuring. However these times are higher than those obtained in the blank experiment, showing the heparin present on PU film maintains anticoagulant activity. The reptilase clotting time shows that the increase of clotting times (APTT and TT) is not due to fibrinogen modifications. Moreover samples of PU films, either with N_2LL grafted on the surface or with heparin linked to the N_2LL grafted PU, were then examined by Scanning Electron Microscope (SEM) (Figure 14-15). The results showed that considerably thick deposit of platelets and fibrinogen were present on the N_2LL grafted PU. In contrast the heparinized film was practically clear.

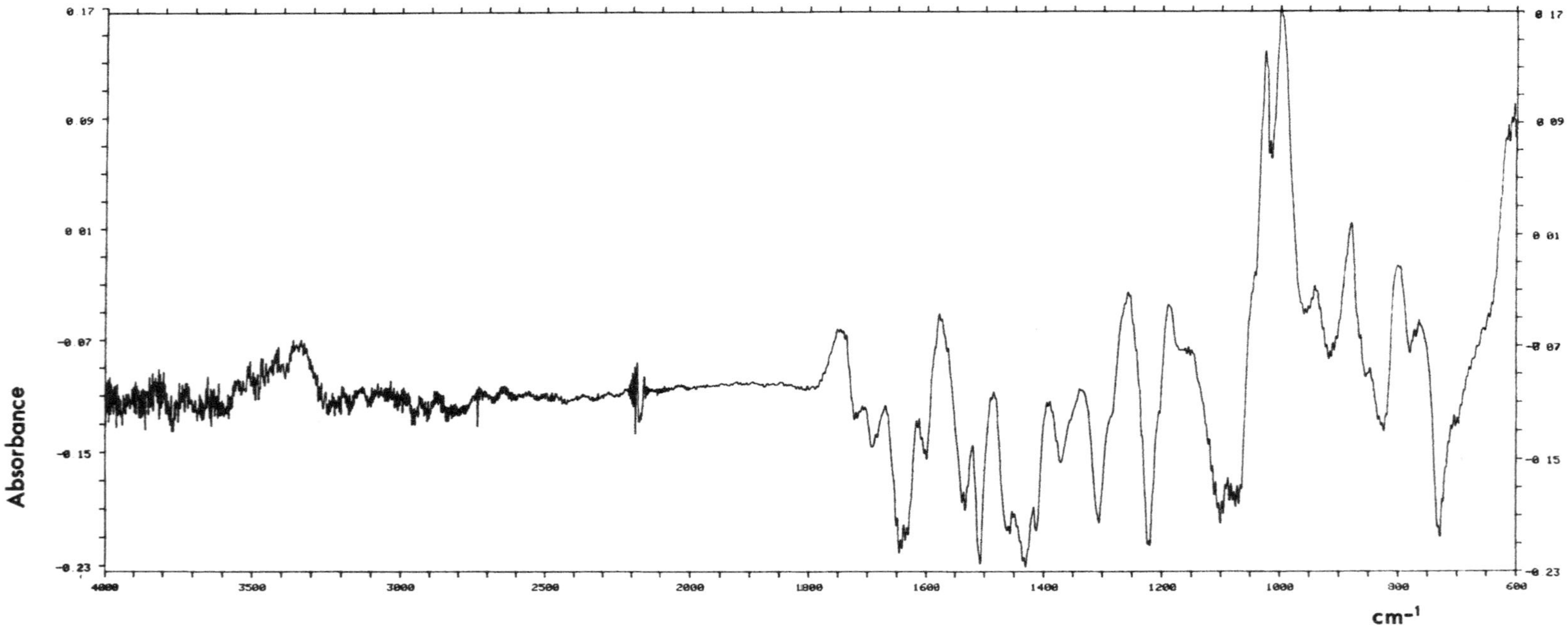

Figure 13. ATR-FTIR difference spectrum of heparin obtained by subtracting the spectrum of N_2LL grafted PU from that of heparinized N_2LL grafted PU.

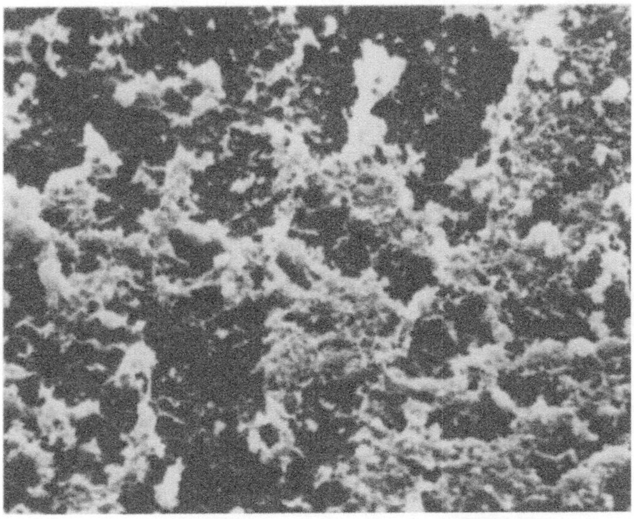

Figure 14. SEM of the surface of N_2LL grafted PU after exposition to PRP.

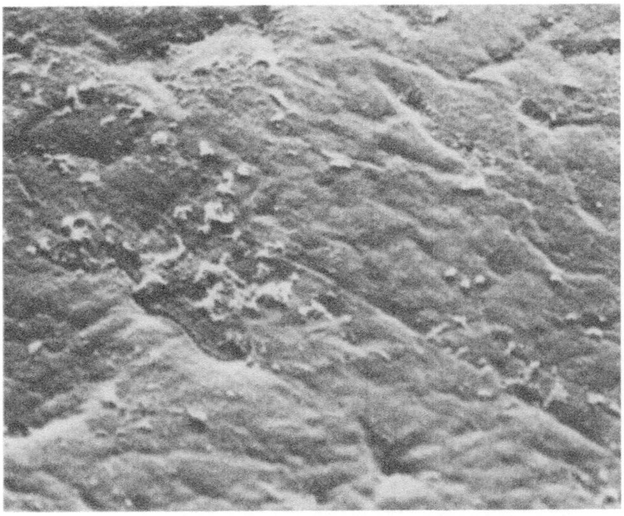

Figure 15. SEM of the surface of heparinized PU after exposition to PRP.

REFERENCES

1. J. D. Andrade, Med. Instrum., 7, 110-120 (1973).
   R. Barbucci, G. Casini, P. Ferruti and F. Tempesti, Polymer, 26, 1349-1352 (1985).
2. S. W. Kim, C. D. Ebert, J. Y. Lin and J. C. McRea, ASAIO, 6, 76-87 (1983).
   J. C. McRea, C. D. Ebert, and S. W. Kim, Trans. ASAIO, 27, 511-516 (1981).
3. G. Schwer, Trans. ASAIO, 18, 321-324 (1972).
4. P. Ferruti and R. Barbucci, Adv. in Polymer Sci., 58, 55-92 (1984).
5. P. Ferruti, M. A. Marchisio and R. Barbucci, Polymer, 26, 1336-1348 (1985).
6. F. Danusso and P. Ferruti, Polymer, 11, 88-113 (1980).
7. R. Barbucci, M. Casolaro, V. Barone, P. Ferruti and M. Tramontini, Macromalecules, 16, 1159-1164 (1983).
8. R. Barbucci, M. Casolaro, P. Ferruti, V. Barone, F. Lelj and L. Oliva, Macromolecules, 14, 1203-1209 (1981).
9. F. Tempesti and I Wadso, preliminary study.
10. T. E. Jensen, (1977) Ph. D. Dissertation, Brigham Young University, Provo, Utah.
11. V. Crescenzi, C. Airoldi, M. Dentini, L. Pietrelli and R. Rizzo, Makromol. Chem., 182, 219-223 (1981).
12. R. Barbucci, M. Benvenuti, G. Casini, P. Ferruti and M. Nocentini, Makromol. Chem.,186, 2291-2300 (1985).
13. R. Barbucci, M. Benvenuti, G. Casini, P. Ferruti and F. Tempesti, "Proceedings of the International Congress on Polymer in Medicine", Capri, June 3-7, 1985.

THE SYNTHESIS OF SOME POTENTIALLLY BLOOD COMPATIBLE HEPARIN-LIKE POLYMERIC BIOMATERIALS.

Charles G. Gebelein and Daniel Murphy

Department of Chemistry
Youngstown State University
Youngstown, OH 44555

When artificial polymeric implants are used in the human body, one of the most important factors is the blood compatibility. The polymeric biomaterials reported here were prepared by the cycloaddition reaction of chlorosulfonyl isocyanate (CSI) to unsaturated polymers, such as polybutadiene or polyisoprene, to give a poly(N-chlorosulfonyl beta-lactam) structure which could then be hydrolyzed to give a polyanionic polymer that contains sulfamate and carboxylate groups. Previous researchers have shown that this type of material can be used as an artificial anti-coagulant with activities fairly similar to those of bovine plasma, but somewhat lower than heparin. These polymeric compounds have many of the same functional groups that are found in heparin, and might be expected to behave in a manner similar to heparinized polymers. Unfortunately, most of the previously reported research on these heparinoids has involved water soluble polyelectrolytes which would not be useful as a potential blood vessel prosthesis. Our research has concentrated upon the use of partially reacted, insoluble polymeric systems, and lightly crosslinked polymers which might retain much of the anti-coagulant properties found in the polyelectrolytes, but might also be usable as replacement blood vessels. The main emphasis in this paper is on the synthesis of these insoluble heparinoids in a manner to keep the polymeric system free of the impurities that can readily arise in this reaction.

INTRODUCTION

Whenever a polymeric biomaterial is to be used in an artifical organ, it is almost certain to come into direct contact with the blood and must, therefore, be a blood compatible material. Many attempts have been made to develop polymers with good blood compatibility, but these efforts have not been entirely successful. Essentially all of the common polymers have been examined for blood compatibile applications, but these usually show marginal to very poor results (1-3). Although a detailed review is beyond the scope of this present paper, a brief overview would be of value.

A large number of approaches to achieve blood compatibility have been attempted. Probably the most widely studied approach has been the use of the process termed "heparinization" in which the natural polyanionic, polysaccharide heparin is coated onto the surface of a polymer (or other material) in order to impart coagulation resistance to the material. This has been done using physical adsorption or covalent bonding methodology (4-7). In many cases, the results have shown promise, but the question of desorption and/or deactivation in long term usage has not been fully resolved. Should such a phenomenon occur, the system would revert back to the natural blood compatibility of the polymer. Because polymers, and most unnatural materials in general, are highly noncompatibile with the blood, the loss of the heparin activity would most likely result in a thrombogenic surface. Primarily for this reason, the use of heparinized materials have been limited to extracorporeal applications, such as dialysis tubing.

Researchers at the Cleveland Clinic and elsewhere have found that a process called "biolization", in which the polymer surface is coated with a denatured natural polymer such as albumin, can sometimes give a surface with good thrombo-resistance (8-10). Although the question of long term stability has not been completely resolved, these biolized surfaces have been coated on Hexsyn, which is a synthetic rubber, and are being used in the Cleveland Clinic's model of an artificial heart (11).

Many types of synthetic polymers have been evaluated in blood compatibility applications (12-17). The polymers that are the most widely used, at this time, in long term blood compatibility applications, are the the poly(ether urethane urea)s [PEUUs], of which Biomer (Ethicon) is an example. These materials were developed when it was recognized that the polyurethanes exhibited greater flexibility and durability than most of the other polymers then under consideration. The specific PEUUs commonly used are the results of extensive studies comparing blood interactions as a function of structure. These PEUU materials have been used in coating lead-wires for heart pacers, in the Jarvik 7 total artificial hearts, and some experimental analogs are being studied as possible small-sized blood vessel prostheses (15-17). Unlike natural blood vessel tissues, however, these PEUUs are usually hydrophobic. Their success appears to be due to a preferential adsorption of albumin, rather than globulin or fibrinogen, which confers a degree of blood compatibilty to the surface. In a sense, these PEUUs could be considered as operating by an in situ biolization process. The question of long term effectiveness, however, continues to remain unresolved. Another factor that may be important is that the surfaces of the PEUUs are slippery and incipient blood clots do not tend to adhere well. This is not necessarily an advantage; the small clots could cause severe problems elsewhere as emboli. The fact that nearly all recipients of artificial hearts based on the PEUU polymers have had strokes may possibly arise from this combination of factors.

A more recently recognized problem with the PEUU materials involves calcification of the surface which reduces the elasticity and restricts the opening through the blood can flow. These problems are discussed at greater length in other chapters of this book, but appears to be related to the primary polymer structure behaving like a crown-compound and extracting the Ca^{2+} ions from the blood stream (18). At this time, it appears that the same structural features which promote blood compatibility also promote calcification.

It would appear to be more desireable to have a synthetic polymer that was hydrophilic and which resembled the natural blood vessel tissues more closely. Alternatively, the synthetic polymer could be patterned after some known thrombo-resistant polymer such as heparin or albumin. The extensive research on hydrogels has been one example of the first approach in which a hydrophilic surface was sought (3). Numerous other workers have explored synthetic heparin-like polymers or heparinoids (19-38). Unfortunately, some of these polymers are actually water soluble polyelectrolytes which would not be suitable for prosthetic applications. Many of these polyelectrolytes do, however, show a fair degree of blood compatibility.

Heparin is a natural polymer that contains hydroxyl, sulfamate, sulfonic acid and carboxylic acid groups. Most of the heparinoid polymers contain at least one of these functional groups. Some of the copolymers with sulfonates, etc. are water insoluble and show a degree of blood compatibility (22-26,28-31). One of the simplest heparinoid systems is formed by the reaction of N-chlorosulfonylisocyanate (CSI) with polymers containing a double bond. After hydrolysis, these water soluble polymers contain sulfamate and carboxylic acid groups and are moderately blood compatible (33-38). The anti-thrombogenic results reported for some of these simple polyelectrolytes is promising (38). We are re-examining these polymers for possible use in prosthetic devices. This paper presents our initial results.

EXPERIMENTAL

Materials

In all of the experiments, Aldrich chlorosulfonyl isocyanate (Catalog #14,266-2: b.p. 107°C; m.p. -44°C; density = 1.626; 98+% purity) was used as the source of CSI. Due to the extremly high reactivity of the CSI and the fact that the major source of contamination is SO_3; this was used as supplied with the special further purification noted below.

Procedure

The amount of CSI used in all the synthesis experiments was 10.0 ml (0.115 m.) which was convenient to handle. Because CSI is highly reactive, it was easier to vary the amount of the saturated compounds than the CSI. A small amount of KCl was added to the addition funnel containing the CSI about a half hour prior to each addition. This reacts with the SO_3 according to equation 1, and effectively removes all of this contaminant. In most cases, the reaction solvent used was CH_2Cl_2. This procedure is similar to those in the original literature (39,40).

$$SO_3 + KCl \rightarrow Cl_2 + K_2SO_3 \qquad \text{(Equation 1)}$$

The CSI addition reaction was run on polybutadiene and polyisoprene polymers and with a variety of low molecual weight model compounds including 2,3-dimethyl-2-butene, 2-methyl-1-pentene, 2-methyl-2-pentene, 2,6-dimethyl-1,2,6-octatriene, cyclohexene and cyclooctene. The amount of unsaturated compound used corresponded to 0.093 m. The basic addition reaction is shown below in Equation 2 (39-42).

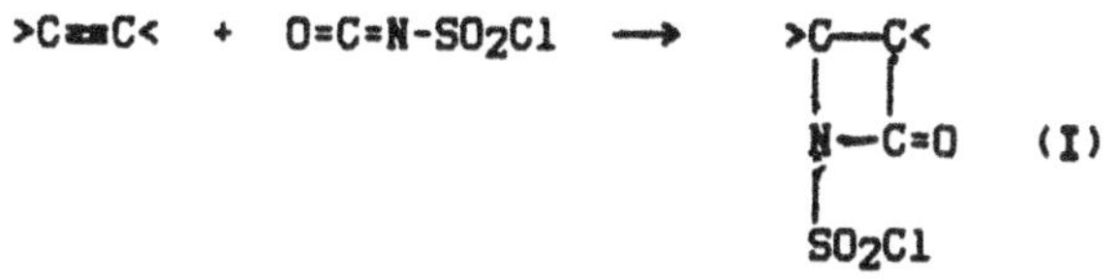

(Equation 2)

The reactions were run in 24/40, 500 ml, three-neck flasks, which were covered with Al foil to exclude light, and were fitted with a mechanical stirrer, an addition funnel and a gas inlet connected to a valve that was used to select either a vacuum system or a nitrogen flow. The unsaturated compound, in CH_2Cl_2 solution, was placed in the flask and the CSI (in CH_2Cl_2) was placed in the addition funnel. Each part of this system was placed under vacum and then sparged with N_2 alternately to insure a complete nitrogen atmosphere. After cooling the reaction mixture in an ice bath, the CSI solution was added at a flow rate of about 1 or 2 drops a second. (The temperature of $0^{o}C$ appears to be the best compromise for the polymer solutions. Below this temperature the phase separation is sometimes excessive; above this temperature the amount of discoloration is too great.) The reaction was allowed to stand for 2 hr. prior to work-up. At the end of that time, the foil was removed and the off-white product was isolated by removal of the solvent. Yields were, in general, nearly theoretical.

The lactams (Structure I, Equation 2) were hydrolzed by stirring the sample with an excess of NaOH solution until the entire sample dissolved (for fully reacted double bond systems) or until IR analysis showed the hydrolysis was complete (for partially reacted double bond systems). This reaction proceeds according to Equation 3.

$$(I) \quad + \quad NaOH \quad \longrightarrow \quad \begin{array}{ll} >C- & C< \\ \;\;| & \;| \\ NH & CO_2^-Na^+ \\ \;\;| & \\ SO_3^-Na^+ & \end{array} \quad (II) \qquad \text{(Equation 3)}$$

Selected samples were tested for elemental analysis. The FTIR spectra of the samples was determined using an IBM IR/32. Although essentially all the products corresponding to Structure (II) were water soluble for the fully reacted double bond systems, many of the partially reacted polymeric materials were insoluble in water.

DISCUSSION

The reaction of chlorosulfonyl isocyanate (CSI) to unsaturated compounds does proceed smoothly if precautions are taken to remove any SO_3 present and to exclude light from the reaction until at least after the CSI has added to the double bond. The reaction does proceed more readily with most of the low molecular weight materials than with the polymers studied, however. The major problems encountered were the purification of the isolated polymers and the hydrolysis of the lactam intermediate (I) under the hetergeneous reaction conditions employed. With care, however, the modified polymers can be obtained whose spectra closely resemble those of the low molecular weight samples.

Figure 1 shows the FTIR spectra, (B), of a polyisoprene that was reacted with 50 mole% of of chlorosulfonyl isocyanate, [CSI], to form the adduct outlined in Equation 2, and the final reaction product, (C), after alkaline hydrolysis according to Equation 3. The FTIR spectrum, (A), of the original polyisoprene is included for comparison. The spectra of these individual materials are clearly different and can be distinguished from each other readily.

The most difficult problem with the polymeric materials arises in attempting to remove any impurities that might arise because of some side

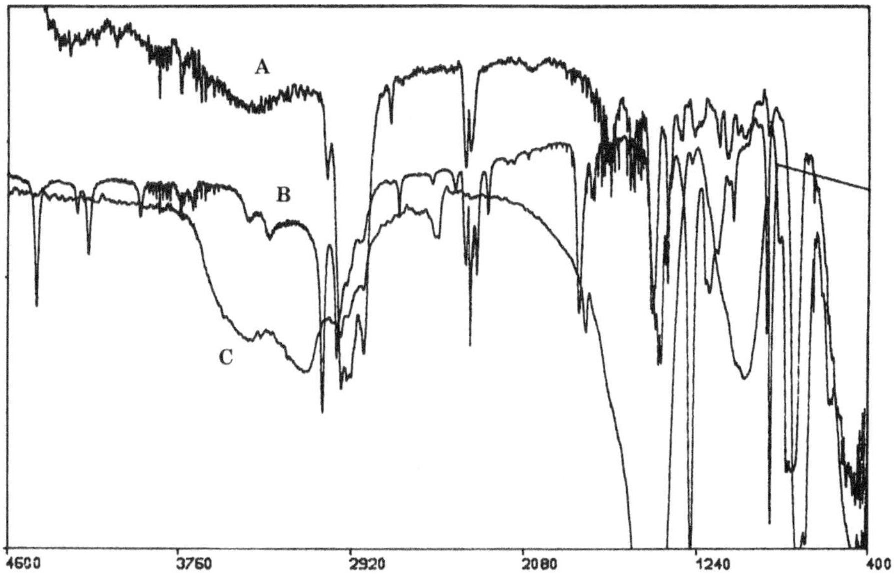

Figure 1. The FTIR spectra of: [A], polyisoprene; [B], a 50 mole% CSI adduct; and [C], the final hydrolyzed reaction product of this adduct.

reaction. Most of these side reactions cause colored impurities. Although these side reactions do occur with low molecular weight compounds, if care is not taken to exclude light and to keep the reaction temperature low, these impurities are readily removed by recrystallization or chromatographic techniques. These procedures do not work well with the polymeric materials and it is essential to prevent their formation or the final reaction products will remain contaminated with these impurities. No attempt was made in this present study to determine the nature of these impurities, but we can speculate that they may involve adducts of SO_3, or reaction products involving this compound and the modified polymer. The only successful way to prepare good CSI adducts is to prevent these side reactions.

Our studies with the low molecular weight model compounds showed that less colored by-products were formed when the temperature was kept low. This is probably the second most important factor in controlling this reaction, but there is a practical limit that can be used with the polymeric systems because the polymers become insoluble at low temperatures. (The major factors involve keeping the system free of moisture and removing the SO_3.) Exclusion of light also helps to prevent the formation of colored products. Even with the low molecular weight compounds, the use of an inert atmosphere aided in reducing colored products, but this is essential with the polymeric systems. Much of the prior literature does not emphasize these factors adequately. To summarize, the only way to prepare low-color, by-product free CSI adducts of polymers is to run the reactions under an inert atmosphere (N_2), at as low a temperature as possible, maintain rigorously dry reaction conditions, remove any residual SO_3, and to exclude light completely.

Although the fully substituted polybutadiene or polyisoprene CSI adducts are water soluble, after hydrolysis to form the product of struc-

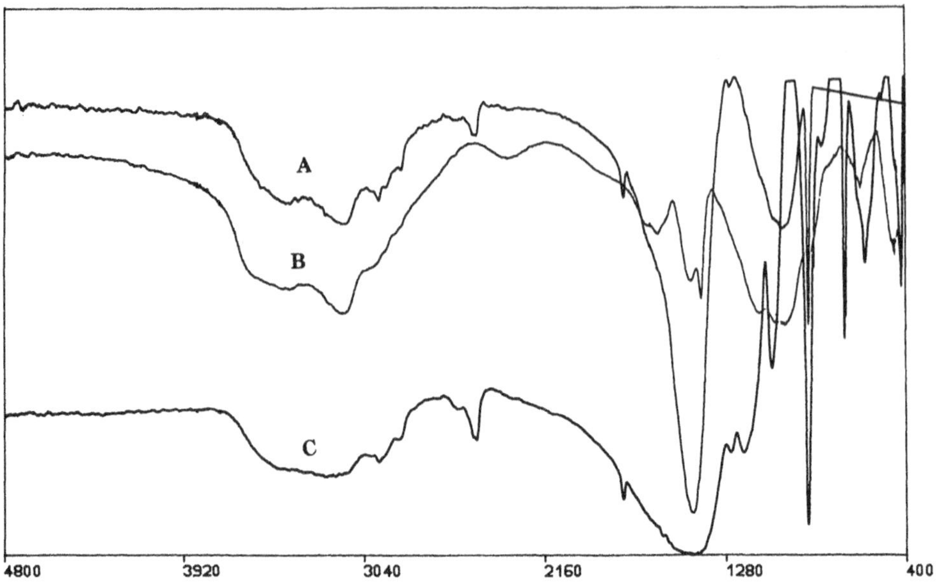

Figure 2. The FTIR spectra of polyisoprene that has been reacted with: [A], 50 mole%; [B], 37.5 mole%; or [C], 100 mole% chlorosulfonyl isocyanate.

Table I. The effect of the level of CSI substitution on the solubility of the final hydrolysis products of the polyisoprene adducts.

	Percent CSI Substitution		
Solvent	37.5	50	100
Water	I	Swells	Sol
Methyl ethyl ketone	I	I	I
Dimethyl sulfoxide	I	I	I
Toluene	I	I	I
Methanol	I	I	I
Butyrolactone	I	I	I
Acetone	I	I	I
Chloroform	I	I	I
Heptane	I	I	I

Sol = soluble; I = insoluble.

ture (II), the polymers with under 50% substitution were not water soluble and might have some potential utility in blood contacting prostheses. Figure 2 shows the FTIR spectra of a series of polyisoprene-CSI adducts that were 37.5, 50, or 100 mole% reacted. All of these materials can be hydrolyzed completely, but only the 100% CSI adduct was water soluble. The solubility characteristics of the polyisoprene:CSI adducts are summarized in Table I. The CSI adducts were essentially insoluble in all the solvents studied, although the 50% adduct did swell in water and the 100% adduct was completely water soluble. Crosslinking does not appear to be the cause of this insolubility because the most reacted adduct is soluble in water. The low solubility is probably due to the fact that the adduct portion of the molecule would be very hydrophilic whereas the non-reacted portions would be highly hydrophobic.

The partial reaction of crosslinked polyisoprene and related polymers gives modifications that are completely insoluble. Highly substituted adducts, however, will swell in water to some extent, depending on the degree of substitution and the extent of crosslinking. Further studies are in progress on these systems and will these be discussed in more detail in a later paper. By comparison with related materials in the literature, these modified polymers could show blood compatibility. Unlike heparin coated systems, however, this compatibility should not be subject to potential desorption effects leading to clotting.

ACKNOWLEDGEMENTS

This research was supported, in part, by a summer research grant (for DM) from PPG, Inc., and was abstracted, in part, from a Master's Thesis of DM which will be presented to Youngstown State University.

REFERENCES

1. C. G. Gebelein in: "Kirk-Othmer Encyclopedia of Chemical Technology", Vol. 19, 3rd Ed., John Wiley & Sons, New York, 1982, pp. 275-313.
2. C. G. Gebelein in: "Polymeric Materials and Artificial Organs", C. G. Gebelein, ed., ACS Symposium Series No. 256, American Chemical Society, Washington, DC, 1984, p. 1.
3. A. S. Hoffman in: "Polymeric Materials and Artificial Organs", C. G. Gebelein, ed., ACS Symp. No 256, Am. Chem. Soc., Washington, DC, 1984, p. 13.
4. R. I. Leininger in: "Biomedical and Dental Applications of Polymers", C. G. Gebelein & F. F. Koblitz, eds., Plenum Press, New York, 1981, p. 99.
5. S. W. Kim, MD&DI, 99-103, August, 1984.
6. T. C. Lin, J. H. Joist & R. E. Sparks, Trans. Biomat. Soc., 7, 289 (1984).
7. W. P. Ip, M. V. Sefton & W. Zingg, Trans. Biomat. Soc., 7, 292 (1984).
8. S. Murabayashi & Y. Nose in: "Biomedical and Dental Applications of Polymers", C. G. Gebelein & F. F. Koblitz, eds., Plenum Press, New York, 1981, p. 111.
9. D. E. Gregonis, D. E. Buerger, R. A. Van Wagenen, S. K. Hunter & J. D. Andrade, Trans. Biomat. Soc., 7, 266 (1984).
10. S. Winters, D. E. Gregonis, D. Buerger & J. D. Andrade, Trans. Biomat. Soc., 7, 290 (1984).
11. H. Kambic, S. Murabayashi & Y. Nose, Chemical & Engineering News, April, 14 , 1986, pp. 31-48.
12. T. Akutsu, N. Yamamoto, M. A. Serrato, J. Denning & M. A. Drummond

in: Biomedical and Dental Applications of Polymers", C. G. Gebelein & F. F. Koblitz, eds., Plenum Press, New York, 1981, p. 119.
13. L. L. Hench & E. C. Ethridge, "Biomedical. An Interfacial Approach", Academic Press, New York, 1982, Chapter 9.
14. Y. Ikada in: "Advances in Polymer Science", 57, 103-140 (1984).
15. D. J. Lyman & K. Knutson in: "Biomedical Polymers. Polymeric Materials and Pharmaceuticals for Biomedical Use", E. P. Goldberg & A. Nakajima, eds., Academic Press, New York, 1980, p. 1.
16. D. J. Lyman, K. B. Seifert, H. Knowlton & D. Albo, Jr., in: "Biomedical & Dental Applications of Polymers", C. G. Gebelein & F. F. Koblitz, eds., Plenum Press, New York, 1981, p. 163.
17. K. Knutson & D. J. Lyman in: "Biomedical & Dental Applications of Polymers", Plenum Press, New York, 1981, p. 173.
18. R. J. Thoma, T. Q. Hung, E. Nyilas, A. D. Haubold & R. E. Phillips, This book, pp. 131-145
19. J. Ehrlich & S. S. Stivala, J. Pharmaceutical Sci., 62, 517 (1973).
20. H. P. Gregor in: "Biomedical Applications of Polymers", H. P. Gregor, ed., Plenum Press, New York, 1975, p. 51.
21. M. Sorm & S. Nespurek, Eur. Polymer J., 14, 977 (1978).
22. M. Sorm, S. Nespurek, L. Mrkvickova & J. Kalal, J. Polymer Sci., Symp. 66, 349 (1979).
23. S. J. Whicher & J. L. Brash, Trans. Soc. Biomaterials, 4, 123 (1981).
24. I. Kaneko, T. Takamatsu & A. Eisenberg, Trans. Soc. Biomaterials, 5, 12 (1982).
25. C. V. Baquey, C. Fougnot, B. Basse-Cathalinat, V. Migonney, B. Masson, S. Winnock, M. Jozefowicz & D. Ducassou, Trans. Soc. Biomaterials, 6, 19 (1983).
26. N. Muramatsu & T. Kondo, J. Biomed. Mater. Res., 17, 959 (1983).
27. J. M. Nigretto, E. Corretge, M. Jozefowicz & M. Mauzac, Trans. Soc. Biomaterials, 7, 205 (1984).
28. J. Jozefonvicz, M. Mauzac, A. M. Fischer, J. Tapon-Bretaudiere & S. Beguin, Trans. Soc. Biomaterials, 7, 206 (1984).
29. M. Jozefonvicz, D. Labarre, H. Serne, C. Fougnot, C. Douzon & F. M. Kanmangne, Trans. Soc. Biomaterials, 7, 207 (1984).
30. F. Maillet, M. Mauzac, J. Jozefonvicz & M. D. Kazatchkine, Trans. Soc. Biomaterials, 7, 208 (1984).
31. V. Migonney, M. Jozefowicz & C. Fougnot, Trans. Soc. Biomaterials, 7, 288 (1984).
32. D. Gulino, C. Boisson & J. Jozefonvicz, Trans. Soc. Biomaterials, 7, 291 (1984).
33. L. van der Does, J. Hofman & T. E. van Uttern, Polymer Letters, 11, 169 (1973).
34. C. P. Pinazzi, P. Noireaux & D. Reyx, Makromol. Chem., 175, 2849 (1974).
35. P. M. van der Velden, M. H. V. Mulder, L. van der Does & C. A. Smolders, Polymer Letters, 14, 5 (1976).
36. L. van der Does, T. Bevgeling, P. E. Froehling & A. Bantjes, J. Polymer Sci., Symp. 66, 337 (1979).
37. L. C. Sederel, L. van der Does, J. F. van Duijl, T. Beugeling & A. Bantjes, J. Biomed. Mater. Res., 15, 819 (1981).
38. C. P. Sharma & A. K. Nair, Trans. Soc. Biomaterials, 7, 100 (1984).

39. R. Graf, Ann. Chem., 661, 111 (1963).
40. R. Graf, Angew. Chem. Internat. Edit., 7, 172 (1968).
41. E. J. Moriconi & W. C. Meyer, J. Org. Chem., 36, 2841 (1971).
42. E. J. Moriconi & Y. Shimakawa, J. Org. Chem., 37, 196 (1972).

CHITIN AND CHITOSAN FOR USE AS A NOVEL BIOMEDICAL MATERIAL

Shigehiro Hirano, Yasuharu Noishiki, Junko Kinugawa, Hitoshi Higashijima and Takatoshi Hayashi

Department of Agricultural Biochemistry
Tottori University
Tottori 680, Japan

Chitin [a (1->4)-linked 2-acetamido-2-deoxy-β-D-glucan], and chitosan (its N-deacetylated compound) were usable as a novel biomedical material: (1) N-Acetylated hybrid gels of chitosan and collagen were transparent and elastic. The lyophilized gels adsorbed rabbit blood or distilled water (up to 20 mL/g). (2) A segment of polyester polyfilament suture coated with a thin material to be examined was directly inserted into the lumen of dog's peripheral veins, and the *in vivo* blood compatibility was macroscopically determined from the blood coagulum formed on the suture surface, indicating that chitosan was thrombogenic, and N-fatty acyl (C_6-C_{12}) chitosans, and N-acetylated hybrids of chitosan and collagen were nonthrombogenic. (3) The anticoagulant activity [units/mg] was in the sequence: O-sulfated N-acetylchitosan [331-379]> N,O-sulfated chitosan [190-280]. heparin [174]. sulfated O-carboxymethyl-chitosan [27-31] with respect to activated partial thromboplastin time. (4) The lipolytic activity (meg/L) was in the sequence: N,O-sulfated chitosan [3200] heparin [950]. sulfated O-carboxymethylchitosan [300]. O-sulfated N-acetylchitosan [100] with respect to lipoprotein lipase.

INTRODUCTION

Chitosan is a linear (1->4)-linked 2-amino-2-deoxy-β-D-glucan, and is chemically prepared from naturally occuring chitin (its N-acetyl product) by treatment with alkali at elevated temperature. Chitin and chitosan are the main structural element in the cuticles of crustacea, insects and molluscan organs, and in the cell walls of fungi (1). These polysaccharides are biodegradable, and almost nontoxic, LD_{50} 16g/Kg for chitosan (2), naturally abundant, and have reactive hydroxyl and amino groups, which can be modified chemically. Chitin and chitosan have been used in the biomedical field: powders and tablets for the controlled release and bioavailability of drugs and nutrients (3), sheets and membranes for wound-healings and dressings (4), and dialysis membranes (5), fibers for digestible suture (6), liposome stabilization (7), antibacterial (8), antivirus (9) and antitumor (10) agents, carriers for drugs to a target organ (11), hemostatics (12), hypocholesterlemic and hypolipidemic agents (13), immunoadjuvants (14), and orthopedic materials (15).

In order to develop the other usages of chitin and chitosan in the biomedical field, we now report some properties of their hybrids with collagen, their blood compatibility, and blood anticoagulant and lipolytic activities of their sulfated derivatives.

EXPERIMENTAL

Materials

Chitin and chitosan

Crab shell chitin (d.s. 0.98 for NAc, $[\alpha]_D^{22}$, -7° (c 0.9, 2% aqueous acetic acid)) was supplied from Katakura Chikkarin Co., and chitosan (Flonac-N, commercial chitosan of crab shell, d.s. 0.23 for NAc, $[\alpha]_D^{26}$, -5° (c 0.9, 2% aqueous acetic acid)) was supplied from Kyowa Yushi Co. Flonic-N was treated with aqueous 40% NaOH containing $NaBH_4$ (0.1 g/500 mL) at 110° for 5 h to afford a purified product (d.s., 0.05 for NAc), which had $[\alpha]_D^{26}$, -10° (c 0.8, 2% aqueous acetic acid).

N-Acetylated hybrids of chitosan and collagen

Chitosan (0.16 g, d.s., 0.05 for NAc) and atelocollagen (3.2-32 mg, Koken Co.) were dissolved in aqueous 2% acetic acid (19 mL), and the solution was diluted with methanol (938 mL), and acetic anhydride (0.47 mL, 5 mol eq.) was added. The mixture solution was kept at room temperature overnight to afford the title gels, which were soaked in distilled water overnight and lyophilized.

Sulfated derivatives

Chitosan was N-acylated (16) with a carboxylic anhydride in aqueous 2% acetic acid-methanol (1:2), and O-carboxymethylchitosan (d.s. 0.61) was prepared from O-carboxymethyl-N-acetylchitosan (17). Sulfation was effected with N, N-dimethylformamide-sulfur trioxide (18), and the product was isolated as the sodium salt. N-Desulfation was effected (19) with 0.04M HCl at 100° for 2.5 h.

Methods

General

Hardness of gels was determined with an Iio M-302 curdmeter. IR spectra were recorded with a Hitachi 215 spectrometer, and optical rotations with a JASCO Dip-180 polarimeter. Sulfate content was determined by the barium chloranolate method (20).

Molecular weights

The molecular weights of sulfated derivatives were estimated by gel chromatography on either a column (1.6x46 cm) of Bio-Gel P-300 fine or a column (1.6x74 cm) of Toyopearl HW 65 fine. Each column was equilibrated and eluted with 0.3M NaCl at room temperature. Each calibration curve was obtained by analysis of relationship between Ve/Vo and log mol. wt. of standard pullulans [mol. wts. (x 10^3) 853, 186, 100, and 12.2, Shodex Co.] and dextrans [mol. wts. (x10^3) 40.1, 25 and 17, gifts from Professor K. Kawahara, Nagasaki University]. The elutes were monitored by the phenol-sulfuric acid method (21) for pullulans and dextrans, the orcinol method (22) for heparin, and u.v. absorption at 194 nm for the sulfated derivatives of chitosan.

An in vivo test for the blood compatibility

A segment of pliable polyester polyfilament suture [No. 2-0 (USP)], ca. 10 cm long, was dipped in a 1% solution of chitosan in aqueous 2% acetic acid, and the suture was taken out and dehydrated by dipping in 60% ethanol at room temperature overnight. The suture, which was coated with chitosan without dehydrating in 60% ethanol as described above, was dipped in a methanol solution (50 mL) containing acetic anhydride (0.5g) or a fatty acid anhydride (1.0 g) at room temperature for 3 h. The suture surface was coated with a thin layer of the corresponding N-acylchitosan gel, and it was sterilized by dipping in 60% ethanol at room temperature overnight, and stored in a sterilized physiological saline.

As reported in our previous paper (23), an 18-guage needle was inserted into the peripheral veins under general anesthesia, and four segments of the suture, which were coated with a thin membrane of chitosan, and N-acetyl-, N-long fatty acyl (C_6-C_{12}) chitosans, were introduced through the needle into the lumen of dog's jugular (two samples) and femoral (two samples) veins. After the needle was withdrawn, the end of the suture was fixed near the area of the vessel puncture. Total four samples were examined in the same dog, and the test was duplicated in the alternative veins of two dogs with three samples and one control. After 2 h or two days, the dog was sacrificed by acute exanguination under general anesthesia and heparin administration (2-5 mg/Kg). The vein in which the suture was inserted was opened, and blood clotting on the suture surface was macroscopically examined. The results were compared with S-aminoethylkeratin-heparin crosslinked with glutaraldehyde and Avocothan H-RSD (Toray Co.) which are known nonthrombogenic materials, as reported previously (23).

Anticoagulant activity

The activity was determined from human plasma with respect to activated partial thromboplastin time [APTT](24). The activity was expressed by units/mg in relation to that of porcine intestinal heparin (174 units/mg, Lot no. CH-108M, Mitsui Seiyaku Kogyo Co.) as a standard.

Lipolytic activity

Blood specimens (5 mL) collected from male rabbits (2.7-3 Kg), were centrifuged at 3,000 r.p.m. at 4°C for 20 min to afford serum specimens. A mixture of serum specimen-coconut oil emulsion-0.07M phosphate buffer (pH 8.0)-physiological saline (1.0:0.1:1.0:1.0) was incubated at 37°C for 30 min, and used as the substrate (25).

After 14-15 h, a solution of a sulfated derivative (0.1 or 1.0 mg/Kg) in physiological saline (5 ml) was injected intravenously into each of three rabbits (2.7-3.0 Kg). After 10 min, a blood specimen (5 ml) was collected, mixed with 0.1M sodium oxalate (0.5 ml), and centrifuged at 3,000 r.p.m. at 4°C for 20 min. The resulting heparinoid plasma specimen (1.0 ml) was throughly mixed with the substrate (1.5 mL) prepared above, and the mixture was incubated at 37°C for 15, 30, and 60 min. Free fatty acids were extracted with 2-propanol:n-heptane:0.5M H_2SO_4 (40:10:1), and the heptane layer was titrated with 0.05M NaOH in the presence of Thymol Blue as the indicator (26). The lipolytic activity was calculated as follows: meq=[free fatty acids (mmol/L) after incubation] - [free fatty acids (mmol/L) before incubation]. The results were compared with those for porcine intestinal heparin as a standard.

RESULTS

The following compounds were prepared (Figure 1): N-acetylchitosan [**1**](16), N-hexanoylchitosan [**2**](16), N-octanoylchitosan [**3**](16), N-decanoylchitosan [**4**](16), N-lauroylchitosan [**5**](16), N,O-sulfated chitosans [**6** and **7**], O-sulfated N-acetylchitosans [**8** and **9**], O-sulfated N-hexanoylchitosan [**10**], O-sulfated chitosan [**11**], N-sulfated O-carboxymethylchitosan [**12** and **13**], and N-acetylated hybrids of chitosan and collagen. The compounds **8** (mol.wt. 26 x 10^3) and **9** (150 x 10^3), and **12** (540 x 10^3) and **13** (245 x 10^3) have similar structures each other, but different molecular weights. The average molecular weights of these derivatives were in the range of 21-26 x 10^3 [**6-8**, **10**, and **11**] and 150-540 x 10^3 [**9**, **12**, and **13**] due to the different molecular weights of chitosan used (Flonac-N or its purified compound). The d.s. for sulfate was 2.6, 1.7, 2.0, 1.8, and 0.7 in **6-11** and 0.6 and 1.0 in **12** and **13**, respectively. The latter two compounds has d.s. 0.9 and 0.6 for O-carboxymethyl group, respectively. Table I summerizes the yields, specific rotations, and elemental analyses of **6-13**. Each sulfate derivative had i.r. absorptions at 1240-1250 (S=O) and 800-810 cm^{-1} (eq. C-O-S). The d.s. for N-acyl was 1.0 in **1-5** and N-acetylated hybrids of chitosan and collagen. These structures were supported by the elemental analyses (Table I). Compounds **1-5**, and the N-acetylated hybrids were insoluble in water, and **6-13** were soluble in water.

Collagen was almost completely immobilized in N-acetylated hybrid gels. The hardness (dyne/cm^2) of gels was 11.8x10^5 on N-acetylchitosan, and 7.2x10^5 on N-acetylated hybrids (chitosan-collagen,1:0.2) at concentration 0.28 for chitosan. The hardness was decreased by addition of collagen (Figure 2). The hybrid gels were transparent, heat-irreversible, and elastic. Their lypophilized gels were sponge-like and adsorbed rabbit blood and distilled water (up to 20 mL/g).

Blood compatibility

As shown in Figure 3, an intense thick blood coagulum formed on chitosan, and the suture surface became red and its diameter was ca. 1.7 times thicker than the original even at 2 h. A thin blood coagulum formed on N-acetylchitosan, but no blood coagulum formed on N-fatty acyl (C_6-C_{12}) chitosans, and the N-acetylated hybrids on two days. The nonthrombogenic property was similar to that of the known nonthrombogenic materials (23). The same results were found in duplicated tests.

Anticoagulant activity

As shown in Figure 4, a straight line was observed by plotting APTT (log) versus each of the sulfated compounds (g/mL). APTT was significantly prolonged with **8**, and its activity at 5-15 μg/mL was 1.9-2.2 times the activity of heparin. APTT was also prolonged with **6**, and its activity was 1.1 (5 g.mL), 1.2 (10 μg/mL), and 1.6 (15 μg/mL) times that of heparin. APTT was slightly prolonged with **12**, and its activity at 100-300 g/mL was 0.2-0.3 times that of heparin. The anticoagulant activity (units/mg) with respect to APTT was 331-379 at 5-15 μg/mL for **8**, 190-287 at 5-15 μg/mL for **6**, and 21-31 at 100-300 μg/mL for **12**. The plotted line of APTT for **8** was almost parallel to that for heparin, but the plotted lines of APTT for **6** and **12** were not parallel.

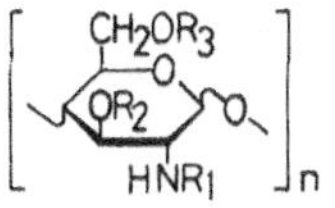

Figure 1. Structures of the repeating unit of N-acylchitosans and sulfated derivatives of chitosan.

Legend of Figure 1.

Chitosan	R^1	R^2	R^3
N-acetyl [1]	H	H	H
N-Hexanoyl [2]	$CO(CH_2)_4CH_3$	H	H
N-Octanoyl [3]	$CO(CH_2)_6CH_3$	H	H
N-Decanoyl [4]	$CO(CH_2)_8CH_3$	H	H
N-Lauroyl [5]	$CO(CH_2)_{10}CH_3$	H	H
N,O-Sulfated [6][(a)]	SO_3Na	H, SO_3Na	SO_3Na
[7][(a)]	SO_3Na	H	H, SO_3Na
O-Sulfated N-acetyl [8][(b)]	$COCH_3$	SO_3Na	SO_3Na
[9][(b)]	$COCH_3$	SO_3Na	SO_3Na
O-Sulfated N-hexanoyl [10]	$CO(CH_2)_4CH_3$	SO_3Na	SO_3Na
O-Sulfated [11]	H	H	H, SO_3Na
Sulfated O-carboxyl-methyl [12][(b)]	SO_3Na	H	$CH_2CH_2CO_2Na$, SO_3Na
- [13][(b)]	SO_3Na	H	$CH_2CH_2CO_2Na$, SO_3Na

(a) Differed in molecular weight.
(b) Differed in d.s. for sulfate.

Lipolytic activity

Lipoprotein lipase activity of rabbit plasma after the intravenous injection of 6-13 varied mainly with dose, incubation time, molecular weight, and chemical structure (Table II). The maximum activity for a dose 0.1 mg/Kg was 3.4 times the activity of heparin for 7 at 15 min, and 0.7 and 0.1 times that of heparin for 13 and 6, respectively, at 30 min (the maximum activity of heparin occured at 30 min); 9-11 and 13 were inactive (Table II).

DISCUSSION

The characteristic blood compatibility of these materials may mainly depend on their ultrastructures and chemical structures. Free amino groups are present on chitosan. Acidic groups of the cellular elements of blood may form polyelectrolyte complexes (27,28) with free amino groups present on the chitosan membrane surface. Malette et al (12) found that

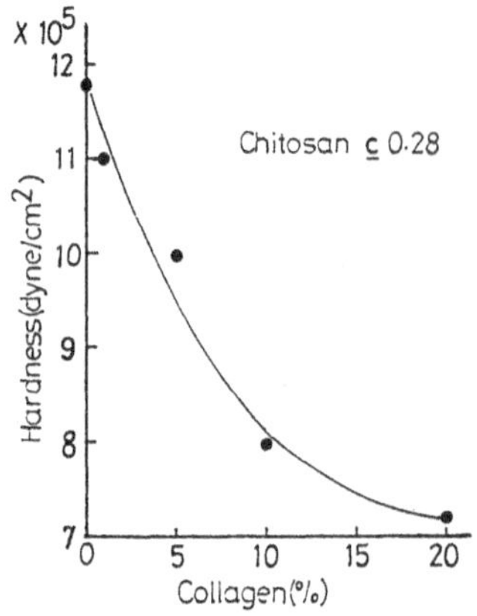

Figure 2. Hardness of N-acetylated hybrid gels of chitosan and collagen with respect to collagen content (0-20%) against chitosan) at conc. 0.28 for chitosan.

an intense thick coagulum formed rapidly with defibrinated blood, heparinized blood, and washed red blood cells in contact with chitosan. The chitosan has a rough surface consisting of many undefined fibrils, but N-acylchitosan membrane surfaces have a smooth surface consisting of fine fibrils, as observed by SEM (29). Kaifu and Komai (30) reported that O-diacetylchitosan membrane was more nonthrombogenic than chitin membrane. The fatty acyl group of N-acylchitosans adsorbed specific proteins without any denaturation (31). This kind of affinity adsorption differs from the denatured polyelectrolte aggregation of proteins on blood coagulum on the surface. In fact, we detected a thin layer on the nonthrombogenic material surface by SEM at a cross section of the treated suture (23).

An O-sulfate group or acidic group at position 6 in the hexosamine moiety is a main active site for anticoagulant activity, and there is no activity when it is absent, and the activity is maintained more or less when position 6 is carboxyl (32,33) or carboxymethyl group. An O-sulfate at position 3 in the hexosamine moiety promotes the activity when position 6 is an acidic group, but it is no essential. The important contribution of N-sulfate group to the anticoagulant activity has been pointed out, but its presence is not a prerequisite for the activity. In fact, dextran sulfate, which has no N-sulfate, has the anticoagulant activity, and the N-acetyl group is a minor component of heparin. On the other hand, an N-sulfate group at position 2 in the hexosamine moiety is a main active site for lipolytic activity, and there is no activity when it is absent. However, the weak activity is maintained when position 2 is N-acetyl. Therefore, these sulfate derivatives may have different active sites for anticoagulant and lipolytic activities.

→

Figure 3. Macroscopic views of the blood coagulum formed on the sutured surface, which was coated with a thin material to be examined. The thick blood coagulum formed on chitosan at 2 h (1); the thin blood coagulum formed on N-acetylated hybrids of chitosan and collagen on 2 days (3), N-hexanoylchitosan on 2 days (4), N-octanoylchitosan on 2 days (5), N-decanoylchitosan on 2 days (6), and N-lauroylchitosan (7) on 2 days.

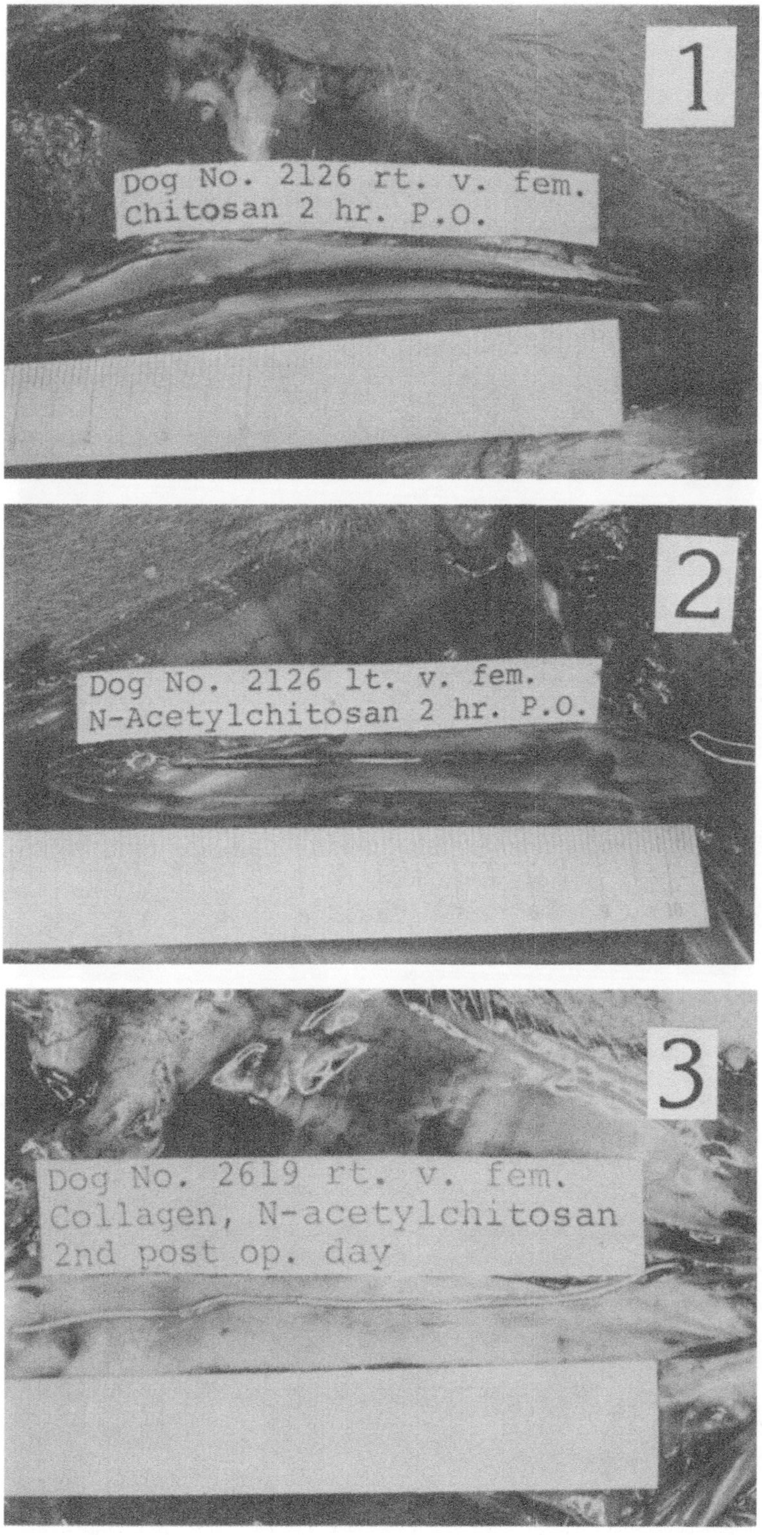
1
Dog No. 2126 rt. v. fem.
Chitosan 2 hr. P.O.
2
Dog No. 2126 lt. v. fem.
N-Acetylchitosan 2 hr. P.O.
3
Dog No. 2619 rt. v. fem.
Collagen, N-acetylchitosan
2nd post op. day

Figure 3.

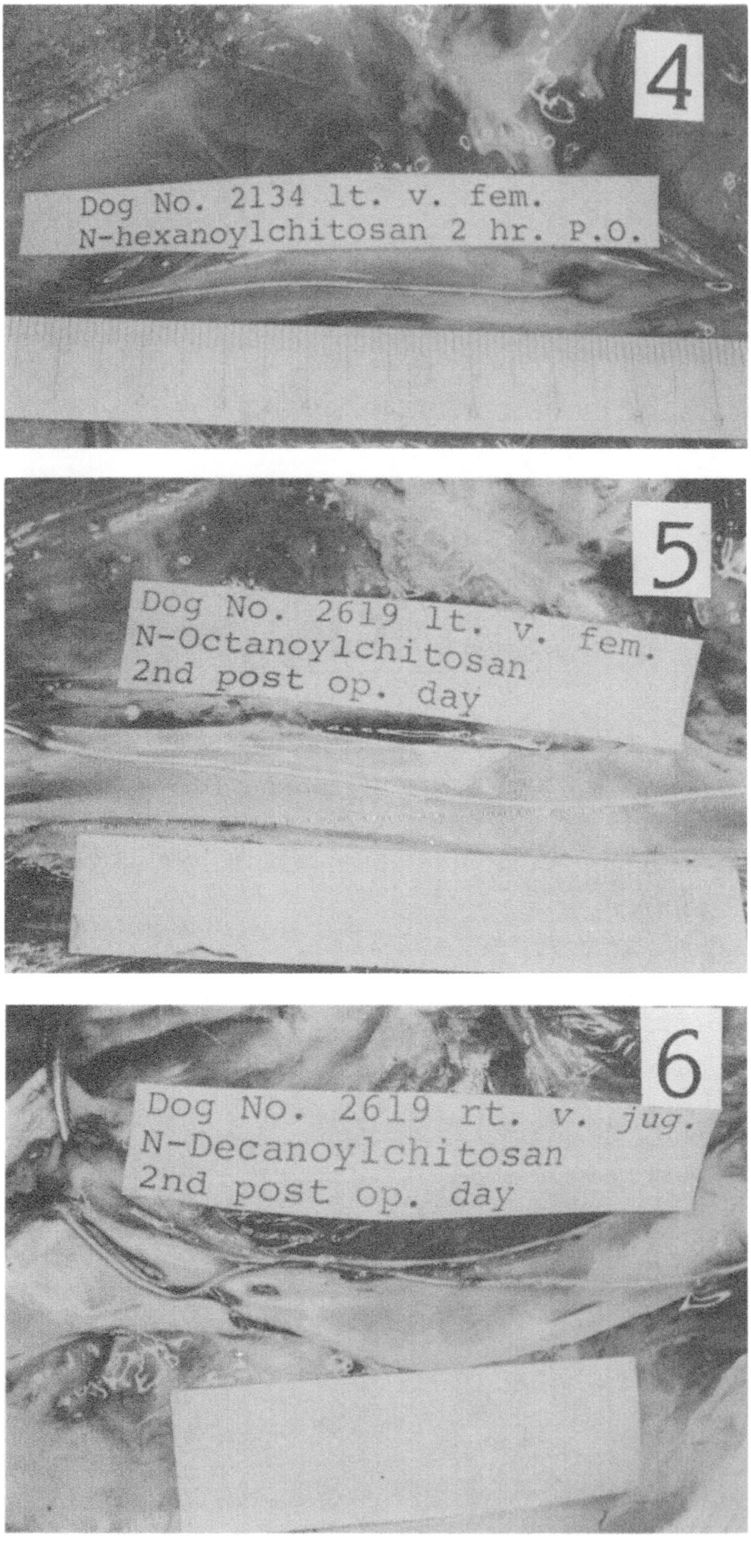

Figure 3. Continued.

Figure 3. Concluded.

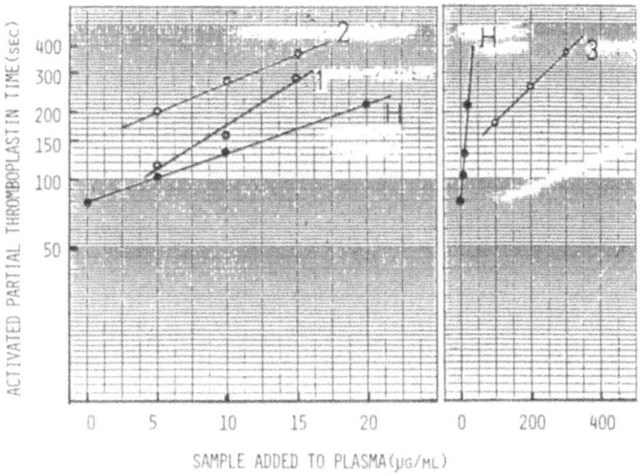

Figure 4. Anticoagulant activities of some sulfated derivatives of chitosan with reference to heparin (H) and APTT.

Table I. Data for some Sulfated Derivatives of Chitosan (a).

Compound	Yield %	$[\alpha]_D$ (c, °C)	Calc.(%)/ Found(%) C	H	N	S
6	79	-11 (0.5, 17)	15.72 15.44	2.54 2.69	3.06 3.10	18.40 18.7
7	80	-11 (0.5, 17)	18.36 18.16	3.77 3.91	3.55 3.70	13.80 14.1
8	86	-31 (1.2, 18)	22.00 22.02	3.66 3.98	3.21 3.18	14.70 14.7
9	75	-16 (0.8, 17)	21.04 21.04	3.64 3.71	3.07 2.93	14.04 13.7
10	83	-22 (0.9, 21)	31.11 31.10	4.75 4.75	3.02 3.00	12.25 12.2
11	83	- 9 (0.8, 17)	28.56 28.44	4.78 4.79	5.55 5.85	9.40 9.4
12	75	-13 (0.5, 14)	30.71 30.90	4.74 5.04	4.61 4.65	6.13 6.4
13	54	- 7 (0.8, 17)	24.21 24.22	4.43 4.43	3.91 3.88	8.95 8.9

(a) The repeat unit is:

6 $C_6H_8NO_4(SO_3Na)_{2.63}(H)_{0.37}(H_2O)_{1.60}$

7 $C_6H_8NO_8(SO_3Na)_{1.69}(H_2O)_{3.34}$

8 $C_8H_{11}NO_{11}S_2Na(H)_{0.45}(Na)_{0.54}(H_2O)2.20$

9 $C_8H_{11}NO_{11}S_2Na_2(H_2O)2.24$

10 $C_{12}H_{19}NO_8SNa(SO_3Na)_{0.77}(H)_{0.23}(H_2O1.30$

11 $C_6H_{10}NO_4(SO_3Na)_{0.74}(H)_{0.26}(H_2O)0.87$

12 $C_6H_8NO_4(C_2H_2O_2Na)_{0.88}(SO_3Na)_{0.58}(H)_{1.54}(H_2O)1.49$

13 $C_6H_9NO_7SNa(C_2H_2O_2Na)_{0.61}(H)_{0.39}(H_2O)2.56$

(b) In water.
(c) In aqueous 0.5 NaOH.

It is also clear that molecular weight is an important factor for both the activities as reflected by the sequence of activities: 8 (mol. wt. 26 x $10^3$0 > 12 (540 x 10^3) with respect to APTT, and 7 (22 x 10^3) >

Table II. Anticoagulant and Lipolytic activities of some sulfated derivatives.

Compound	Mol. wt.[a] (x10^3)	d.s. for sulfate	Lipolytic activity[b] (meq/L)	Anticoagulant activity[c] (units/mg)
Heparin	21	--	950 (1.0)[d]	174 (1.0)
6	12	2.6	n.d.[e]	190-287 (1.1-1.6)
7	22	1.7	3200 (3.4)[f]	n.d.
8[(g)]	26	2.0	100 (0.1)[d]	331-379 (1.9-2.2)
9	150	2.0	i.a.[(h)]	n.d.
10	27	1.8	i.a.	n.d.
11	22	0.7	i.a.	n.d.
12[(i)]	540	0.6	n.d.	21-31 (0.2-0.3)
13	245	1.0	700 (0.7)[d]	n.d.

(a) Estimated by gel filtration (see experimental); the average molecular weight is shown.
(b) The average maximum value to two or three experiments for dose 0.1 mg/Kg is shown, and the relative value to the activity of heparin is shown in the parentheses.
(c) With respect to APTT.
(d) At 30 min incubation.
(e) Not determined.
(f) At 15 min incubation.
(g) The lipolytic activity 2600 (2.7) was observed with the dose 1.0 mg/Kg.
(h) Inactive.
(i) Compounds **12** and **13** contained d.s. 0.9 and 0.6, respectively, for carboxymethyl group.

heparin (21 x 10^3) > **13** (245 x 10^3) > **8** (26 x 10^3) with respect to LPL. Although possessing similar structure, **8** (mol. wt. 26x10^3) had lipolytic activity, but **9** (mol. wt. 150 x 10^3) did not, reflecting the difference in molecular weight (34,35). However, **13** (mol. Wt. 245 x 10^3) had 0.7 times the activity of heparin, in spite of its relatively high molecular weight, because of the presence of carboxymethyl groups at position 6. The N-sulfate compounds **6** and **13** showed activity, but the N-hexanoyl and NH_2 compounds **10** and **11** showed no activity.

LPL catalyses the hydrolysis of the triacylglycerol moiety of lipoproteins to afford free fatty acids, which generally go into solution as complexes with plasma albumin. However, abnormally high concentration of free fatty acids in plasma can produce several serious pathological effects (36), especially in hemodialysis with heparin (37). The sulfate derivatives of chitin and chitosan possess low toxicity (LD_{50}1.25-3.25 g/Kg)(38), which is similar to that of heparin (LD_{50}1.5-2.0 g/Kg)(39). Therefore, it is significant to note that **8** possesses 1.9-2.2 times that activity of heparin with respect to APTT, and only 0.1 time the activity of heparin with respect to LPL.

REFERENCES

1. D. Knor, Food Technol., 85-97 (1984).
2. K. Arai, T. Kinumaki, and T. Fujita, Bull. Tokai Reg. Fish. Res. Lab., 56, 89-94 (1968).
3. Y. Sawayanagi, N. Nambu, and N. Nagai, Chem. Pharm. Bull., 30, 4213-4215 (1982).
4. L. L. Balassa, GB, 1,252,373 (1971).
5. S. Mima, S. Yoshikawa, and M. Mima, JP Tokkyo Koho, 130,870 (1975).
6. M. Nakajima, K. Atsumi, and K. Kifune, in "Chitin, Chitosan, and Related Enzymes", J. P. Zikkakis, ed., Academic Press, Orlando, 1984, p. 407-410.
7. A. Kato, M. Arakawa, and M. Kondo, J. Microcapsulation, 1, 105-112 (1984).
8. K. Suzuki, Y. Okawa, K. Hashimoto, S. Suzuki, and M. Suzuki, Micobiol. Immunol., 28, 903-912 (1984).
9. B. Alarcon, J. C. Lacal, J. M. Fernandes-Sousa, and L. Carrasco, Antivival Res., 4s, 234-244 (1984).
10. S. Suzuki, Y. Okawa, Y. Ohura, K. Hashimoto, and M. Suzuki, Proc. 2nd, Internat. Conf. Chitin/Chitosan, S. Hirano and S. Tokura, eds., Japan. Soc. Chitin/Chitosan, 1982, p. 210-212.
11. S. Hirano and Y. Ohe, Carbohydr. Polym., 4, 15-22 (1984).
12. W. G. Malette, H. J. Quigley, Jr., R. D. Gaines, N. D. Johnson, and W. G. Rainer, Ann. Thoracic. Sug., 36, 55-58 (1983).
13. M. Sugano, T. Fujikawa, Y. Hiratsuji, and Y. Hasegawa, Nutr. Rep. Int., 18, 531-537 (1978).
14. K. Nishioka, S. Nishimura, N. Nishi, I. Saiki, S. Tokura, and I. Azuma, Vaccine, 2, 93-99 (1984).
15. M. Maeda, Y. Inoue, H. Iwasa, and K. Kifune, in "Chitin, Chitosan, and Related Enzymes", J. P. Zikakis, ed., Academic Press, Orlando, 1984, p. 411-415.
16. S. Hirano, Y. Ohe, and H. Ono, Carbohydr. Res., 47, 315-320 (1976).
17. R. Trujillo, Carbohydr. Res., 7, 483-485 (1968).
18. R. C. Schweiger, Carbohydr. Res., 21, 219-228 (1972).
19. T. Danishefsky, Methods Carbohydr. Chem., 5, 407-409 (1965).
20. R. J. Bertolacini and J. E. Barney, Anal. Chem., 29, 281-283 (1957).
21. J. E. Hodge and B. T. Hofreiter, Methods Carbohydr. Chem., 1, 388-389 (1962).
22. J. X. Khym and D. Doherty, J. Am. Chem. Soc., 74, 3199-3200 (1952).
23. Y. Noishiki, Jpn. J. Artif. Organs, 11, 794-797 (1982).
24. R. R. Proctor and S. I. Rapaport, Am. J. Clin. Pathol., 36, 212-219 (1961).
25. Z. Yosizawa and M. Matsuno, in "Comprehensive Science of Polysaccharides", (Japanese), T. Harada and G. Koizumi, eds., Kodansha Scientific Co., 1973, p. 609-621.
26. V. P. Dole, J. Clin. Invest., 35, 150-154 (1956).
27. K. Shinoda and A. Nakajima, Bull. Inst. Chem. Res. Kyoto Univ., 53, 400-408 (1975).
28. S. Hirano, C. Mizutani, R. Yamaguchi, and O. Miura, Biopolymers, 17, 805-810 (1978).
29. S. Hirano, K. Tobetto, and Y. Noishiki, J. Biomed. Mater. Res., 15, 903-911 (1981).
30. K. Kaifu, and T. Komai, J. Biomed. Mater. Res., 16, 757-766 (1982).
31. T. Baba, R. Yamaguchi, Y. Arai, and T. Ito, Carbohydr. Res., 86, 161-164 (1980).
32. D. Horton and E. K. Just, Carbohydr. Res., 29, 173-179 (1973).
33. R. J. Whistler and M. Kosik, Arch. Biochem. Biophys., 142, 106-110 (1971).
34. H. E. Edward and J. C. Allen, Int. J. Pharm., 18, 39-45 (1984).
35. B. Nijmeyer, L. O. Anderson, and E. Homer, Blood, 63, 836-842 (1984).

36. T. Wessel-Aas and B. Christophersen, Clin. Nephrol., 18, 135-140 (1982).
37. F. Ogata, S. Takahashi, K. Imai, and Y. Hirasawa, Tosekikaishi, 17, 353-357 (1984).
38. J. Doczi, A. Fishman, and J. A. King, J. Am. Chem. Soc., 75, 1512-1513 (1953).
39. J. Seifter and A. J. Begany, Am. J. Med. Soc., 216, 234-241 (1984).

A STUDY OF LIPOSOME-TYPE ARTIFICIAL RED BLOOD CELLS STABILIZED WITH CARBOXYMETHYL CHITIN

Atsushi Kato and Tamotsu Kondo

Faculty of Pharmaceutical Sciences
Science University of Tokyo
Shinjuku-ku, Tokyo 162, Japan

Liposome-type artificial red blood cells (ARBC) stabilized with carboxymethyl chitin containing human hemolysate were prepared by a two-step emulsification technique. Electron microscopic observations using various techniques revealed that the ARBC have a mean diameter of about 310 nm and a multilamellar structure. Oxygen binding abilities of the ARBC and human hemolysate were measured by the use of a Clark-type oxygen electrode. They exibited a striking similarity in the shape of oxygenation curve. Disintegration tests on the ARBC using the enzymes (lysozyme, chitinase and phospholipase C), that can digest the components of the ARBC membrane, suggested that the membrane has a structure in which the phospholipid layer is covered by a mesh of CM chitin molecules in an aqueous phase. The acute toxicity of the ARBC to male mice was examined and LD_{50} values of the ARBC for 2 ml of intravenous injection was evaluated to be 13.8 ml/kg by the Litchfield-Wilcoxon method. The effect of the injected ARBC suspensions on the weights of the organs (liver, spleen and kidneys) were investigated and the differences in organ weight between the injected groups and the control ones were evaluated by a t-test.

INTRODUCTION

Liposome technology has been developed by many investigators in the hope that liposomes can be used as carriers of such substances as drugs, enzymes and hormones, because they are expected to carry these substances to the targeted organs or to deliver them in the whole body more effectively than conventional injection of these substances (1). On the other hand, many studies on red blood cell substitutes have been carried out to get rid of the chronic shortage of blood for transfusion owing to the increase of surgical operations (2,3,4). In addition, the necessity to develop such blood substitutes has become increasingly important, because it is inferred from the viewpoint of immunology that man-to-man transfusion may be one of the causes of many immunologically infectious diseases, e.g. AIDS.

Therefore, we attempted to develop a red blood cell substitute by liposome technology, that is, we tried to prepare lecithin liposomes containing human hemolysate, so-called liposome-type artificial red blood cells, by the various conventional methods. However, the liposomes obtained were quite unstable as was mentioned by Kim and Martin (5). Such unstable liposomes containing human hemolysate are likely to be disintegrated by the actions of various plasma components immediately after they are injected into the blood stream.

Hence, we attempted to reinforce the liposomal membrane with a certain biopolymer that has biocompatible properties. The biopolymer chosen in this preparation of the liposomes containing human hemolsate is carboxymethyl chitin (CM chitin), a derivative of poly β(1-4) linked N-acetyl-D-glucosamine, chitin, which is widely distributed in nature as the major exoskelton material of crustaceans such as crab. Chitin and their derivatives are expected to be usable for biomaterials in future because of their non-toxic and exzymatically biodegradable properties (6,7). Then, we succeeded in preparing liposome-type artificial red blood cells (ARBC) containing human hemolysate using lecithin and CM chitin by a two-step emulsification technique.

Although several physicochemical properties of the ARBC stabilized with CM chitin such as stability (8) and flow properties (9) were revealed, there are still many uncertainties about the cells. For example, oxygen binding ability, which is the most important function of the vesicles as a red blood cell substitute, membrane structure, toxicity to animals and so on. We intend to report the new findings concerning these properties of the liposome-type artificial red blood cells stabilized with carboxymethyl chitin in this paper.

EXPERIMENTAL DETAILS

Preparation of Liposome-type Artificial Red Blood Cells Stabilized with CM Chitin (ARBC)

The ARBC were prepared in the following way (8). Purified egg yolk lecithin (Asahi Kasei Chemical Industries, Ltd., Tokyo) was dissolved in dichloromethane at 50 mg/ml under nitrogen stream. CM chitin was dissolved in a phosphate buffer solution (PBS; pH 7.4, ionic strength 0.154) at a concentration of 0.2%(w/v). To 10 ml of human hemolysate was added an equal volume of the lecithin solution, and the mixture was vigorously agitated by a voltex mixer for 30 sec to give a W/O emulsion (first emulsification). The emulsion obtained was quickly added with stirring to 100 ml of the CM chitin solution to yield a W/O/W complex emulsion (second emulsification). After 10 min stirring, another 100 ml of the CM chitin solution was added to the complex emulsion under stirring and the stirring was further continued until dichloromethane was completely evaporated out. In this way, liposome-type artificial red blood cells stabilized with CM chitin(ARBC) were obtained.

The aqueous dispersion of the ARBC was centrifuged at 2,000 g for 15 min and then at 15,000 g for 30 min to collect the cells. The collected cells were washed several times with the PBS solution on the centrifuge and redispersed in the same medium.

Evaporation of Oxygen Binding Ability of the ARBC

The oxygen binding ability of the ARBC in suspension and human hemolysate were measured by a Gilson Model 5/6H Oxygraph equipped with an OXY module and a Clark-type oxygen electrode immersed in a water-jacketed closed cell. A polarizing voltage of -0.8 V was applied to the electrode.

A suspension of the ARBC at 17.6% (v/v) particle concentration, human hemolysate at 50 mg/ml hemoglobin concentration and PBS solution (pH 7.4), each saturated with air at 25°C by means of airing, were used as samples for measurements. The air-saturated PBS solution was used as the standard solution having a constant partial pressure of oxygen at 25°C. A gas mixture consisting of nitrogen and air was introduced into 4.0 ml of the standard PBS solution in the cell with stirring to give a PBS solution having a given partial pressure of oxygen. The mixing ratio of nitrogen and air was varied in as wide a range as possible. A 0.5 ml volume of the air-saturated sample (ARBC suspension or human hemolysate) was added to 4.0 ml of each of PBS solutions with various partial pressure of oxygen, and then the amount of liberated oxygen from the sample was measured by the electrode.

The degree of oxygen saturation (%) for the hemoglobin in the ARBC or human hemolysate was calculated from the data obtained and oxygenation curves for the ARBC and human hemolysate were determined by plotting the degree of oxygen saturation against partial pressure of oxygen.

At the same time, the hemoglobin concentrations in the ARBC and human hemolysate used in the above measurements were determined spectrophotometrically using a wavelength of 555 nm. The ARBC suspension at 17.6% (v/v) particle concentration and human hemolysate at 50 mg/ml hemoglobin concentration were diluted 90 times with PBS solution. A volume of the diluted ARBC suspension was mixed with an equal volume of sodium 1-dodecyl sulfate (SDS) solution at 0.04% (w/v) concentration and the mixture was allowed to stand until the ARBC were completely disintegrated by the action of the surfactant. Similarly, the human hemolysate was treated with the SDS solution. The hemoglobin concentration in these mixtures were determined spectrophotometrically.

<u>Evaluation of Enzymatic Disintegration of the ARBC</u>

Three times recrystalized egg-white lysozyme (E.C.3.2.1.17) and phospholipase C from <u>clostridium perfringes</u> (E.C.3.1.4.3) were purchased from Sigma. Chitinase from <u>Streptomyces</u> (E.C.3.2.1.14) was supplied by Calbiochem-Behrings (La Jolla, Ca.). These enzymes were used without further purification. All assays were performed in the PBS solution at 37°C.

The enzymatic disintegration of the ARBC was evaluated by measuring the hemoglobin concentration leaked from the ARBC. A 2.5 ml volume of ARBC suspension at 0.5% (v/v) particle concentration in the PBS solution was mixed with an equal volume of enzyme solution in the same buffer. After incubation for 2 hrs, the mixture was centrifuged at 15,000 g for 30 min to remove the unbroken ARBC and the supernatant containing leaked hemoglobin was collected as the sample. To one ml of SDS solution at 0.1% (w/v) was added an equal volume of the sample, and the concentration of denatured hemoglobin in this mixture was measured by a spectrophotometer at a wavelength of 555 nm. In the same way, the total hemoglobin concentration in 2.5 ml of the ARBC suspension at the same particle concentration as above was determined and it was used as the standard (100% disintegration) for calculating the degree of disintegration of the ARBC.

In the cases of lysozyme and chitinase, the degree of enzymatic hydrolysis of CM chitin was evaluated by measuring the increase in reducing power of CM chitin according to modified Schales' procedure (11). Four ml of the supernatant obtained above was allowed to stand in the boiling water for 15 min to stop the enzymatic reaction and to make the hemoglobin molecules in the solution coagulate. After cooling, the mixture was centrifuged at 15,000 g for 15 min to remove the coagulated hemoglobin molecules and the increase in reducing power of CM chitin was determined as follows: the coloring reagent solution was made by dissolving 0.5 g potassium ferricyanide in 1,000 ml of 0.5M sodium carbonate and stored in a brown glass bottle. Four ml of the coloring reagent solution was mixed with 3 ml of the sample solution and the mixture was incubated in boiling water for 15 min in a test tube stoppered with an alminum foil. After cooling the mixture for 30 min, the optical density at 420 nm was measured and the increase in reducing power of CM chitin was calculated from the decrease in the optical density.

Evaluation of Stability of the Enzyme-Treated ARBC

The stability of the ARBC treated with lysozyme or chitinase was evaluated by determining the degree of disintegration by SDS of the cells. Five ml of ARBC suspension at 0.5% (v/v) particle concentration in the PBS solution was mixed with an equal volume of enzyme solution at 100 g/ml in the PBS solution and the mixture was incubated for 2 hrs at 37°C. After incubation, the enzyme-treated ARBC were settled in a centrifuge at 15,000 g for 30 min. The supernatant was removed and the settled ARBC were redispersed gently into 5 ml of the PBS solution. Five ml each of SDS solutions at various concentrations was added to an equal volume of the suspension of redispersed ARBC and the mixtures were incubated for 2 hrs at 37°C. After incubation, the mixtures were centrifuged at 15,000 g for 30 min to remove the unbroken ARBC, and the concentrations of leaked hemoglobin in the supernatants were determined spectrophotometrically. The degree of disintegration of the ARBC was calculated and plotted against the SDS concentration.

Evaluation of Toxicity of the ARBC

The ARBC were washed three times with physiological saline solution on the centrifuge after preparation and were redispersed in the same medium. Male mice (BALB/c, 20-25 g) were used in this study and one animal group consisted of five mice. Various volumes (1,2 and 3 ml) of suspensions of the ARBC at 5, 10, 15 and 20%(v/v) particle concentration in physiological saline solution were injected into the animals of each group from the tail vein, respectively. As the control, the same volumes of physiological saline solution were injected into the animals of the control groups in the same manner. The toxicity of the ARBC was determined by the death ratio of the animals two weeks after injection. The death ratio was plotted against the logarithm of the particle volume of the ARBC. LD_{50} value of the ARBC was evaluated by the Litchfield-Wilcoxon method on the basis of the data obtained and expressed in particle volume per body weight (ml/kg).

Effect of Injected ARBC Suspension on Organ Weight

One ml each of the ARBC suspensions at 5, 10 and 15%(v/v) particle concentrations was injected into the male mice groups consisting of five animals in the same way as described above, respectively. At the same time, the same volume of the physiological saline solution was injected to the animals of the control group. Two weeks later, these animals were sacrificed to measure the weights of their organs (liver, spleen and kidneys), and then the difference in organ weight between the ARBC injected groups and the control ones were evaluated by a t-test with a computer.

RESULTS AND DISCUSSION

Figure 1 shows a scanning electron micrograph of the liposome-type ARBC stabilized with CM chitin. It was found that the ARBC are spherical

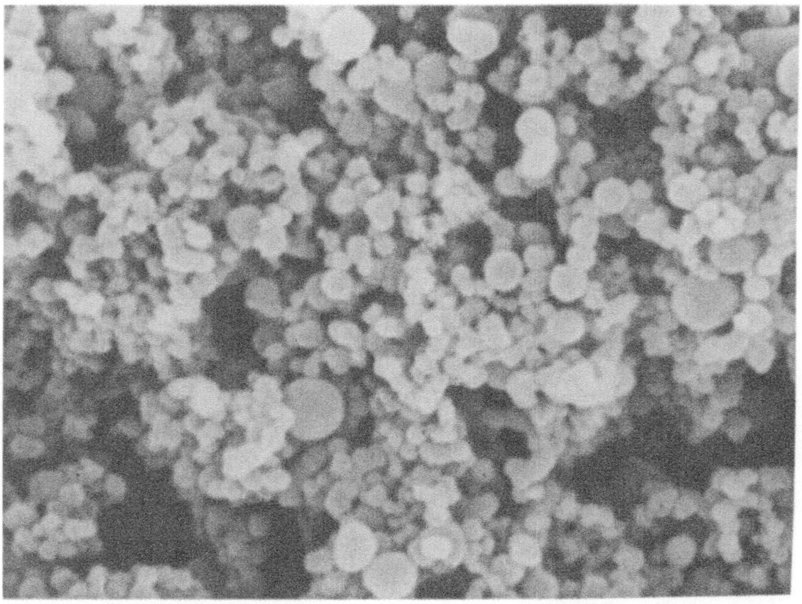

Figure 1. Scanning electron micrograph of liposome-type artificial red blood cells (ARBC) stabilized with carboxymethyl chitin (8).

in shape with a mean diameter of about 310 nm. The mean diameter of the cells seems to be larger than that of the conventional liposomes. Nevertheless, the cells can be expected to pass through capillary vessels with a mean diameter of a few micrometer.

Figure 2 shows a freeze-fracture image of the ARBC. This picture reveals that the ARBC have a multilamellar structure and many granulated structures on their surface. In this preparation of the ARBC, two kinds of vesicles, large and small ones (not shown here), were obtained. According to our previous electron microscopic study (10), it was found that the former have a multilamellar structure and the latter a unilamellar one, and that the portion of the former is dominant while that of latter is insignificant. The granulated structures that can be seen on the surface of the vesicles seems to be CM chitin molecules which contribute to the stabilization of the liposomal ARBC membrane. As this picture shows, CM chitin does not yield a continuous layer but rather occurs in patches on the membrane surface, probably forming a mesh-like layer on the entire ARBC surface.

Figure 3 shows oxygenation curves for the ARBC suspension and human hemolysate solution measured with the Clark-type oxygen electrode. The ordinate gives the degree of oxygen saturation and abscissa partial pressure of oxygen. The results suggest that the ARBC have an oxygen binding ability quite similar to that of human hemolysate, though a careful inspection of the figure reveals that there exists small differences in oxygenation curve between them. That is, the degree of oxygen saturation is slightly lower for the ARBC than for human hemolysate at partial pressures of oxygen between 50 and 100 mmHg while the shape of oxygenation curve and the value of $P(O_2)_{50}$ are quite similar for both of them. This seems to be caused by the fact that a portion of the hemoglobin molecules encapsulated in the ARBC may undergo certain changes. The relationship between the hemoglobin concentration and the partial pres-

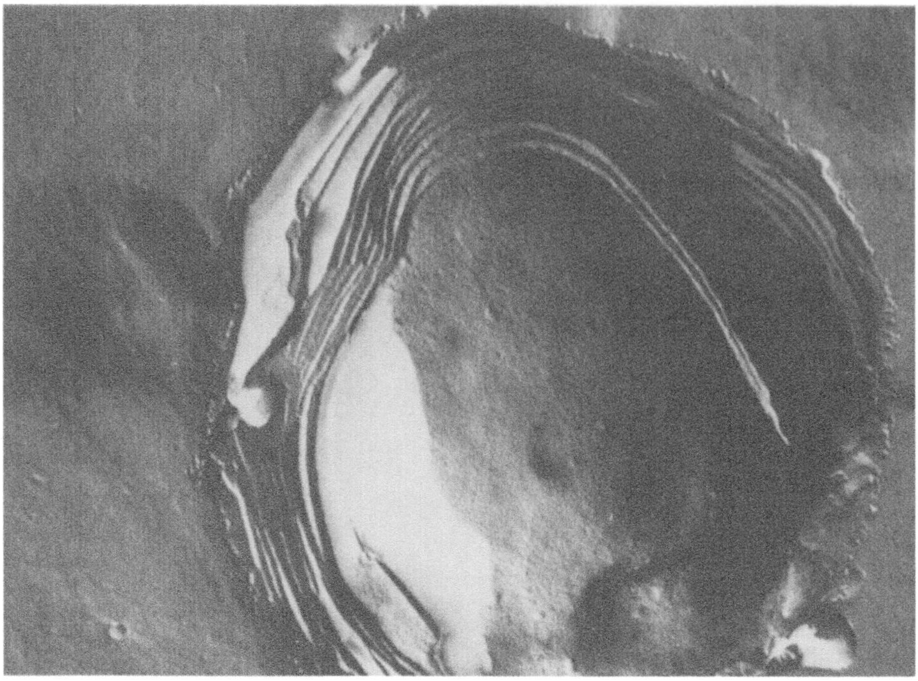

Figure 2. Freeze-fracture electron micrograph of the ARBC (10).

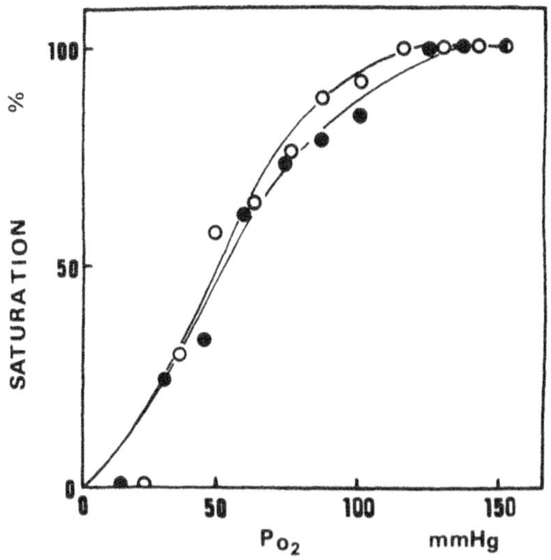

Figure 3. Oxygenation curves of the ARBC (●) and human hemolysate (o) as determined by a Clark-type oxygen electrode at 25°C (12).

sure of oxygen liberated in oxygen free condition for the ARBC and human hemolysate was then investigated in order to know the amount of denatured hemoglobin molecules in the ARBC.

Table I shows the hemoglobin concentrations in the ARBC and human hemolysate that were used in the oxygenation measurements, and the partial pressures of oxygen liberated from the air-saturated samples which were added to the PBS solution saturated with nitrogen. In this table, the hemoglobin concentrations are those of hemoglobin denatured by SDS because the ARBC stabilized with CM chitin were disintegrated completely by the action of SDS molecules (8). As shown in Table I, the ratios of the hemoglobin concentration in the ARBC to that in the hemolysate and the partial pressure of oxygen liberated from the ARBC to that from the hemolysate are 0.432 and 0.361, respectively. If the hemoglobin molecules in the ARBC should retain the same oxygen binding ability as those in the hemolysate, in other words, if all of the hemoglobin molecules trapped in the ARBC were intact, it would be expected that the above ratios have an identical value. However, the ratio obtained from the measurements of liberated oxygen is lower than that from the measurements of the hemoglobin concentration. That is, the amount of oxygen liberated from the ARBC is 84.6% of the value predicted from the hemoglobin concentration, suggesting that 15.4% of the hemoglobin molecules trapped in the ARBC might have been denatured and lost their oxygen binding ability.

Two possible causes for the reduced oxygen binding ability are given below. [1] When the w/o emulsion is formed in the first step of the preparation of the ARBC, some of the hemoglobin molecules in the aqueous phase of the emulsion may be denatured through contact with dichloromethane molecules at the interface of the aqueous and organic phases, and then they may lose their oxygen binding ability. [2] When dichloromethane molecules are evaporated out of the w/o/w complex emulsion in the second step of the preparation, a certain number of hemoglobin molecules may leak from the water droplets in the emulsion to adsorb on the surface of the ARBC formed and then remain there as denatured hemoglobin molecules. The presence of this sort of denatured hemoglobin on the surface of the ARBC was suggested in the previous electron microscopic observations (10).

Lysozyme and chitinase are able to hydrolyze CM chitin molecules accompanied with an increase of reducing power of the polymer, and phospholipase C can digest lecithin molecules. As both of the enzymes at a

Table I. Partial pressure of oxygen liberated from the ARBC and human hemolysate under oxygen free condition at 25°C and the total hemoglobin concentrations in the samples (12).

	Hemoglobin conc. (mg/ml)	Liberated oxygen (mmHg)
ARBC	9.77	4.61
HEMOLYSATE	22.60	12.77
ARBC/HEMOLYSATE	0.432	0.361

concentration of 50 μg/ml produced almost the same degrees of increase in reducing power of CM chitin molecules dissolved in the PBS solution (not shown here), these enzymes at this concentration were used for the disintegration tests of the ARBC and for measurements of increase in reducing power of the CM chitin molecules on the surface of the ARBC. Figure 4 shows the results of the disintegration tests of the ARBC with lysozyme and chitinase, indicating that the ARBC were disintegrated by the action of the chitinase to a slight extent, whereas no disintegration of the vesicles was found with lysozyme. At the same time, we measured the increase of reducing power of the CM chitin molecules on the surface of the vesicles on the sample used in the disintegration tests by the hydrolytic action of the enzymes. Although both enzymes at this concentration have the same hydrolytic activity on the CM chitin solution in the PBS as mentioned above, no increase of reducing power of CM chitin molecules was found for lysozyme, while the reducing power of CM chitin molecules increased with an increase of chitinase concentration as shown in Figure 5. This seems to result from the difference in the mechanism of hydrolytic action on CM chitin between the two enzymes. Namely, as is reported by many investigators, when lysozyme undergoes hydrolytic action on a substrate, the six active subsites of the enzyme have to be bound simultaneously by six glucosamine residues of the substrate (6). This seems to be a very strict condition for the enzyme to exhibit its hydrolytic activity on CM chitin adsorbed on the surface of the ARBC. On the other hand, no such severe restriction has been reported in the case of chitinase though the detailed mechanism of its hydrolytic action is still unknown. Therefore, it can be presumed that CM chitin molecules should adsorb on the surface of the ARBC in the process of the ARBC preparation without exposing their residues cleavable by the hydrolytic action of lysozyme to the bulk water phase and/or they should form a certain structure on the surface of the ARBC to restrict their molecular motions.

Although neither lysozyme nor chitinase could disintegrate the ARBC significantly as shown in Figure 4, it can be expected that the enzymes, chitinase in particular, must have degraded CM chitin molecules on the surface of the ARBC to some degree to lower the stability of the ARBC, because the increase of reducing power of the CM chitin molecules was observed by the action of chitinase. Then, we investigated the stability to SDS, an anionic surfactant, of the ARBC treated with lysozyme and chitinase. Figure 6 shows the relationship between the degree of disintegration of the ARBC treated with the enzymes and SDS concentration. Actually, it was revealed that the concentration of the SDS needed to disintegrate the ARBC is increased in the following order: non-treated > lysozyme-treated > chitinase-treated. This indicates that the stability

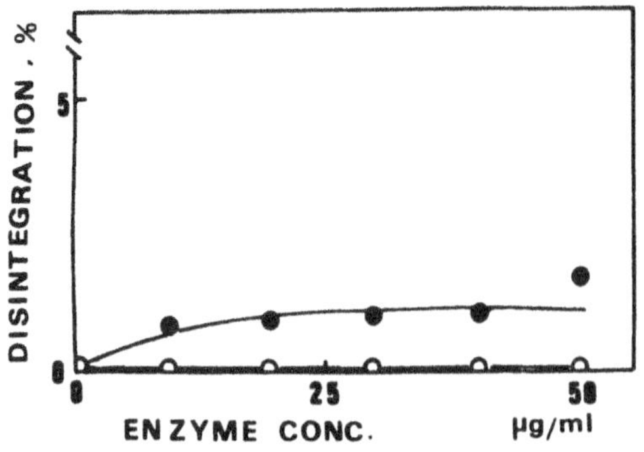

Figure 4. Disintegration of the ARBC at 0.25% (v/v) particle concentration by lysozyme (o) and chitinase (●) at 37°C after 2 hr incubation (12).

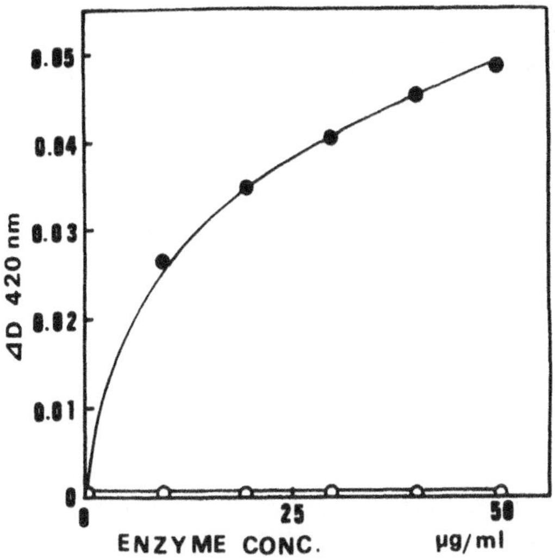

Figure 5. Enzymatic digestion of the CM chitin molecules on the surface of the ARBC dispersed in the PBS solution (pH 7.4) at 0.25% (v/v) particle concentration by lysozyme (o) and chitinase (●) at 37°C after 2 hr incubation (12).

of the ARBC is reduced by treatment with chitinase and lysozyme, the former is more effective than the latter in reducing the stability to SDS of the ARBC. In addition, the stability of the ARBC treated with lysozyme was lower than that of the non-treated cells in spite of the fact that no increase of reducing power of the CM chitin molecules was observed with the enzyme as shown in Figure 5, suggesting that lysozyme should hydrolyze the CM chitin molecules on the surface of the ARBC in a very little amount, probably too little to be detected by the modified Schales' procedure.

Furthermore, in order to elucidate the form in which CM chitin molecules are attached to the surface of the ARBC, we carried out disintegra-

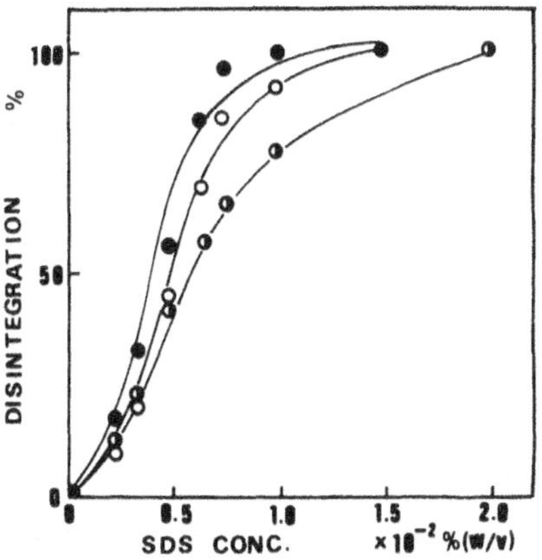

Figure 6. Disintegration by SDS of the ARBC treated with lysozyme (o) and chitinase (●), and PBS (◐) at 37°C after 2 hr. incubation.

tion tests of the ARBC by the action of phospholipase C, which can digest lecithin molecules. The results are shown in Figure 7. The degree of disintegration of the ARBC increased linearly with increase in the enzyme concentration and it was higher than that caused by chitinase, indicating that the lecithin molecules constituting the ARBC membrane should be digested by the action of the enzyme. Hence, the CM chitin layers on the surface of the ARBC should have openings large enough to allow the enzyme to make access to and to act on the ARBC surface composed of lecithin molecules. Consequently, CM chitin molecules are quite likely to form a mesh-like structure on the surface of the ARBC as presumed by the electron microscopic observations.

Finally, we carried out <u>in vivo</u> toxicity tests of the ARBC suspensions using mice (BALB/c) by means of intravenous injection from tail vein without bloodletting. Figure 8 shows the relation between the death ratio of the mouse and the logarithm of the total particle volume of the injected ARBC at three suspension volumes (1, 2 and 3 ml). The death ratio increased steeply with an increase in the particle volume in all cases and was independent of injected suspension volume. Since no animal was found dead in the control groups even by injection of the largest (3 ml) volume of physiological saline solution, it can be mentioned that the death ratio is mainly determined by the total particle volume of the injected ARBC. Based on the data in this figure, the value of LD_{50} was estimated by the Litchfield-Wilcoxon method to be 13.8 for 2 ml injection in terms of the total particle volume of the injected ARBC per kilogram of body weight of the animal (ml/kg). Hence, if 10 per cent of the LD_{50} value is allowed for a margin of safety, it would be estimated that 966 ml of the ARBC suspension at 10% (v/v) particle concentration can be used in transfusion for a recipient who weighs 70 kg. Although this value is of a presumptive nature to estimate the toxicity of the ARBC expediently, the value seems to be slightly low for transfusion.

Then, we investigated the effect of ARBC injection on several organs (liver, spleen and kidneys). That is, we evaluated changes in weights of the organs by the t-test. One ml each of the ARBC suspensions at various particle concentrations was injected to the mice from tail vein and the

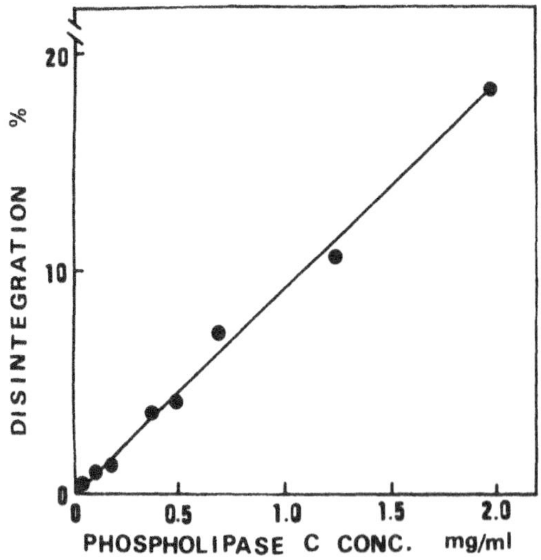

Figure 7. Disintegration of the ARBC at 0.25% (v/v) particle concentration by phospholipase C at 37°C after 2 hr incubation (12).

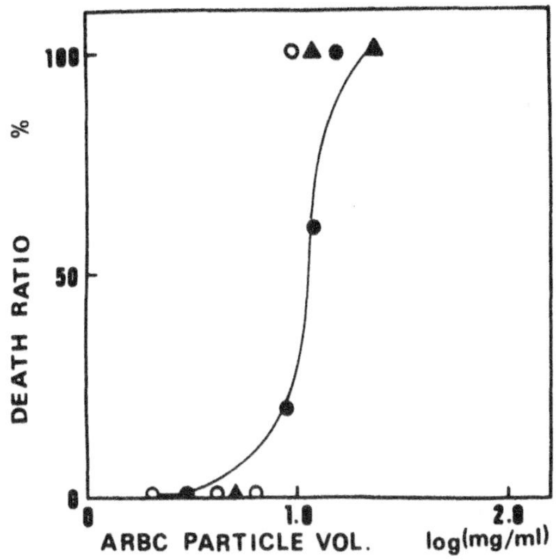

Figure 8. Relation between the death ratio of mice (BALB/c) and the logarithm of particle volume of the injected ARBC (12). Injected volume of the ARBC suspension: 1 ml (o), 2 ml (●) and 3 ml (▲).

animals were sacrificed to remove the organs two weeks later. The weights were then compared with those of the control animals that were injected the same volume of the physiological saline solution. The results obtained are shown in Table II, in which an increase in weight is found in the spleen at Mode III (highest particle concentration) and in the liver at Mode I (lowest particle concentration). In the case of kidneys, the change in weight of the organs decreased with increase of the particle concentration of the injected ARBC suspension. Thus, no distinct tendency of change in weight was found for any of the organs in this test and we cannot discuss the relation between the cause of toxicity of the ARBC suspension and the effect of ARBC injection on changes in the weights of the organs based on such a small amount of information. Hence, further efforts should be made to seek the cause of this rather high toxicity of the ARBC in future.

Table II. Effect of injected suspensions of the ARBC on the weights of liver, spleen and kidneys after a time lapse of two weeks.

Mode	Liver	Spleen	Kidneys
I	19.70%[a]	--	22.94%[a]
II	--	--	17.36%[a]
III	--	25.19%[a]	--

(a) ($p < 0.01$). Injected ARBC particle concentration, % (v/v); Mode I: 5%, Mode II: 10%, MOde III: 15%.

REFERENCES

1. C. G. Knight, ed., "Liposomes: from Physical Structure to Therapeutic Applications", Elsevier, Amsterdam, 1981.
2. L. Djordjevich and I. F. Miller, Exp. Hemat., 8, 584-592 (1980).
3. B. P. Gaber, P. Yager, P. Scheridan and E. L. Chang, FEBS LETT., 153, 258-288 (1983).
4. M. Arakawa and T. Kondo, J. Pharm. Sci., 70, 354-357 (1981).
5. S. Kim and G. M. Martin, Biochim. Biophys. Acta, 645, 1-9 (1981).
6. S. H. Pangburn, P. V. Tresconey and J. Heller, Biomaterials, 3, 105-108 (1981).
7. P. R. Austin, C. J. Brine, J. E. Castle and J. P. Zikakis, Science, 212, 749-753 (1981).
8. A. Kato, M. Arakawa and T. Kondo, J. Microencap., 1, 105-112 (1984).
9. A. Kato, M. Arakawa and T. Kondo, Bioeology, 20, 593-601 (1983).
10. E. Wehrli, T. Kondo, A. Kato, H. Izawa and J. Kreuter, J. Microencap., 1, 329-333 (1984).
11. T. Imoto and Y. Yagishita, Agr. Biol. Chem., 35, 1154-1156 (1971).
12. A. Kato, I. Tanaka, M. Arakawa and T. Kondo, Biomat. Med. Dev. Art. Org., 13, 61-82 (1985).

POLYMERIC PLATINOL DERIVATIVE OF METHOTREXATE FOR THE TREATMENT OF VIRALLY RELATED JUVENILE DIABETES

Charles E. Carraher, Jr.[a,b], Isabel Lopez[b] and David J. Giron[c]

(a) Department of Chemistry
Florida Atlantic University
Boca Raton, Florida 33431
Departments of (b) Chemistry and
(c) Microbiology and Immunology
Wright State University
Dayton, Ohio 45435

The polyamine derived from tetrachloroplatinate and methotrexate successfully inhibits the growth of the diabetic encephalomyocarditis, EMC-D, virus in male mice and female mice treated with testosterone employing a vaccine like protocol where the Pt-MX polymer is administered prior to infection by the EMC-D virus. Infection by the EMC-D virus is successfully halted if the Pt-MX polymer is administered within one day after viral infection. Thus the Pt-MX polymer is effective at minimizing or eliminating a virus that produces juvenile diabete-like symptoms.

INTRODUCTION AND HISTORY

Diabetes

In 1975 a National Commission on Diabetes reported on the status of diabetes as being that the "problem (was) severe, therapy (was) inadequate" (1). The disease now affects 5% of the USA population and is projected to double every 15 years. Thus the average newly born has a better than 20% chance of developing diabetes. The disease accounts for about 2% of all deaths in the USA making it the third largest killer after heart and cancer related diseases (1-3). Furthermore, it is the leading cause of blindness in adults.

Blank (4) reports that in the last twenty years that no new drug types have been added to the clinical arsenal for the treatment of diabetes although advancement has been made in a number of medical aspects including (a) diagnosis of diabetes, (b) better understanding of the biochemical processes associated with diabetes, (c) synthesis and secretion of insulin and (d) insulin-receptor interaction.

There are two major forms of diabetes; mellitus, once referred to as maturity onset and the other juvenile onset. While about eighty percent of all diabetics have the adult type, it is the juvenile variety that is the most dangerous. For juvenile diabetes, the symptoms often occur suddenly. While it largely affects young people, it can occur in adults of any age.

Juvenile diabetes, also known as brittle or unstable diabetes, is believed to be largely if not wholly virally related and may be epidemically related (4-8). Thus development of a vaccine may prevent the development of juvenile diabetes.

A diabetic is unable to metabolize sugar properly bringing about a buildup of sugar. The kidneys of a diabetic normally discharge some of this excess sugar into the urine. Thus, urine sugar analyses are typically employed as a measure of the extent of diabetes. The buildup in sugar is a result of a diabetic's inability to produce enough insulin which is needed to metabolize and utilize saccharides. Insulin is a hormone produced by beta cells of the pancreas and is part of a larger peptide known as proinsulin. The virus responsible for the onset of juvenile diabetes attack and kill these beta cells. One family of viruses known to produce onset of juvenile diabetes in humans is the Picorna or little RNA viruses. Coxsakies virus, known to cause diabetes in humans, and Encephalomyocarditis, EMC virus, suspected to cause diabetes in humans, are the two most used viruses to produce juvenile diabetes in the laboratory and both are members of the Picorna virus family. Because of ease of handling, the EMC virus is by far the most employed virus.

Juvenile diabetics are typically lean and have an absolute requirement of needing insulin. As noted above, juvenile diabetics are not able to metabolize saccharides. Since saccarides are a major foodstuff serving as major sources of metabolic intermediates and energy, the body reverts to usage of proteins and lipids as alternative sources of energy and metabolics. These changes result in polyuria (loss of salts and water), polyphasin (excessive hunger) and polydipsin (excessive thirst). Clinical tests show hyperglycemia and glucosuria (excessive amounts of glucose in the blood and urine) and ketosis and ketonemia (ketones and fatty acids in the blood, and urine).

On the other hand, the symptoms of maturity onset diabetes are more subtle ranging from overt, readily detectable hyperglycemia to a prediabetic condition. Maturity diabetics are often overweight.

Juvenile diabetics must be treated with tightly monitored diets and exogenously administered insulin. Maturity onset diabetics may require only dietary restrictions and there are a number of drugs that may be used, although insulin remains the drug of choice for exaggerated cases.

Treatment with insulin is currently under study. Banting and Best, in 1921, isolated insulin and demonstrated its ability to lower blood glucose and to ameliorate diabetes symptoms (9). This ushered in the notion that a cure for diabetics was at hand. Only in the late 1940's and early 1950's, when it was found that a disproportionate number of insulin-treated diabetics suffered from strokes, renal and cardiac problems and retinopathics, did it become evident that insulin alone was not the answer. Even so, insulin remains the sole drug employed in the treatment of juvenile diabetes.

<u>Encephalomyocarditis Virus</u>

A murine model of insulin-dependent diabetes mellitus was developed by Craighead and colleagues in 1978, using the M variant of an encephalomyocarditis virus (EMC-M). EMC is a picorna virus which is a small lytic

RNA virus in the cardivirus family. The virus was recovered from swine that developed fatal myocarditis. This virus, passed once in mice, exhibited specific diabetogenic properties. These properties suggested that "wild" viruses which have specific beta-cell tropism and diabetogenic potential may be found in nature (10). Using the EMC-M for infection, beta cell degranulation, coagulation necrosis, and alteration of the islet cells were noted in susceptible mice (11). The degree of beta cell damage correlated with the degree of hyperglycemia seen in the mice (12). In 1979, Yoon and colleagues discovered that EMC-M was a mixture of at least two viruses, the diabetogenic D variant, (known as EMC-D) and the nondiabetogenic B variant (EMC-B) (13). The induction of diabetes by EMC-D was inhibited by EMC-B during co-infection experiments. This demonstrated that the ratio of EMC-B:EMC-D in the inoculation determined the incidence of diabetes in the susceptible SJL/ mouse.

The two variants of EMC-M exhibit different characteristics. The D variant of EMC-M produces large diffuse plaques on mouse L929 cells and does not induce interferon (IFN)d; in contrast, the B variant produces small distinct plaques on mouse L929 cells and induces large amounts of interferon (14). Yoon noted that the passage history of EMC-M influenced its diabetogenicity (13). After passage in mouse kidney cell cultures, the diabetogenic capacity of the virus was markedly diminished but was restored after one passage in mice (15). No demonstrable time difference was seen in the appearance of neutralizing antibody to either the B or D variants (16). It was further noted, that antibody that acted against one variant neutralized the infectivity of the other variant to the same extent (13).

Differences in the two variants were also identified by Ray, et al. who analyzed the two genomes by nucleic acid hybridization and RNA fingerprinting (17). The ribonuclease T-digested RNA's, fingerprinted on two-dimensional polyacrylamide gels, revealed at least one oligonucleotide (20 to 25 nucleotides in length) present in the EMC-D RNA which was not present in EMC-B (17). Gould and colleagues demonstrated differences in tissue tropism and replication of the two variants in ICR Swiss mice (14). For the present study the EMC-D variant was employed to cause the onset of juvenile diabetes.

Methotrexate

Folic acid (N-[p-[(2-amino-4-hydroxy-6-peteridinyl) methyl] amino]-benzoyl]-L-glutamic acid) must be reduced to 1, L-tetrahydrofolic acid (Figure 1) before it can serve as a coenzyme. The reduction occurs stepwise by way of dihydrofolic acid, and both steps are generally thought to be carried out by the same enzyme, called both folic reductase and dihydrofolic reductase, though separation of two enzymic activities has been reported (18). Folic acid coenzymes participate at three places in the nucleic acid biosynthetic pathway. N^5, N^{10}-Methenyletrahydrofolic acid supplies the formyl group for the conversion of glycinamide ribonucleotide; N^{10}-methylenetetrahydrofolic acid donates the methodyl group necessary for the conversion of 2^I-deoxyuridylic acid to thymdylic acid (19-22).

A number of folic acid antagonists with structures similar to folic acid have been synthesized and partially characterized. The N^{10}-methyl derivative of aminopterin, N-[p[[2,4-diamino-6-pteridinyl) methyl] amino]benzoyl] -L-glutamic acid (also called methotrexate, MX, and amethopterin) is a potent antifolate with a high degree of anticancer activity. Several studies with purified dihydrofolic reductase have shown that the pH dependent binding of methotrexate is stoichiometric with excess enzyme, but reversible with equivalent or excess amount of the

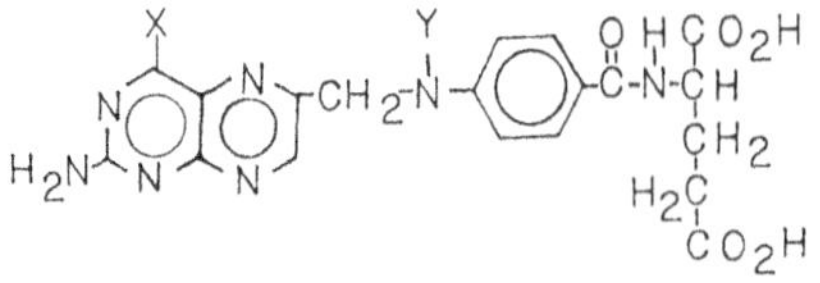

Figure 1. The structures of folic acid and related compounds.
Where X = OH, Y = H for folic acid
X = NH_2, Y = H for aminopterin
and X = NH_2, Y = CH_3 for methotrexate

inhibitors (23-27). Thus while the binding is reversible and dependent on the conditions under which it is measured, it is extremely tight and as such, the binding has been termed "pseudoirreversible" (24), binding more tightly to the dihydrofolic reductase than folic acid itself. Zakrzewski proposed that methotrexate (and aminopterin) bind through hydrogen binding between the enzyme and the 2,4-diminopyrimidine portion. Further, from thermodynamic studies, it was found that folic acid, which normally exists in the keto form, must tautomerize to the enol form to bind to the enzyme and that this event is associated with an increase in entropy (28,29). Baker proposes that the tight binding is due to the increased basicity of methotrexate which results in its protonation and then the protonated inhibitor binds to anionic groups in the active site of the enzyme (30,31). Further, it is found that the presence of alkyl groups on the 2-and 4-amino groups of aminopterin destroys biological activity probably because they sterically interfere with binding to the dihydrofolic reductase (32,33).

Regardless of the true mechanism(s) of activity, the biological activity appears principally due to methotrexates ability to prevent cells from reducing folic and dihydrofolic acid to tetrahydrofolic acid, although they may also interfere with one carbon transfers (34-36).

Methotrexate, MX, has been used in the treatment of arthritis (37), though its wider application as a general purpose drug is discouraged by such serious side effects as oral and gastrointestinal ulceration, liver damage and hepatic damage (38). Methotrexate is slightly less toxic than aminopterin and is more widely used in such autoimmune diseases as lupus erthematosus and psoriasis (39). The toxic effects of methotrexate can reversed if the immunosuppressive dose is followed by folinic acid (5-formyltetrahydrofolic acid), leaving the immunosuppressive capacity unaltered (40). This led to the discovery of the "rescue technique" in therapy.

Platinum Polyamines

Cancer is the second leading cause of death in the U.S.A., accounting for 21.9% of the total deaths. A great variety of compounds have been investigated and a few have proven sufficiently useful in the clinical treatment of human neoplasms at acceptable levels of toxicity to deserve the designation of chemotherapeutic agents. Among these are cis-diamminodichloroplatinum (II), C-DDP, and methotrexate.

In 1965 Rosenberg and coworkers discovered that in the presence of platinum electrodes E. coli failed to replicate but continued growing, giving filamentous cells (41). Later it was demonstrated that inhibition of cell division also occured in the presence of platinum compounds. The

```
 Cl     Cl
   \   /
    \ /
     Pt
    / \
   /   \
 NH3   NH3
```

C-DDP

Figure 2. The structure of C-DDP.

most effective of these compounds was found to be c-diamminodichloroplatinum (II), C-DDP (**2**) (Figure 2).

Since then studies have been done on a number of cis-platinum (II) derivatives and certain of these substances have demonstrated antineoplastic activity in many animal tumor screening systems. Among the compounds tested c-DDP (also called cis-platinum and Platinol) exhibited the greatest antineoplastic activity and was most effective in treating a wide variety of tumors. For example, this compound was active against animal transplantable tumors such as Sarcome 180, L1210 luekemia, Walker 256 carcinosarcoma, Dunning ascitic leukemia and 7, 12-dimethylbenzanthracene (DMBA) a chemically induced mammary tumor (for instance 42,43). Ultimately cis-DDP was chosen for clinical assessment.

Clinical trials using cis-DDP began in 1971. Although the initial usefulness of this compound was set by its nephrotoxicity, it was found to show prominent antitumoral activity against a wide variety of human neoplasms. Cis-DDP has shown great activity, and is currently being used alone or in combination with other drugs, in the treatment of testical tumors (in combination with Vinblastin and Bleomycin), ovarian carcinoma (in combination with doxorubicin), osteosarcoma (44), carcinoma of the bladder, head and neck, lung, renal and endometrial carcinomas, hepatoma, leukemia and non-Hodgkin's lymphoma (for instance 44-46).

The antineoplastic activity of cis-platinum and its derivatives appears to be related to the chemical nature of the species attached to the platinum atom and the position of the ligands relative to one another. The requirements are: that the compounds be electrically neutral, containing two inert ligands and two labile ligands which are oriented cis to each other. Derivatives of C-DDP not having these necessary features typically do not exhibit antineoplastic properties.

Unfortunatly, the potential of c-DDP or its derivatives as antineoplastic agents has been greatly resticted by a wide range of undesirable side effects. Some of these include gastrointestinal side effects such as nausea, vomiting and diarrhea, occasional liver dysfunction, myelosuppression, leukopenia and thrombocytopenia, neutrotoxicity, neophrotoxicity, etc. (for instance 47-49). The most important of the dose limiting side effects of cis-platinum is nephrotoxicity. This results in cumulative and eventually irreversible damage. Acute tubular necrosis is a prominent feature of cis-platinum nephrotoxicity. Clinical signs include elevated BUN, serum creatinine proteinuria and hyperuricema (44-50). There are structural analogues of cis-platinum which do not exhibit nephrotoxicity, some which are currently on clinical trial (51). Several approaches have been examined for protective effects against cis-platinum induced nephrotoxicity. It has been reported that hydration and diuresis with mannitol or furosemide ameliorated the toxicity in some cases (49,51,52).

Attempts to Minimize Toxicity

Several approaches to minimize cis-platinum induced nephrotoxicity have been described (53-55). The synthesis of chemically modified derivatives based on cis-DDP needs special mention. Modifications can range from simple ligand substitutions to the incorporation of the drug into complex high molecular weight polymer systems. Second generation platinum compound analogs or structural derivatives of cis-DDP which, within the limits of necessary structural requirements, may show not only reduced toxicities but in many cases, also enhances antitumoral activity have been synthesized. Carraher and coworkers first synthesized some platinum polyamines that are polymic derivatives of c-DDP in 1976 (56).

The polymeric derivatives of c-DDP generally show equal to greater activity towards all tested cell lines compared to c-DDP itself (for instance 57,58). For most of the polymers derived from diamines, inhibition of tumor cell lines begins at a polymer concentration above 20 micrograms per milliliter and is complete by 50 micrograms per milliliter. C-DDP shows only 50 percentage inhibition at 50 micrograms per milliliter for HeLa cells and 50 at 50 micrograms per milliliter for L929 cells (c.f. Table I).

The platinum polyamines are also active with RNA virus at concentrations below where cell damage occurs (compare results from Tables I and II). About two-thirds of the polyamines are found to inhibit viral growth. The platinum polyamines are found to be much less toxic than c-DDP itself. Thus mice were injected with 10 to 100 times the lethal dosage of polyamine based on c-DDP once every other day for a month without mouse death (59).

Cell differentiation values based on transformed cells were obtained for a number of polyamines (60). A cell differentiation value, CD, is the ratio of cell death for transformed, cancer-like cells compared to the cell death of healthy cells. CP values greater than one were found for about two-thirds of the polyamines. Live animal tests were conducted

Table I. Inhibition of Tumor Cell Lines By Selected Platinum Polyamines[a]

	Concentration (Micrograms per Milliliter)							
	10		20		30		50	
Diamine Reactant, PtX_4	HeLa	L929	HeLa	L929	HeLa	L929	HeLa	L929
Hexanediamine, $PtCl_4$	0	0	0	0	25	25	100	100
p-Phenylenediamine, $PtCl_4$	0	0	0	0	25	25	100	100
Hexanediamine, PtI_4	0	0	0	0	25	25	100	100

(a) Percentage cytopathic effect after 24 hours of treatment.

Table II. Effects of Selected Platinum Polyamines on Poliovirus Type I(a)

	Concentration (Micrograms per Milliliter)					
	0		10		20	
Reactants	Pfu/ml	%-Control	Pfu/ml	%-Control	Pfu/ml	%-Control
p-Phenylene-diamine, $PtCl_4$	$3.2x10^7$	---	$6.5x10^7$	200	$3.2x10^7$	100
Hexanediamine, PtI_4	$3.2x10^7$	---	$3.2x10^7$	100	$3.2x10^7$	100
Hexanediamine, $PtCl_4$	$3.2x10_4$	---	$3.4x10^6$	10	$1.1x10^7$	34

(a) Plaque forming units per milliliter.

where mice were injected with lethal dosages of cancer and a comparison is made to their life expectancy compared with mice treated with a potential drug. The polymer from tetrachloroplatinate and methotrexate has showed superior results in many tests. For instance, it has a CD ratio between 3 and 4 and has extended the average life expectancy of cancerous mice 150 to 200% (59,60). Thus this polymer was picked for further study, in particular as an antiviral agent.

Summary

The viruses tested thus far are mainly of the RNA variety. There are a number of RNA virally related maladies. Some, such as polio Type I, already have effective methods of control. Thus we looked for RNA virally related diseases where no good prevention control approaches are practiced. One such malady is juvenile diabetes. This is the initial report of some of our efforts aimed at the control of juvenile diabetes employing product 3 (Figure 3).

EXPERIMENTAL

Methotrexate (MX, 5.42 mmoles) was dissolved in 150 ml of distilled deionized water, DDW. The solubility of MX is poor at room temperature so the solution was heated to about 60°C to dissolve the MX, then the solution was cooled. A solution of K_2PtCl_4 (J & J Materials, Neptune City, NJ) 5.42 mmoles in 15 ml DDW was added and the mixture allowed to stir at 25°C for 48 hours at which time the product was filtered, washed with hot DDW and allowed to dry in air. Product yield was 91%.

Infrared spectra were obtained employing Perkin-Elmer 457 and 1130 Infrared Spectrophotometers and a Digilab FTS-20 C/D FT-IR. Thermal analysis was obtained employing a DuPont 900 DSC cell attached to a DuPont 950 Thermal Analyzer Console and a DuPont 990 TGA. Mass spectral analysis was performed using a direct insertion probe in a Kratos MS-50

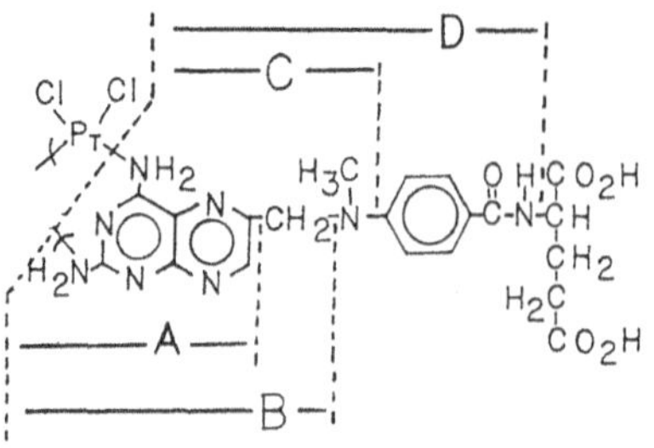

Figure 3. Structure of polymer that is being examined for controlling juvenile diabetes.

mass spectrometer, operating in the EI mode, 8 KV acceleration with a probe temperature of about 350°C. Molecular weight determinations were performed using a Brice-Phoenix Model BP-3000 Universal Light Scattering Photometer. Refractive index increments, dn/dc, were determined using a Bausch and Lomb Abbe Refractometer, Model 3-L. Elemental analysis for Cl and Pt were determined using wet analysis techniques for Cl and thermodecomposition for Pt. Elemental analysis for C, H and N were carried out employing a Perkin-Elmer, 240 Elemental Analyser.

In Vivo Pre-treatment study Female Mice

Eight to nine week old Swiss ICR female mice were employed and were treated as per usual protocol. The mice were divided into three groups, each containing 18 mice. A stock solution of polymer was made by dissolving the polymer in DMSO and then 0.9% saline solution added dropwise forming a 1:1 DMSO/Saline (V/V) solution. Mice received intraperitionially (IP) injections employing the following regimen:

Day 0: All groups received 0.5 cc IP of polymer solution (6.4 mg of polymer/kg mouse weight).

Day 7: Each mouse in groups I and II was treated with 0.80 mg testosterone (treatment with testosterone is necessary with female mice to permit them to develop diabetes in response to virus infection).

Day 8: Group I received a second 0.5 cc IP inoculation of the polymer solution.

Day 9: All groups received 1×10^4 pfu (plaque forming units) of EMC-D virus.

Day 17: A one hour glucose tolerance test was done on each animal.

In Vivo Pre Treatment/Study Male Mice

Eight to nine week old Swiss ICR male mice were employed and were treated as per usual protocol. The mice were divided into five groups, each containing ten mice. A stock solution of polymer was added dropwise to form a 1:3 DMSO/Saline (V/V) solution. Mice received IP injections employing the following regimen:

Day 0: Groups III, IV and V were inoculated with 0.5 cc of K_2PtCl_4 (group III), MX (group IV) and polymer (group V) (6.4 mg of test material/kg of mouse weight).

Day 7: Groups III-V were again inoculated as per day 0.

Day 8: Groups II-V were inoculated with 0.2 cc (1×10^4 pfu) of EMC-D virus.

Day 16: A one hour glucose tolerance test was done on each animal.

In Vivo Study Male Mice Pre-Drug Treatment

Eight to nine week old Swiss ICR male mice were employed. The mice were divided into 4 groups each containing 10 mice. A stock solution of polymer was made (as described above). Mice received IP injections employing the following regimen:

Day 0: Group I, II and III were injected with 0.2 cc (1×10^4 pfu) of EMC-D virus.

Day 1: Group II was injected with 0.5 cc of Pt-Methotrexate polymer (6.4 mg test material/kg mouse weight).

Day 8: Group I and II were injected as per day 1.

Day 16: A one hour glucose tolerance test was done on each animal.

RESULTS AND DISCUSSION

Structural

Carraher and coworkers (56) first synthesized polymeric derivatives of cis-DDP in 1976. Initial synthesis involved the use of inexpensive diamines as 1,6-hexanediamine and p-phenylenediamine. More recently Carraher and Manek (61-63) employed additional nitrogen-containing reactants including hydrazines, amides and thioamides. Cell culture and live animal test results are consistent with improved anticancer activity being found for platinum containing products derived from biologically active diamines. Thus the synthesis of biologically active nitrogen containing reactants was undertaken.

The trans effect conveniently leads to the desired cis products and the observed much greater reactivity of nitrogen containing Lewis bases compared with other Lewis bases (water, alcohols, ketones, aldehydes and acids) give reactions only at nitrogen sites. The following is a brief summary of the structural analyses carried out on the polymer employed in the biological testing described in this chapter.

While the products synthesized can be described structurally as derivatives of c-DDP, those derived from multinitrogen containing reactants as methotrexate probable exist as a combination of products with reaction occuring at all of the nitrogens. Even so, the products are linear, based on solubility, and contain one platinum atom, two chlorides and one methotrexate per repeat unit, based on elemental analysis.

From infrared spectoscopy, the appearance of two bands at 450 and 415 cm^1 are attributed to the Pt-Cl stretch. The carbonyl stretching band is found about $1710 cm^{-1}$. Other bands characteristic of the presence of the MX moiety are also present.

From the mass spectra of the MX-Pt product ion fragments are characteristics of the presence of the MX moiety ion fragments at (all given in m/e) 367(MX-2CO_2), 322(D-cf.3), 204(C), 176-178(B), 163(A), 160(B-NH_2), 148-151($H_3CNC_6H_4CONH$), 137($C_5H_7N_5$), 136($C_5H_6N_5$), 134 and 135(NC_6H_4CONH), 118-121(C_6H_5CON and NC_6H_4CO), 108-107(H_3CNC_4), 103-104(C_6H_4CO), 90-93(M-

C_6H_4), 83-87(C_3H_6NHCO), 76-79(C_6H_5), 71(NCH_2CO_2), 68-70(C_2H_4NHCO).

The particular polymer employed for the biological studies has a weight average degree of polymerization from light scattering photometry of 300.

Biological Results

Biological testing of the polymer for the treatment of juvenile diabetes began about five years ago. Results of three experiments are reported here. The initial experiment involves use of female mice. Treatment with EMC-D virus results in a diabetes like condition when a number of mouse strains are treated, but typically only with male mice or female mice treated with testosterone. Three groups of mice were treated initially with the MX-Pt polymer, thereafter referred to as simply the polymer. A week later groups I and II were treated with testosterone, making groups I and II susceptible to virus induced diabetes. All groups of animals received the diabetogenic virus. The amount of blood glucose (mg/dl) for healthy mice varies depending on the protocol, mouse strains, etc., but is typically about 200 mg/dl (64). This particular experiment was designed to test several items. First, are untreated female mice subject to contracting virally induced diabetes. Group II constitutes a group of untreated female mice and in agreement with literature, the blood glucose level of about 180 mg/dl is consistent with the mice not contrating diabetes. Second, does the use of testosterone cause diabetes in female mice. Group II constitutes a group of testosterone treated mice with a blood glucose average level indicative of diabetes. Third, does the use of polymer prevent diabetes in susceptible mice. Group I is testosteron treated mice inoculated with polymer just prior (one day) to infection by virus. The glucose level is about that for the nondiabetic mice, group III, and significantly below the average for the diabetic mice, group II (Table III). For this experiment, it appears that the polymer has effectively blocked the diabetogenic effects of the virus.

The second experiment is similar to the first with the exception that male mice were employed and the test compounds includes the two reactants employed in the synthesis of the polymer. As can be seen by comparing the mean GT values of each group (Table IV), the polymer shows a greater

Table III. Results of Glucose Tolerance Test for _in vivo_ study (Female mice) for Preinfection Treatment.

	Group I (Treated)	Group II	Group III (Control)*
(Blood) Glucose mg/dl	194	326	180
Diabetic, % (a)	7	56	0
Hypoglycemic, % (b)	0	0	0
Normal, %	93	44	100

(*) 3 SD for control is 51 mg/dl
(a) Glucose values 3SD above control group
(b) Glucose values 3SD below control group

Table IV. Results of Glucose Tolerance Test for in vivo Study (Male Mice) for Preinfection Treatment.

	Control	Virus	K_2PtCl_4	MX	Polymer
(Blood) Glucose mg/dl	215*	465	331	342	261
Diabetic, %(a)	0	100	40	50	20
Hypoglycemic, %(b)	0	0	0	0	10
Normal, %	100	0	60	50	70

(*) 3SD for control is 75 mg/dl
(a) Glucose values 3SD above control group
(b) Glucose values 3SD below control group

positive effect on the control of diabetes than either of the two reactants, registering blood glucose values of near that of the control group.

These two experiments are related to generation of a vaccine that can be employed to prevent onset of beta-cell damage by RNA viruses. The results show that the Pt-MX polymer can effectively minimize beta-cell damage, halting onset of juvenile-like conditions.

In the third experiment the virus is injected 24 hours prior to the first administration of the test compound (group II), whereas group I was injected 1 week later. As noted by comparing (Table V) the mean glucose tolerance test values of each group injection of the polymer 24 hours after injecting the virus results in the percentage diabetis for this group to be 0%. Also the average blood glucose value is very near that of the control group for the group treated with the polymer after only one day and 64% normal with polymer treatment occuring one week after viral infection.

The third experiment is related to viral treatment subsequent to viral infection. While the 100% cure rate for mice treated by polymer one

Table V. Results of Glucose Tolerance Test for in vivo Study (Male Mice) for Postinfection Treatment.

	Group I (one week)	Group II (one day)	Group III Control *
(Blood) Glucose mg/dl	273	193	199
Diabetic, % (a)	36	0	0

(*) 3SD for control is 78 mg/dl
(a) Glucose values 3SD above control group

day after viral infection is striking, it must be noted that it takes several days for the virus to grow to epidemic proportions. The test for antiviral activity by addition of the polymer one week after viral infection is a truer test of the polymers potential to control a viral attach. The results show a modest response but further work must be done before the antiviral, postpolymer treatment can be accepted as a real alternative to a vaccine approach. Further, it must be remembered that with today's technology, once the insulin producing beta cells are destroyed they cannot be regenerated.

The precise mechanism of action of the polymer is unknown but may involve the inhibition of EMC-D virus at any one of the following sites: Inhibition of penetration of the virus into the cell, transcription, translation or maturation, preventing destruction of pancreatic beta cells.

ACKNOWLEDGEMENT

We gratefully acknowledge a gift of the methotrexate from Monsanto Chemical Company, Dayton, Ohio.

REFERENCES

1. T. H. Maugh, Science, 191, 272 (1976).
2. P. J. Palumbo, L. R. Elveback, C.-P. Chu, D. C. Connolly and L. T. Kurland, Diabetes, 25, 566 (1976).
3. D. L. Rimoin, Calif. Med., 119, 14 (1973).
4. B. Blank, "Burger's Medicinal Chemistry, Part II", 4th Ed., M. E. Wolff, Ed., Chapter 31, Wiley-Interscience, NY, 1979.
5. J. E. Craighead and M. F. McLane, Science, 162, 913 (1968).
6. D. W. Boucher and A. L. Notkins, J. Exp. Med., 137, 1226 (1973).
7. T. H. Maugh, Science, 188, 347 and 436 (1975).
8. C. E. Hunt, J. R. Lindsey and S. U. Walkley, Fed. Proc., 35, 1206 (1976).
9. F. G. Banting and C. H. Best, J. Lab. Clin. Med., 7, 251 (1922).
10. J. Craighead, N. Eng. J. Med., 299, 1439 (1978).
11. E. Rayfield and Y. Seto, Diabetes, 27, 1126 (1978).
12. K. Hayashi, D. Boucher and A. Hotkins, Amer. J. Path., 75, (1974).
13. H. Whitcomb, V. Merluzzi and S. Copperband, Cell. Immunol., 21, 272 (1976).
14. C. L. Gould, M. Trombley, N. Bigley, K. McMannamor and D. J. Giron, Proc. Soc. Exp. Biol. Med., 175, 449 (1984).
15. J. Yoon, T. Onodera and A. Notkins, J. Gen. Virol., 37, 225 (1977).
16. D. J. Giron, R. Tu, F. Hemphil, F. Pindak and J. Schmidt, Proc. Soc. Exp. Biol. Med., 163, 142 (1980).
17. U. Ray, G. Aulaakh, M. Schubert, P. McClintock, J. Yoon and A. Notkins, J. Gen. Virol., 64, 947 (1983).
18. U. K. Kenkare and B. M. Braganca, Biochem. J., 86, 160 (1963).
19. F. M. Heunnekens and M. J. Osborn, Advan. Enzymol., 21, 369 (1959).
20. J. C. Rabinowitz, "The Enzymes", 2nd Ed., Vol. 2, Part A, P. D. Boyer, H. Lardy and K. Myrback, Eds., Academic Press, NY, 1960, p. 185.
21. J. S. O'Brien, Cancer Res., 22, 267 (1962).
22. M. Friedkin, Ann. Rev. Biochem., 32, 185 (1963).
23. J. R. Bertino, J. P. Perkins and D. G. Johns, Biochemistry, 4, 839 (1965).
24. W. C. Werkheiser, J. Bio. Chem., 236, 888 (1961).
25. W. C. Werkheiser, Cancer Res., 25, 1608 (1965).
26. J. R. Bertino, B. A. Booth, A. L. Beiber, A. Cashmore and A. C.

Sartorelli, J. Bio. Chem., 239, 479 (1964).
27. A. W. Schrecker and F. M. Huennakens, Biochem. Pharmacol., 13, 731 (1964).
28. S. F. Zakrzewski, J. Bio. Chem., 238, 1485 (1963).
29. S. F. Zakrzewski, J, Bio. Chem., 238, 4002 (1963).
30. B. R. Baker, Cancer Chemother. Rep., 4, 1 (1959).
31. B. R. Baker and J. H. Jordan, J. Pharm. Sci., 54, 1740 (1965).
32. R. Roth, J. M. Smith and M. E. Hultquist, J. Amer. Chem. Soc., 72, 1914 (1950).
33. R. Roth, J. M. Smith and M. E. Hultquist, J. Amer. Chem. Soc., 73, 2864 (1951).
34. C. A. Nichol and A. D. Welch, Proc. Soc. Exp. Bio. Med., 74, 403 (1950).
35. H. H. Jukes and H. P. Broquist, "Metabolic Inhibitors", Vol. 1, R. M. Hochester and J. H. Quastral, Eds., Academic Press, NY, 1963, pp. 481-534.
36. C. G. Hitchings and J. J. Burchall, "Advances in Enzymology", Vol. 27, F. F. Nord, Ed., Wiley-Interscience, NY, 1965, pp. 417-468.
37. T. Y. Shen, "Burger's Medicinal Chemistry", 4th Ed., Part III, M. E. Wolff, Ed., Wiley-Interscience, NY, 1981, p.1254.
38. H. H. Roenigk, H. J. Mailback and G. D. Weinstein, Arch. Dermatol, 105, 363 (1972).
39. R. L. Black, W. M. O'Brien, E. J. Van Scott, R. Auerbach, A. Z. Eisen and J. J. Bunim, J. Amer. Med. Assoc., 189, 743 (1964).
40. M. C. Berenbaum, Lancet, 2, 1363 (1964).
41. B. Rosenberg, L. Van Camp and T. Krigas, Nature, 205, 698 (1965).
42. C. Litterest, A. Leroy and A. Guarino, Cancer Treat. Rep., 63, 1485 (1979).
43. H. Vandenberg, H. Fravaled and J. Roberts, J. Clin. Hem. and Onc., 7, 349 (1977).
44. M. Rosenwieg, D. VonHoff and M. Slavik, Ann. Int. Med., 86, 803 (1977).
45. L. Einhorn and J. Donahue, J. Urol, 117, 65 (1977).
46. D. Calvo, Y. Patt and S. Wallace, Cancer, 45, 1278 (1980).
47. D. VonHoff, R. Schilsky and C. Reichert, Cancer Treat. Rep., 63, 1527 (1979).
48. I. Krakoff, Cancer Treat. Rep., 63, 1523 (19790.
49. C. Litterest, I. Torres and A. Guarino, J. Clin. Hem. and Onc., 7, 357 (1973).
50. J. Kovach, C. Moretel and A. Schutt, Cancer Chemother. Rep., 51, 357 (1973).
51. C, Carraher, D. Giron and W. Scott, "Bioactive Polymeric Systems", C. G. Gebelein and C. Carraher, Eds., Plenum Press, NY, 1985, Chapter 20.
52. C. Cvitkovic, J. Spaulding and V. Bethune, Cancer, 39, 1357 (1977).
53. J. Hill, E. Loeb and A. MacLellan, Cancer Chemother. Rep., 59, 647 (1975).
54. S. D. Williams and L. H. Einhorn, Adv. Int. Med., 27, 531 (1982).
55. C. E. Merrin, Cancer Treat. Rep., 63, 1579 (1979).
56. C. Carraher, W. Scott, D. Giron and J. Schroeder, J. Macromol. Sci., A-1, A15(4), 725 (1981).
57. C. Carraher, D. Giron, I. Lopez, D. R. Cerutis and W. J. Scott, Organic Coatings and Plastic Chemistry, 44, 120 (1981).
58. C. Carraher, W. J. Scott, J. A. Schroeder and D. Giron, J. Macromol. Sco. Chem., A15(4), 625 (1981).
59. C. Carraher, D. Giron and coworkers, Unpublished results.
60. D. Giron. M. Espy, C. Carraher, I. Lopez and C. Turner, Polymeric Materials, 51, 312 (1984).
61. C. Carraher, T. Manek, D. Giron, M. Trombley, G. Hess, K. Casberg and W. Scott, "New Monomers and Polymers", Eds. W. Culberson and C. Pittman, Plenum Publ. Co., NY (1984).

62. C. Carraher, T. Manek, M. Trombley, G. Hess and D. Giron, Polymer Preprints, 24(1), 69 (1983).
63. C. Carraher and T. Manek, Polymer Preprints, 23(2), 77 (1982).
64. J. Yoon and A. Notkins, Metabolism, 32(7), 37 (1983).

ANTIBACTERIAL AND ANTITUMORAL ACTIVITY OF SELECTED POLYPHOSPHONOANHYDRIDES AND POLYPHOSPHOROANHYDRIDES

Charles E. Carraher, Jr.[a], Raymond J. Linville[b], Isabel Lopez[c] and David J. Giron[d]

(a) Florida Atlantic University, Department of Chemistry
Boca Raton, Florida 33431
(b) Air Force Wright Aeronautical Laboratories (AFWAL/MLPJ)
Wright-Patterson AFB, Ohio 45433
(c) Departments of Chemistry and
(d) Microbiology and Immunology, Wright State University
Dayton, Ohio 45435

A variety of polyphosphoanhydrides and polyphosohoranhydrides were tested for biological activity. Most of the phosphorus-containing polymers inhibited several tested bacteria and about one half inhibited most of the tested bacteria. Preliminary tests involving cancer cell lines indicated that the polymers inhibited these cell lines at concentrations equal to or less than required for the cancer drug Platinol.

INTRODUCTION

Phosphate and phosphonate-like monomeric and polymeric products show a wide spectrum of biological activities and serve as the basic structure for the biologically significant nucleic acids. We have previously reported the synthesis of a wide variety of related phosphorus containing polymers including polyphosphate and polyphoaphonate esters and polyphosphoramides, (for instance 1-8).

We have reported the initial synthesis of a wide variety of phosphorus containing polymers based on the condensation of organophosphorus diacid chlorides with salts of dicarboxyic acids (1; 9, 10) (Equation 1). Our purposes for desiring the synthesis of such products include the following. First, these polymers had not been previously synthesized. Their synthesis would allow comparisons to be made with other similar polymers. Second, such products may show useful physical and biological properties. The corresponding polyesters are employed as flame retarders for textiles, used as specialty adhesives for glass and inhibit a wide range of organisms including maggots. Nucleic acids are natural examples of phosphorus polyesters.

Third, this is the final step towards the reaction of organophosphorus (V) dichlorides with amino acids allowing the synthesis of a phosphate-like and amino acid-like polymer- a phosphorus polyaminoanhydride.

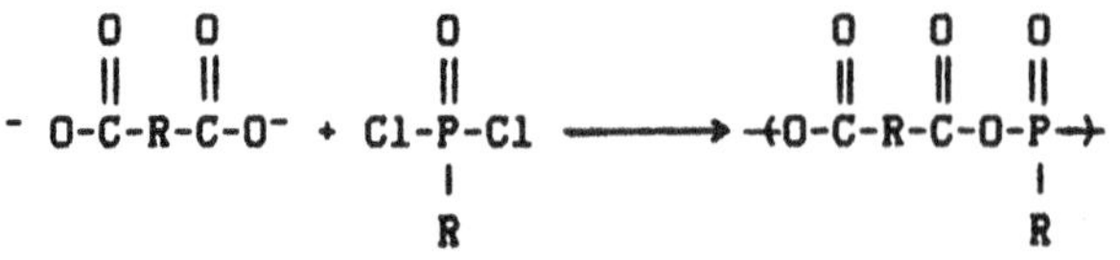

(Equation 1)

1

Specifically, the aminoacids may be of the synthetic variety such as the biologically important para-aminobenzoic acid (PABA; **2**) (Equation 2) or alpha-aminoacids, preferably dipeptides and greater since alpha-amino acids may lead to five-membered cyclic products (**3**) (Equation 3) rather than the desired linear products (**4**) (Equation 4).

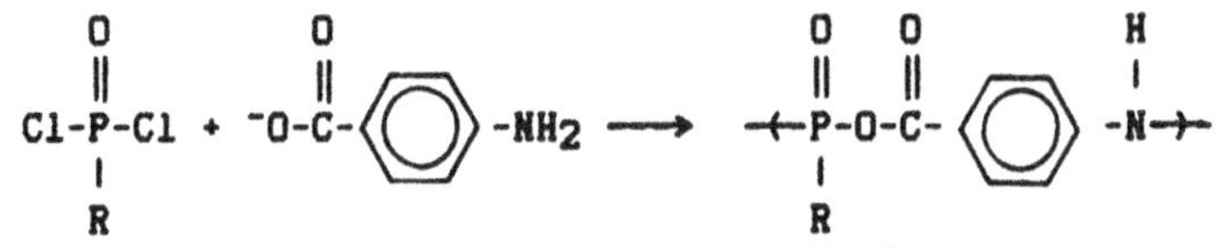

(Equation 2)

2

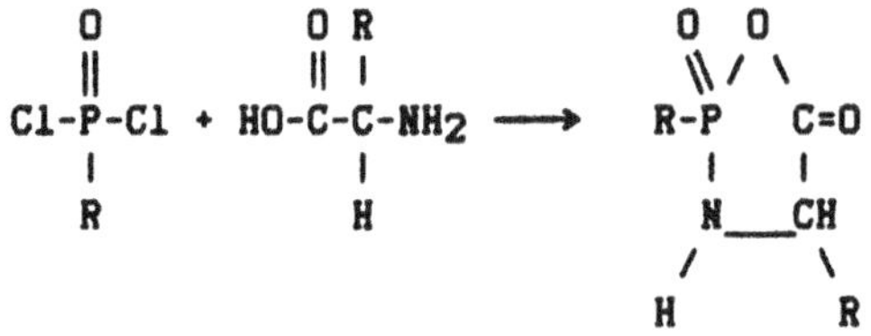

(Equation 3)

3

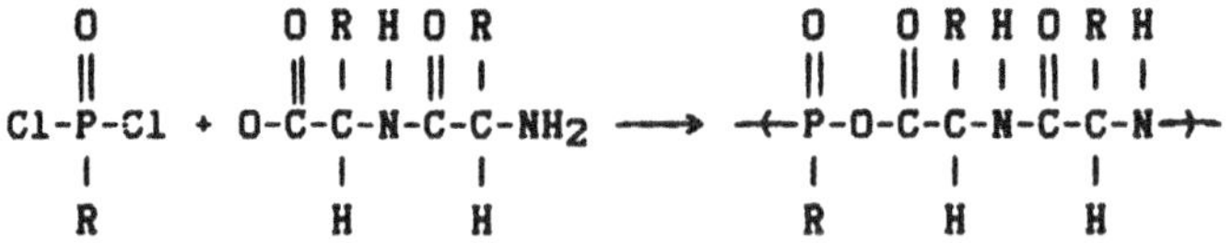

(Equation 4)

4

Fourth, we found that the nature of the end group could be varied depending on the identity of the employed added base. This would allow the study of biological activity as a function of identity of the end group.

EXPERIMENTAL

Synthetic and physical characterization techniques are described in detail elsewhere (10). Briefly, solutions of the organophosphorus compound in an organic solvent (usually chloroform) were added to a Waring Blendor. Rapid stirring was begun and an aqueous solution of a diacid, which also contained an added base, sodium hydroxide, NaOH, or triethylamine, Et_3N; a phase transfer catalyst or other agent, was added. Reaction time began when the organic phase was fully added to the stirred aqueous phase.

The products of each reaction were recovered employing suction filtration, with the collected solid product washed with water and chloroform. The product was then transferred to a preweighed petri dish and permitted to dry.

Molecular weight determination were obtained using a Brice-Phoenix Model BP-3000 Universal Light Scattering Photometer. Refractive Index Increments, dn/dc, were determined using a Bausch and Lomb Abbe Refractometer Model #3-L.

Infrared spectra were obtained employing postassium bromide pellets using a Perkin-Elmer 457 Grating Spectrophotometer. Electron Impact Mass Spectroscopy analyses were performed using a Finnigan 4021 MS coupled with a Finnigan 9610GC system. Direct Insertion Chemical Ionization Mass Spectroscopy was done using a DuPont 21-491 MS coupled with a Hewlett-Packard HP-2216C.

Preliminary toxicology determinations were accomplished using both bacteria and tumor tissue cultures. Organisms were chosen to give a wide range of bacterial species. The procedure involved streaking Tryptic Soy Agar (TSA) plates with the organism, then using a stock solution of polymer at a concentration of 8000 ug/ml of DSMO. Paper discs (1/2 cm diameter) were dipped into the stock polymer solutins and placed on the TSA plates. The plates were then incubated for 24 hours at $37^{0}C$. After this time period, the plates were checked for inhibition which is indicated by a clear zone on the plate. Three cell lines were chosen for the polymer studies. Dimethylsulfoxide (DSMO) was the solvent of choice for the polymers since it is not appreciably toxic to man (for instance 11) and the phosphorus containing polymers are soluble in it.

Various concentrations of the polymer were prepared by diluting the stock solution with Dulbecco's Modified Eagle Medium supplimented with 10% fetal calf serum and 1% Penicillin/streptomycin(DMEM). Confluent cultures of the cell lines L929 (Mouse Connective Tissue Tumor Line), BHK-21 (Baby Hamster Kidney), and HeLa (Epitheloid Carcinoma, Cervic Human) in flasks (75 cm^3) were trypsinized, suspended in DMEM and plated onto Corning plates (1 ml per well). The appropriate concentrations of DMEM/ polymer were then added to each well in volumes of 1 ml each. The plates were incubated overnight at $37^{0}C$ 5% CO_2 to allow the cells to form a monolayer. The cells were then microscopically inspected for cytopathic effects (CPE) to obtain an approximate ratio of dead to live cells (0 for 0% cell death, 1 for 25% cell death, 2 for 50% cell death, 3 for 75% cell death and 4 for 100% cell death).

A staining procedure was employed after the polymers had been on the cells for 48 hours to determine cell death more accurately. The DMEM in each well was aspirated, and a 5% Eosin-methylene blue solution (which stains the dead cells) was diluted to 1:10 with DMEM. One ml of dye solution was placed in each of the wells. The plate was incubated for 1/2 hour and the cells were assigned a 48 hour cell death rating. The rating of 0 to 4 as before, according to the approximate ratio of dead (stained cells) to live cells.

RESULTS AND DISCUSSION

Nature of End-Group

The full discussion related to the evidence related to the nature of the end-group is given elsewhere (10). Here is described briefly results of only two analysis techniques, namely infrared spectroscopy and mass spectral analysis.

Previous condensations carried out employing triethylamine as the added base have often resulted in the presence of triethylamine associated endgroups (for instance 12-15). Ready identification techniques for identification of the presence or absence of the triethylamine moiety have proven to be infrared and mass spectral analyses.

Products that contain the triethylamine moiety exhibit infrared bands within the 2800 to 2400 cm^{-1} region. Thus, the products from phenylphosphonic dichloride and disodium adipate (Figure 1, upper incomplete spectrum) shows no bands in the 2800 to 2400 cm^{-1} region. The analogous product (Figure 1, lower complete spectrum) produced from ditriethylamine adipate shows bands at 2760, 2740, 2665, 2630, 2480 and 2465 cm^{-1}; all characteristic of the presence of the triethylamine moiety.

Mass spectral analyses of analogus products synthesized using triethylamine or sodium hydroxide showed analogous results. Table I gives results compared with the mass spectrum of triethylamine itself. Thus, it is possible to identify products that contain appreciable triethylamine associated endgroups.

Biological Activities

The phosphorus anhydrides were tested in solutions (8×10^3 microgram per milliliter) against a variety of bacteria. Results appear in Tables II and III. Characteristics of the bacteria tested are given in Table IV. Three show no inhibition to any of the tested bacteria. The majority show some inhibition. The products from phenylphorphonic dichloride and tere-

Table I. Mass Spectral ion fragments derived from the product of n-butylphosphonic dichloride and ditriethylamine terephtalate (Product A) and cyclohexylphosphonic dichloride and ditriethylamine terephthalic (Product B).

Triethylamine							
m/e	86	58	101	56	100	87	
John Wiley	1000	202	157	85	66	56	
Product A	1000	276	230	80	86	74	
Product B	1000	380	370	430	100	100	
Terephthalic Acid							
m/e	149	67	50	44	51	76	77
John Wiley	1000	420	340	220	220	180	170
Product A	1000	13	240	260	190	120	187
Product B	1000	89	190	a	140	99	78

(a) MS run from m/e > 50

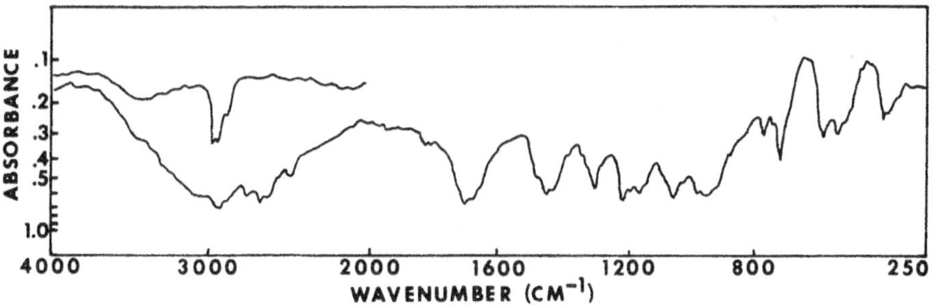

Figure 1. Infrared Spectra of condensation products of phenylphosphonic dichloride and disodium adipate (top, to 2000 cm^{-1}) and ditriethylamine adipate (bottom).

phthalic acid and 2,2-dithiosalicylic acid show complete or partial inhibition towards all tested bacteria. Others showing good inhibition (complete or partial) are products from phenylphosphonic dichloride with adipic acid and 1,3-bis(4-carboxyphenyl)propane acid and terephthalic acid with cyclohexylphosphonic dichloride. The fact that about one half (5 out of 12) showed good broad bacterial inhibition indicates that analogous products might be canidates for futher study as broad spectrum antibacterial agents.

Table II. Results of the bacteriological study of selected products (a).

Diacid (b)	E. coli	Alcal	Staph e.	Staph a.	Kleb	Bran	Actin
2, 5-Dimethyl terepthalic	N	N	N	N	N	N	N
Tetramethyl-terepthalic	P	P	N	P	P	C	P
Adipic	P	N	P	P	P	P	P
2,2-Dithio-salicylic	P	P	P	P	P	C	P
1,3-Bis(4-Carboxy-phenyl)-propane	P	P	N	P	P	P	P
N,N'-Terephthalyidene-bis-anthranilic	N	N	N	N	N	N	N
Terephthalic	C	P	P	P	C	C	C

(a) N = no inhibition
P = partial inhibition
C = complete inhibition
(b) Derived from condensation with phenylphosphonic dichloride.

Table III. Results of the bacteriological study of selected products (a).

Phosphorus Acid Chloride (b)	E. coli	Alcal	Staph E	Staph A	Kleb	Bran	Actin
4-Chlorophenyl-phosphoro-dichloride	N	N	N	N	P	C	B
Cyclohexylphosphonic-dichloride	P	P	P	P	P	N	P
Phenylphosphoro-dichloride	N	N	N	N	N	N	N
Phenylphosphonothio-dichloride	N	N	N	N	N	N	N
n-Butylphosphoro-dichloridate	C	N	N	P	N	C	P

(a) N = no inhibition
P = partial inhibition
C = complete inhibition
(b) Derived from condensation with neutralized terephthalic acid.

Several of the phosphorous anhydrides were evaluated as to their ability to inhibit tumor cells. Results appear in Tables V to VII. Cell inhibition is moderate for all three compounds. For comparison, results of cell inhibition studies for cis-dichlorodiaminoplatimun II, (cis-DDP) licensed as a wide ranging anticancer drug under the name Platinol, are given and are generally the same as those of the phosphorus containing products. Further, the biological activities of the two products derived from employing different added bases are similar indicating that for at least these products, endgroup nature is not critically related to their antitumoral activity.

The biological activities of these products is not unexpected due to their structural similarity to many industrially important monomeric products employed as toxics for home, farm, and lawn uses.

Further biological tests are underway and others are planned. Results will be published in the future.

These biological analyses are meant to evaluate only briefly the potential inhibitory nature of the compounds. As noted, many of the organophosphorus polymers studied, exhibited inhibition to some degree.

A current study involves studying the biological activity of selected phosphorus containing polymers as a function of the nature of the end-group, namely either containing or not containing triethylamine associated endgroups. Results from such studies should assist in defining, in general terms, the importance of selected structural parameters in rela-

Table IV. Characteristics of employed bacteria.

<u>Acinetobacter calcoaceticus</u>

Gram-negative rod
Can be an opportunistic pathogen but is part of the normal flora of the skin and mouth
Associated with conjunctivitis, keratitus, and chronic ear infections

<u>Alcaligens faecalis</u>

Gram-negative rod
Part of the normal intestinal flora
Has been found to cause urinary infections and, in debilitated individuals, septicemia or meningitis

<u>Branhamella catarrhalis</u>

Gram-negative diplococcus
Part of the normal oral and nasopharyngeal flora
Has been associated with mucous membrane inflammations, veneral discharges, meningitis and bacterial endocarditis

<u>Escherichia coli</u>

Gram-negative rod
Part of the normal intestinal flora
Most frequent cause of urinary tract infections
May cause cholecystitus, appendicitis, peritonitis, sinusitis, and summer diarrhea

<u>Klebstella pneumoniae</u>

Gram-negative rod
Part of the normal flora of the nose, mouth and intestines
May cause lung abcess, sinusitis and upper respiratory infections

<u>Staphylococcus epidermidis</u>

Gram positive cocci
Generally causes mild infections but has caused septicemia bacterial endocarditis, and urinary tract infections

<u>Staphylococcus aureau</u>

Gram-positive cocci
Causes "pimples", abscesses, impetigo wound infections, pyelitis, cystitis, "food poisoning", pneumona, meningitis, and enteritis

Table V. Results of selected products effect on the HeLa Cell line (a)

Diacid	Base	150 ug/ml	100 ug/ml	40 ug/ml	16 ug/ml	8 ug/ml
Terephthalic	NaOH	1	1	1	1	1
Terephthalic	Et_3N	±	2	1	±	1
2,2'Dithiosali- cyclic acid	Et_3N	1	1	±	1	±
Cis-DDP	----	3	2	2	---	---

(a) All values are in % cells that die where:
0 = 0% cell death
1 = 25% cell death
3 = 75% cell death
4 = 100% cell death

Table VI. Results of selected products effect on the cell line L929 (a)

Diacid	Base	150 ug/ml	100 ug/ml	40 ug/ml	16 ug/ml	8 ug/ml
Tetrephthalic	NaOH	3	2	---	1	0
Terephthalic	Et_3N	2	2	1	1	+
2,2'-Dithiosali- cyclic acid	Et_3N	2	1	±	1	---
Cis-DDP	---	2	2	0	---	---

(a) Same as in Table V.

Table VII. Results of selected products effect on the cell line BHK-21(a)

Diacid	Base	150 ug/ml	100 ug/ml	40 ug/ml	16 ug/ml	8 ug/ml
Terephthalic	NaOH	2	2	2	2	2
Terephthalic	Et_3N	1	1	1	1	1
2,2"-Dithiosali- cyclic acid	Et_3N	1	1	±	±	±

(a) Same as in Table V.

tion to specific biological activities, thus assisting researchers in the design of polymeric drugs.

REFERENCES

1. C. Carraher and T. Brandt, Makramolekulare Chemie, 126, 66 (1969).
2. C. Carraher and D. Winthers, J. Polymer Sci., 7(A1), 2417 (1969).
3. C. Carraher and D. Posey, J. Polymer Sci., 7(A1), 2436 (1969).
4. C. Carraher and P. Billion, Makromolekulare Chemie, 128, 143 (1969).
5. C. Carraher and D. Burger, Makromolekulare Chemie, 138, 59 (1970).
6. C. Carraher, Inorganic Macromol. Revs., 1(4), 287 (1972).
7. C. Carraher and D. Burger, Angew. Makromolekulare Chemie, 46, 73 (1975).
8. C. Carraher, "Interfacial Synthesis", Vol. II, F. Millich and C. Carraher, eds., Dekker, N.Y., 1977, Chpt. 20.
9. C. Carraher and R. Linville, Polymer Reprints, 25(2), 31 (1984).
10. C. Carraher, R. Linville and H. Blaxall, "Crown Ethers and Phase Transfer Catalysis in Polymer Science", L. Mathias and C. Carraher, eds., Chapter 8, Plenum, N.Y. (1984).
11. "Dimethyl Sulfoxide", Technical Bulletin, Crown Zellerback Corp., Camus, WA, (1966).
12. Y. Naoshina, C. Carraher, G. Hess and M. Kurokawa, "Inorganic and Organometallic Polymers", J. Slat, C. Carraher and G. Pittman, eds., Pelnum, N.Y., 1985.
13. Y. Naoshina, G. Hess and C. Carraher, Polymeric Materials, 49, 215 (1983).
14. C. Carraher, L. Torre and H. M. Molloy, "Interfacial Synthesis", Vol. III, C. Carraher and J. Preston, eds., Plenum, N.Y., 1982.
15. C. Carraher and S. Jorgensen, J. Polymer Sci., A-1, 16, 2965 (1978).

ELASTOMERIC POLYPEPTIDE BIOMATERIALS: INTRODUCTION OF HEXAPEPTIDE REPEATS (HARD SEGMENTS) IN THE POLYPENTAPEPTIDE

D. W. Urry, R. D. Harris, H. Sugano,
M. M. Long and K. U. Prasad

Laboratory of Molecular Biophysics
The University of Alabama at Birmingham
P.O. Box 311/University Station
Birmingham, Alabama 35294

A repeating sequence of elastin, the hexapeptide (Val-Ala-Pro-Gly-Val-Gly), has been variously introduced into matrices of the polypentapeptide of elastin, $(Val^1\text{-}Pro^2\text{-}Gly^3\text{-}Val^4\text{-}Gly^5)_n$, to examine a possible structural role of interlocking of hexapeptide segments of one chain with hexapeptide segments of a second chain. The aggregational properties of oligohexapeptides with n = 1,2,3 and 4, of polyhexapeptides with n = 100, of polypentapeptide with n = 200, of 1:2 mixtures of polyhexapeptide with polypentapeptide, respectively, and of cosequential polyhexapentapeptide with a ratio of hexapeptide to pentapeptide of 1:2 were determined. Samples with 60% water by weight were prepared and γ-irradiation crosslinked at 20 MRAD to obtain elastomeric bands. The introduction of hexapeptide, under conditions where hexapeptide-hexapeptide interactions occured, was shown to be able to increase elastic modulus by as much as an order of magnitude. This demonstrated structural role, of the hexapeptide, its previously demonstrated chemotactic role and the effect of introducing occasional lysine residues in place of Val^4 in the polypentapeptide sequence were considered in a general way in discussing the potentials of elastomeric polypeptide biomaterials.

INTRODUCTION

Durability of the Biological Elastic Fiber and the Essentiality of the Polypentapeptide of Elastin

Biological elastic fibers are comprised of a core of fibrous elastin (derived from the crosslinking of a single precursor protein, tropoelastin) and a fine surface layer of microfibrillar protein. Single biological elastic fibers can last the lifetime of an individual (1,2). This means surviving more than 10^9 stretch-relaxation cycles in the aortic arch and descending thoracic where there is by weight twice as much fibrous elastin as collagen (3) and where the elastic fibers must stretch to damp the pressure surge resulting from left ventricular contractions.

It has been demonstrated by Sandberg and colleagues that the most striking structural feature of this protein is a repeating pentapeptide sequence, $(L\cdot Val^1\text{-}L\cdot Pro^2\text{-}Gly^3\text{-}L\cdot Val^4\text{-}Gly^5)_n$, which repeats some 11 times in pig (4) and 13 times in chick (5). In the evolution of warm blooded animals, birds and mammals are believed to have diverged some 200 million years ago. This means that biology, having settled on the polypentapeptide as an essential component of elastic fibers, has not allowed this repeating sequence to vary in 200 million years of evolution.

Additional repeating sequences have been observed which generally occur for less extensive stretches and with less fidelity. These may be listed in part as the polyhexapeptide $(L\cdot Ala^1\text{-}L\cdot Pro^2\text{-}Gly^3\text{-}L\cdot Val^4\text{-}Gly^5\text{-}L\cdot Val^6)_n$ with n = 5 plus (6), the polytetrapeptide $(L\cdot Val^1\text{-}L\cdot Pro^2\text{-}Gly^3\text{-}Gly^4)_n$ with n = 4 (6), the polynonapeptide $(L\cdot Val^1\text{-}L\cdot Pro^2\text{-}Gly^3\text{-}L\cdot Phe^4\text{-}Gly^5\text{-}L\cdot Val^6\text{-}Gly^7\text{-}L\cdot Ala^8\text{-}Gly^9)_n$ with n = 4 (5,7,8) and, unique so far to chick, is an extensive polytripeptide $(L\cdot Val^1\text{-}L\cdot Pro^2\text{-}Gly^3)_n$ with n = 13 (5). In this laboratory we have synthesized and extensively characterized the polypentapeptide (PPP), the polyhexapeptide (PHP) and the polytetrapeptide (PTP). The crosslinked PTP is elastomeric (9) and the PHP forms a more rigid structure (10) and can associate to form a material with cellophane like properties (11). References to the synthesis and characterization of the sequential polypeptides of elastin will appear to be excessively to our own work. This is because, with the exception of two reports from one other laboratory (12,13), there have been no other laboratories preparing and characterizing the sequential polypeptides of elastin. The reason may in part be that the dominant view of the past several decades has been that fibrous elastin functions by virtue of being a random network of chains (22). The focus of this report is on the PPP with introduction of hexapeptide units as means of modifying the properties of this elastomeric polypentapeptide biomaterial.

Similarity of Polypentapeptide and Elastin Protein

When crosslinked, the PPP is elastomeric and, depending on water content and numbers of crosslinks, can exhibit the same elastic (Young's) modulus as fibrous elastin (14-16). The PPP, α-elastin (a 70,000 dalton chemical fragmentation product of fibrous elastin), and tropoelastin (the soluble precursor protein) - all form similar filamentous aggregates on raising the temperature of diluted solutions to the physiological range (16). Both polypentapeptide and α-elastin exhibit similar temperature dependent dielectric relaxation properties in the 1 MHz to 1 GHz range (17,18). Both PPP (19,20) and fibrous elastin (21) exhibit similar rotational correlation times for their mobile backbone moieties as determined by nuclear magnetic resonance relaxation studies. Both crosslinked PPP and fibrous elastin exhibit similar temperature dependence of elastomeric force in thermoelasticity studies (16). Also, as with fibrous elastin (22), the elastomeric force of crosslinked polypentapeptide (23) in the 40 to 60°C temperature range is almost entirely entropic in origin. Thereby it approximates an ideal elastomer wherein a decrease in number of accessible states on extension, rather than a stressing of bonds, is the origin of the elastomeric force.

Temperature Dependence of Physical Properties of the PPP

Fascination with the PPP, which biological evolution has retained for elastic fibers of warm blooded animals, is heightened by the physical properties that it exhibits as a function of temperature (24). Below 25°C, the PPP is miscible in all proportions with water and exhibits limited order (25).

On going from 25° to 40°C, it exhibits a striking phase and structural transition: (a) it aggregates to become 62% water, 38% peptide by weight at 40°C (25). (This process is reversible and is called coacervation); (b) it self-assembles into anisotropic fibers (14,16); (c) it exhibits an inverse temperature transition demonstrable by nuclear magnetic resonance relaxation studies (20); (d) it develops an intense dielectric relaxation centered near 25 MHz (17); (e) it develops a conformation of regularly repeating β-turns as evidenced by circular dichroism (25-27); and (f) it markedly increases elastomeric force as demonstrated by thermoelasticity studies (16-24).

In the 40° to 60°C range, for a given elongation the elastomeric force divided by temperature (°K) and the conformation and water content remain essentially constant (23-25). Above 60°C, the PPP undergoes a very slow structural and compositional transition. Over a time scale of days to months depending on the size of the PPP-water aggregates, one half of the water content is extruded to become 32% water 68% peptide by weight and the structure disorders (25). These temperature dependances are partially summerized by the phase-structure diagram of Figure 1 (24,25).

Furthermore the temperature of the transitions can be altered by changing solvent composition. For example, on using a solvent mixture, 30% ethylene glycol 70% water by volume, the 25° to 40°C transition is shifted to lower temperatures initiating at 12°C (28) and for fibrous elastin the elastomeric force divided by temperature in a thermoelasticity study becomes constant starting at a lower temperature (22). Also analogs of the PPP not accessible to DNA directed syntheses (e.g. Gly replaced by D.Ala) variously can alter the temperature of the transitions, can increase the elastic modulus of the material (D.Ala^3-PPP) (29)

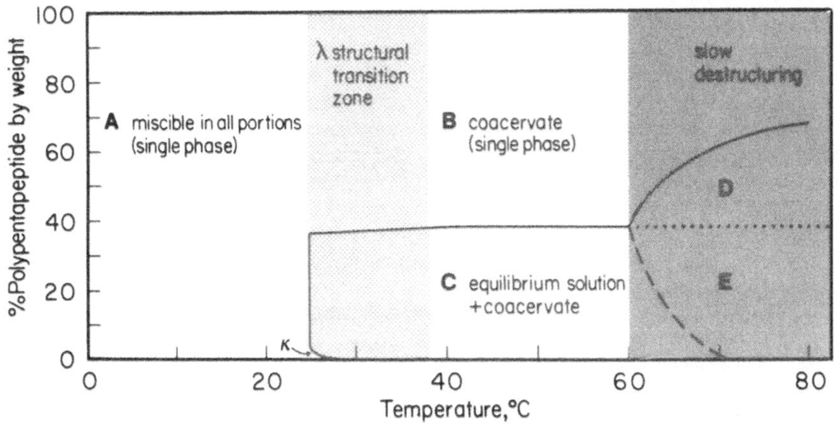

Figure 1. Phase-structure diagram of the polypentapeptide water system. In region A, below 25°C, water and polypentapeptide are miscible in all proportions. Above 25°C when there is less than 37% peptide by weight, a two phase region (c) occurs which is a mixture of a coacervate phase and an aqueous phase (equilibrium solution). The composition of the coacervate is given by the boundary between region B and regions C and D. Above 60°C there is an expulsion of water from the coacervate with a slow destructuring of the polypentapeptide which is a β-spiral in the coacervate state. Adapted with permission from reference (24).

and can decrease the intensity of the dielectric relaxation (D.Ala^5-PPP) (28). Thus the polypentapeptide is a novel elastomeric polymer that undergoes interesting inverse temperature transitions and its properties can be modified by solvent or by carefully selected analogs.

Development of Molecular Structure and a New Basis for Entropic Elastomeric Force

How can this non-random sequential polypeptide be an entropic elastomer? It has been demonstrated by the studies noted above that the order of the PPP increases intermolecularly and intramolecularly as the temperature increases from 20° to 40°C, yet this non-random construct is an entropic elastomer (16,24). Previously findings of a dominantly entropic elastomer have been the basis for concluding a random network of chains as the responsible molecular structure (22). In a departure from this previous perspective, detailed conformational studies on linear and cyclic oligomers of the pentapeptide and on the high molecular weight polypentapeptide (30-33) have resulted in the development of a structure with features required to explain the above properties and have resulted in the proposal of a librational entropy mechanism of elasticity (34).

The derivation of the dynamic β-spiral of the PPP is reviewed elsewhere (16). What follows in this introduction is a brief description. The proposed β-spiral structure is comprised of regularly recurring Type II Pro^2-Gly^3 β-turns (see Figure 2A) which repeat on a helical axis. This is shown schematically in Figure 2B and C and shown in atomic detail in Figure 2D and E. As is apparent in the stereo axis view in Figure 2D, there is space within the loose β-spiral for water. This then would be the water extruded on slow destructuring above 60°C. On raising the temperature above 25°C the chains wrap up into the β-spiral on optimization of interturn hydrophobic contacts and β-spirals associate with supercoiling to form the twisted filaments (see Figures 2F and G) seen in the transmission electron microscope with negative staining (see Figure 4 of reference 16) and these further associate to form the fibrils and fibers of Figure 5 of reference 16. Of special relevance to the elastic mechanism is the L-Val^4-Gly^5-L-Val^1 suspended segment which is most easily seen in Figure 2D. The peptide moieties within this segment are particularly free to rock or to librate with large correlated changes in the flanking i, i + 1 torsion angles. This librational motion provides much configurational entropy to the structure. On stretching this peptide, the librational process is damped (see Figure 15 of reference 16) causing the stretched chain to be at lower entropy. The restoring force of the elastomeric reaction is the lower free energy of the higher entropy, unstretched state. This has been termed the librational entropy mechanism of elasticity.

The dielectic relaxation studies (17) were actually a test of the structure to see if the proposed peptide librational processes ascribed to the dynamic polypentapeptide β-spiral might be observed; the L-Ala^5-PPP (35), the D-Ala^3-PPP (29) and the D-Ala^5-PPP (36) analogs were synthesized and characterized to test the proposed β-spiral with its dynamic suspended segment. These studies have thus far, strengthened the case for the proposed structure and for the perspective utilized here. This first elucidation of a detailed mechanism of biological elasticity demonstrates the requirement of a regular conformation with appropriate dynamic properties and it, therefore, provides the basis and the opportunity to vary systematically the structure to obtain a variety of new biomaterials - a new class of polymers based on a new structural motif.

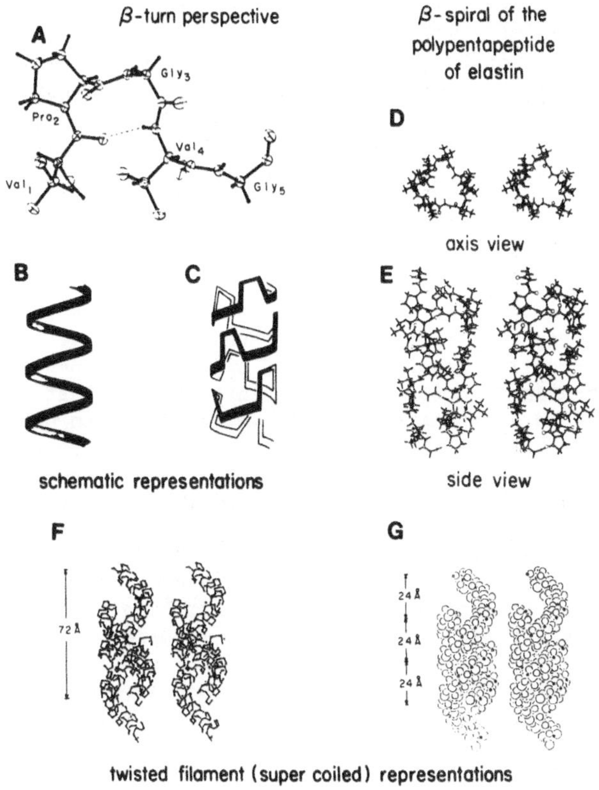

Figure 2. Molecular structure of the polypentapeptide. (A) The β-turn secondary structural fracture with a hydrogen bond between the Val^1 C-O and the Val^4 NH and with Pro^2-Gly^3 at the corners. The orientation of the peptide moiety between residues 2 and 3 is that of a Type II β-turn. (B) Schematic spiral (helical) representation which is modified in (C) to show the β-turns functioning as spacers between the turns of the helix. (D) and (E) stereo perspectives in atomic detail of the β-spiral of the polypentapeptide of elastin. (D) is the spiral axis view showing space for water within the spiral and showing peptide segments, Val^4-Gly^5-Val^1, suspended between the β-turns. It is these suspended segments that have much freedom for the peptide moieties to undergo large rocking motions i.e. to librate. This is the source of the entropy which provides for the favorable free energy of the unstretched state. (E) is a stereo pair side view of the β-spiral showing the β-turns functioning as spacers between turns of the spiral by means of non-restricting hydrophobic contacts. (F) is a stereo pair of three β-spirals, in α-carbon representation, super-coiling to form twisted filaments and resulting in a structure that closely matches the twisted filaments seen in transmission electron micrographs with negative staining. (G) is the same twisted filament as in F but with spheres of different sizes at the α-carbon centers. This allows better viewing of the twisted filaments. (A) adapted with permission from reference 30, (B), (C) and (D) are reproduced with permission from D. W. Urry, Ultrastruct. Pathol., 4, 227-251 (1983)., (E) is from reference 33, and (F) and (G) are from reference 34.

Introducing the Use of Hard and Soft Segments in Analogy to Segmented Polyurethanes

When the polyhexapeptide (PHP), specifically $(VAPGVG)_n$, is dissolved in water at 4°C and the temperature is raised, aggregation occurs over a relatively narrow temperature range and that range shifts to lower temperatures as the concentration is raised (37). The aggregation of the PHP in contrast to that of the PPP is irreversible in water. The aggregation can be redissolved in trifluoroethenol-water mixtures and lyophilized to regain water solubility, that is, the heat elicited aggregation of PHP is not a truly irreversible process. This is thought to be due to the more rigid structure of the PHP (10) wherein on association there is an interlocking of hydrophobic ridges (37). Thus in a cosequential polypeptide comprised of repeating hexapeptides and of repeating pentapeptides, could it occur on heating (to form the associated state) that the hexapeptide repeats would tend to associate selectively, i.e. to cluster, and that these clusters of stiffer hexapeptide units would be separated by softer polypentapeptide segments? This possibility, which would be quite analogous to the hard and soft segments of segmented polyurethanes (38,39), is explored in the present report.

The approach is (a) to prepare and examine oligohexapeptides, with n = 1,2,3, and 4 to determine their heat elicited aggregational properties, (b) to mix PHP and PPP and carry out temperature profile of turbidity studies to see if there is evidence for selective self association, (c) to crosslink a PHP plus PPP mixture to determine if an interesting elastomer is produced, (d) to prepare γ-irradiation crosslinked cosequential polypeptides comprised of randomly incorporated hexapeptide and pentapeptide units and (e) to determine if the crosslinked cosequential polypeptide exhibits altered elastomeric properties on heating at temperatures sufficient to drive the hexapeptide hexapeptide interactions.

EXPERIMENTAL DETAILS

Peptide Synthesis

The synthesis of the sequential polypeptides of the hexapeptide sequence, Boc-$(VAPGVG)_n$-OCH_3 where n = 2, 3, and 4 was carried out by classical solution methods as shown in Scheme I.

Boc-VAPGVG-OCH_3 (I) prepared as described earlier (40), was deblocked with TFA and coupled with Boc-VAPGVG-ONp (II) in the presence of HOBt to obtain Boc-$(VAPGVG)_2$-OCH_3 (III). After deblocking III, it was coupled with Boc-VAPGVG-OH (IV) using the water soluble carbodiimide, EDCI (41), to obtain the next higher homologue V. Similarly Boc-$(VAPGVG)_4$-OCH_3 (VI)

Scheme I. Synthesis of the Sequential Polypeptides of the Hexamer Sequence

```
                 (a) TFA                                  (A) TFA
Boc-VAPGVG-OCH3---------------> Boc-(VAPGVG)2-OCH3-------------------->
     (I)     (b) Boc-VAPGVG-ONp      (III)          (b) Boc-VAPGVG-OH (IV)
                      (II)                                  & EDCI

                    (a) TFA
Boc-(VAPGVG)3-OCH3----------------->Boc-(VAPGVG)4-OCH3
       (V)          (b) (IV) & EDCI        (VI)
```

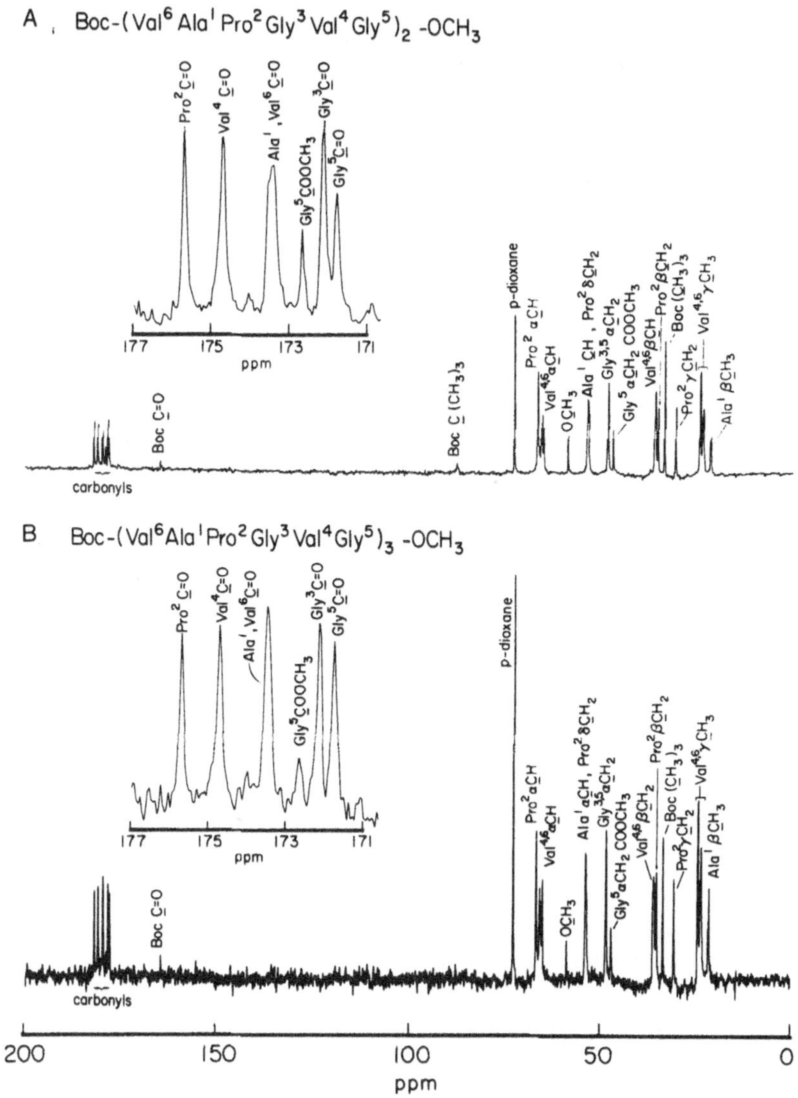

Figure 3. Carbon-13 nuclear magnetic resonance spectra in water of oligohexapeptides, Boc-(Val^{6}-Ala^{1}-Pro^{2}-Gly^{3}-Val^{4}-Gly^{5})$_n$-OCH$_3$ with n = 2, part A, and n = 3, part B. All the resonance are assigned. This and the absence of extraneous resonances is verification of the synthesis. The p-dioxane is added for reference.

was obtained. As a demonstration of the syntheses and verification of purity, the carbon-13 nuclear magnetic resonance spectra are given in Figure 3 for the oligohexapeptides with n = 2 and 3.

The synthesis of the polyhexapeptide, H-(VAPGVG)-V-OCH$_3$ was carried out previously (40) starting from the monomer sequence H-VAPGVG-ONp. Recently, it was observed in the case of the preparation of the PPP, that with the permutation where Pro is present as the C-terminal amino acid in the monomer sequence, as H-GVGVP-ONp instead of H-VPGVG-ONp, a very high molecular weight polymer (about 200,000 dalton) could be obtained (42). Hence a similar approach was taken for the preparation of the polyhexapeptide with Pro as the C-terminal amino acid in the monomer sequence. The synthesis of poly(GVGVAP)-OH is shown in Scheme II. The product may, of course, also be written H-GVG(VAPGVG)$_n$VAP-OH.

Scheme II. Synthesis of H-(GVGVAP)$_n$-OH

```
        Gly      Val       Gly       Val       Ala       Pro
         |        |         |         |         |         |
         |        |         |         |  Boc---|--Oh H---|---OBzl
         |        |         |         |         |         |   i
         |        |         |         |  Boc---|---------|---OBzl (VII)
         |        |         |         |         |         |   i
         |        |         |  Boc--|--OH  H--|---------|---OBzl
         |        |         |         |         |         |   iii
         |        |         |  Boc--|---------|---------|---OBzl (VIII)
         |        |         |         |         |         |   ii
    Boc--|--------|---------|--OH  H--|---------|---------|---OBzl
         |        | (IX)    |         |         |         |   iv
    Boc--|--------|---------|---------|---------|---------|---OBzl (X)
         |        |         |         |         |         |   v
    Boc--|--------|---------|---------|---------|---------|---OH
         |        |         |         |         |         |   vi
    Boc--|--------|---------|---------|---------|---------|---ONp (XI)
         |        |         |         |         |         |   vii
      H--|--------|---------|---------|---------|---------|---ONp
         |        |         |         |         |         |   viii
    H- (---------------------------------------------------)n-OH (XII)
```

(i) IBCF-HOBt; (ii) HCl/Dioxane; (iii) IBCF; (iv) EDCI-HOBt; (v) H_2-Pd/C; (vi) Bis(p-nitrophenyl)carbonate; (vii) TFA; (viii) DMSO-NMM

Boc-AP-OBzl (VII) was prepared by the mixed anhydride method (43) in the presence of HOBt (44). After deblocking, the coupling was carried out with Boc-Val-OH to obtain the tripeptide benzyl ester (VIII) which was deblocked and coupled with Boc-GVG-OH (IX) (44) to obtain the hexapeptide, X. Following hydrogenolysis, the peptide acid was converted to the p-nitrophenyl ester (XI) (45) using bis(p-nitrophenyl)carbonate (46). After removal of the Boc group with TFA, a one molar solution of the active ester in DMSO was polymerized for 14 days in the presence of 1.6 equiv. of NMM. On diluting with water the PHP was dialyzed against water, first using 3,500 molecular weight cut-off dialysis tubing for 7 days and then 50,000 M.Wt. cut-off tubing for 7 more days and finally the retentate was lyophilized.

<u>Random Copolymerization of Hexa and Pentapeptides</u>

The active p-nitrophenyl esters of the hexapeptide, H-GVGVAP-ONp, and the pentapeptide, H-GVGVP-ONp (42), were mixed in the desired ratio (1:2) and polymerized in the usual way. The product is a sequential polypeptide in which in a single chain there is presumably a random mixture of hexapeptides and pentapeptides in a 1:2 ratio. This product will be called a cosequential polyhexapentapeptide and will be abbreviated as P(HP)P. The purity of the intermediate and final compounds was checked by thin layer chromatography, elemental analyses and nuclear magnetic reso-

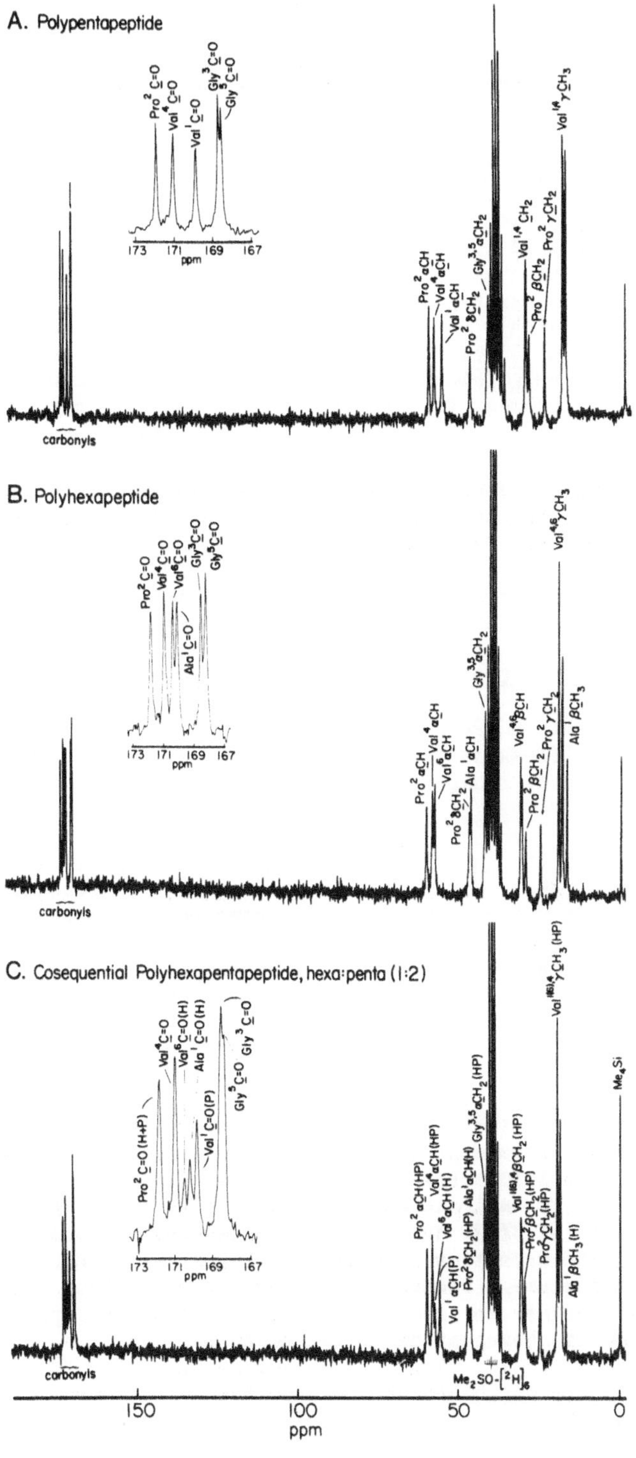

Figure 4. Carbon-13 nuclear magnetic resonance spectra in dimethyl sulfoxide of (A) the polypentapeptide, (B) the polyhexapeptide and (C) the cosequential polyhexapentapeptide with a hexa to penta ratio of 1:2 in a random copolymer. The assignment of all resonances and the absence of extraneous peaks serves as verification of the syntheses.

nance. Demonstration of synthesis and of purity of products is given by the carbon-13 nuclear magnetic resonance spectra in dimethylsulfoxide (DMSO) of the polyhexapeptide, PHP, the cosequential polyhexapentapeptide, P(HP)P, and for comparison of the polypentapeptide, PPP, in Figure 4.

Temperature Profiles of Aggregation

Temperature profiles of aggregation (turbidity profiles) were obtained with a Cary 14 spectrophotometer set at 300 nm to follow light scattering as function of increasing temperature. The heating rate was 30°C per hour and to insure appropriate mixing of the polymers the sample cell was equipped with a 20KHz vibrator. Samples were run in duplicate with fresh material used each time in a 1 mm rectangular cell with a total volume of 0.3 ml. This data is expressed as normalized turbidity to correct for differences in total turbidity due to different sample concentration.

Reversibility of the polypeptides was checked immediately after each temperture run by exposing the sample to an ice bath for 30 minutes. If the aggregation was not reversibile, the sample was further incubated at 4°C for 16 hours. For the oligohexapeptides, reversibility was checked after the sample had been taken to 72°C where n = 3 and to 60°C where n = 4 and then when both had been taken to 80°C.

Preparation of Polyhexapeptide Sheet from Trifluoroethanol

The polyhexapeptide (PHP) was evaporated from trifluoroethanol (TFE) in a round bottom flask after which ether was added, and the resulting PHP sheet was removed. This material was cut into strips for the stress-strain studies.

Preparation of the γ-irradiated crosslinked elastomers: A quantity of 120 mg of dry sample was placed in the bottom of a cryotube to which 204 μl of distilled water were added. The samples were allowed to hydrate at 4°C to form a viscoelastic mass. In each tube a Plexiglas pestle was then inserted which had a 1 mm x 10 mm channel turned in it. As the pestle was inserted, the viscoelastic sample flowed into and filled the channel to form a continuous band with dimensions of 28 mm x 10 mm x 1 mm. The P(HP)P samples within the channel were held at 65°C for 17 hours with slow rotation of the pestle. The samples were then placed in dry ice and taken to the γ-irradiation source. Irradiation crosslinking of the samples was carried out at the Auburn University Nuclear Science Center. The samples were placed at the center of a group of cobalt sources arranged in concentric circles. The configuration gives approximately 0.45 MRAD per hour gamma irradiation. The samples were exposed for 44.19 hours to give a total of 20 MRADs. The elastomeric bands are referred to as X^{20}-PPP for the crosslinked polypentapeptide, X^{20}-(PHP+PPP) for the crosslinked 1:2 mixture, respectively, of polyhexapeptide and polypentapeptide and X^{20}-P(HP)P for the cosequential polyhexapentapeptide with a random mixture of hexapeptide to pentapeptide (1:2) in the original polymerization.

Stress-Strain Studies

A strip of the polyhexapeptide sheet measuring 24 mm x 4.64 mm x 0.28 mm was clamped in the stress-strain apparatus with the distance between the grips being 10 mm. Elongation of the sample was done at a rate of 1.1 mm/min. The axis of the plotter recording the displacement of the moving grip was calibrated to 25.4 mm pen movement for a 0.1 mm grip displacement. The Statham UC-2 transducer with a UL4-2 load cell accessory used in our apparatus has a sensing element displacement of 2.15 x 10^{-4} mm/gram of applied force. This correction was applied in calculating the elastic (Young's) modulus. The cross section of each of the γ-irradiation crosslinked samples was measured prior to and after completion of each run. The elastic modulus of each sample was measured at 40°C in water. The crosslinked cosequential polypeptide, X^{20}-P(HP)P, samples were then allowed to set for 12 hours at 65°C and again the elastic modulus was measured at 40°C. The X^{20}-P(HP)P samples were subsequently swollen at 20°C in water for five hours, heated at 80°C in water for 17 hours and then the elastic moduli were again determined. The samples examined for their elastic moduli are shown in Figure 5. During γ-irradiation crosslinking of samples in Figures 5B, C, and D vacuoles formed, the origin of which appeared to be the cryotube as no such quantity of vacuoles formed when using a glass container. When elastic moduli were determined no correction was made for the cross sectional area of the vacuoles such that the elastic moduli are minimal values and can generally be considered to be about twice the value reported for the X^{20}-PPP and X^{20}-P(HP)P samples. The important element for the latter sample is the increase in elastic modulus on heat treatment at 65°C and 80°C.

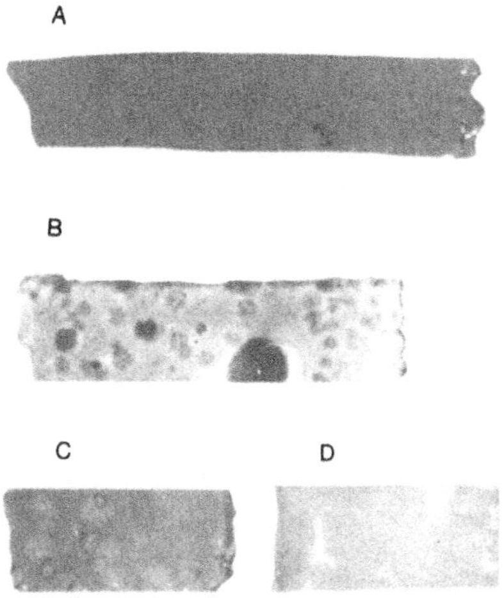

Figure 5. Photographs of the matrices characterized in Figure 7 by stress-strain studies. (A) strip of polyhexapeptide sheet from trifluoroethanol. (B) polypentapeptide coacervate crosslinked by 20 MRAD of γ-irradiation, X^{20}-PPP. (C) cosequential polyhexapentapeptide with a ratio of hexa to penta of 1:2 after crosslinking at 20 MRAD. This is designated as X^{20}-P(HP)P. (D) polyhexapeptide plus polypentapeptide at a weight ratio of 1:2 after crosslinking at 20 MRAD, designated as X^{20}Pr-(PHP+PPP). This product is an elastomeric matrix with a teflon like texture and with an elastic modulus ten times greater than obtained for polypentapeptide alone when prepared under identical conditions.

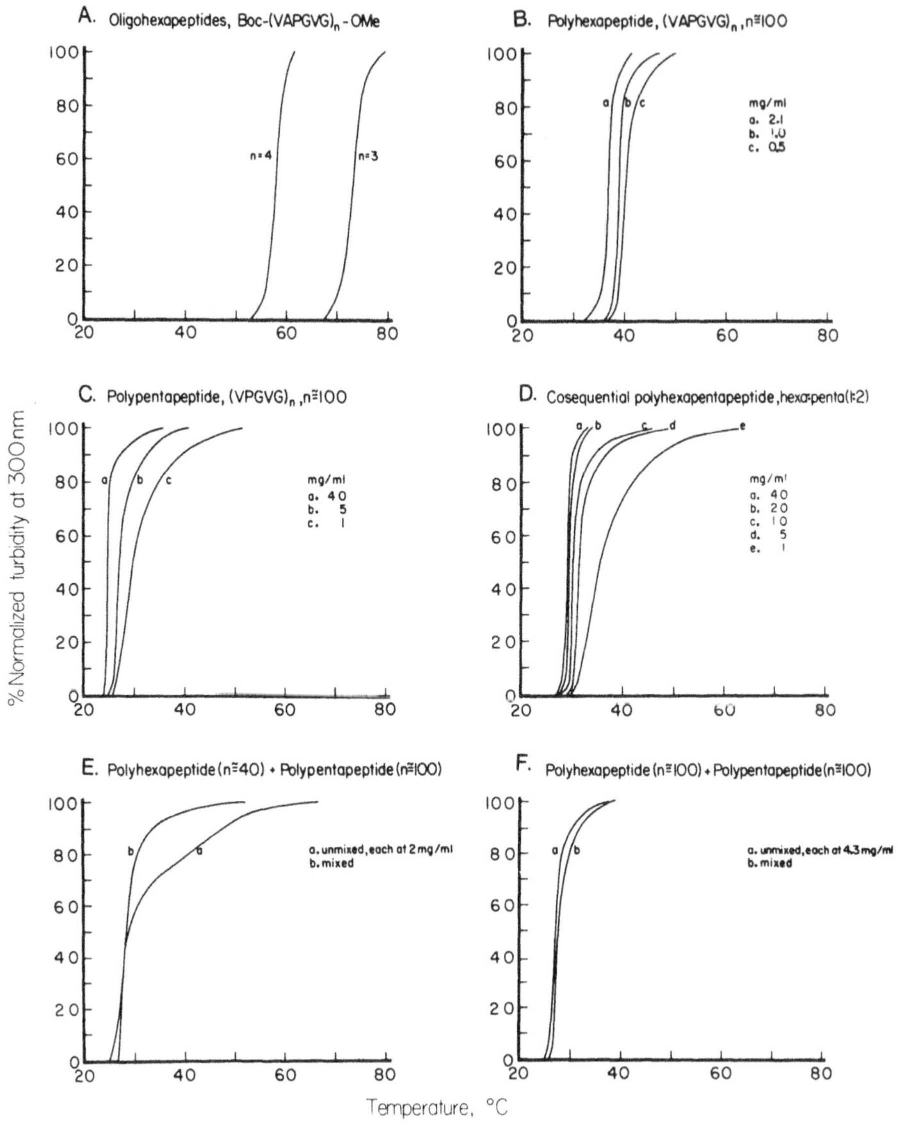

Figure 6. Temperature dependence of aggregation (turbidity profiles) for (A) the oligohexapeptides with n = 3 and 4, (B) the polyhexapeptides (PHP), (C) the polypentapeptide (PPP), (D) the cosequential polyhexapentapeptide, P(HP)P <with a ratio of hexa to penta of 1:2), (E) the mixture of polyhexapeptide (n = 40) plus polypentapeptide (n = 200) and (F) the mixture of polyhexapeptide (n = 100) plus polypentapeptide (n = 200).

RESULTS

Aggregational Properties of Oligohexapeptides

Temperature profiles for association of the oligohexapeptides are included in Figure 6A. When the hexapeptide (n = 1) and the dodecapeptide (n = 2) were examined, they exhibited little if any aggregation on raising the temperature to 80°C. When the octadecapeptide (n = 3) was examined, however, it gave a temperature profile for turbidity with a midpoint just above 70°C. On immediately lowering the temperature the aggregates redissolved, but if the temperature were raised to 80°C and held there for a short period, the association in water did not reverse on standing overnight at 4°C. When the tetraeicosapeptide (n = 4) was examined, the midpoint for the turbidity profile was shifted to a lower temperature, 57°C. This aggregation was found to be irreversible in water. Thus for the oligohexapeptides a repeating sequence of three to four hexapeptides is required before the polyhexapeptide-polyhexapeptide interaction is irreversible. To achieve irreversibility with heating at temperatures less than 80°C an n value of four is required. When considering cosequential polypeptides containing hexapeptide and pentapeptide repeating sequences, the results that there should be continuous runs of uninterrupted pentapeptides in order to have the hexapeptide function in an interlocking role (as in a hard segment of segmented polyurethanes).

The turbidity profiles for high molecular weight polyhexapeptide (n = 100) are shown for several concentrations in Figure 6B. This polyhexapeptide-polyhexapeptide interaction is irreversible in water, and dissolution in trifluoroethanol and lyophilization with increasing amounts of water is required to again attain solubility in water at low temperatures. The turbidity profiles for a series of concentrations of high molecular weight polypentapeptide (n = 200) are given in Figure 6C. This aggregation is immediately reversible on lowering the temperature. In an initial test as to whether hexapeptide sequences might function in an interlocking role when randomly interspersed in a polypeptide chain comprised of mostly pentapeptide, the random cosequential polyhexapentapeptide, P(HP)P, was prepared as outlined in the experimental section. The turbidity profiles for P(HP)P as a function of concentration are given in Figure 6D. The aggregates, formed by heating this random copolymer of hexapeptide and pentapeptide with a ratio of 1:2, respectively, immediately redissolved on lowering the temperature below 25°C. If there are in the P(HP)P sufficiently long runs of uninterrupted repeating hexapeptides to effect some interlocking at elevated temperatures, these associations are separated by the hydration (swelling) of pentamer sequences on lowering the temperature. Whether such association might be present at 40°C can be tested in the crosslinked samples by the stress-strain studies reported below.

Before proceeding to the stress strain studies, however, effects on aggregation of combining separate chains of polyhexapeptide and polypentapeptide are of interest. When a lower molecular weight polyhexapeptide with n = 40 is placed in one chamber of a tandem cell and polypentapeptide in the other, the temperature profile of Figure 6E curve a is obtained. The delayed turbidity change is due to aggregation of the lower molecular weight polyhexapeptide. On mixing the separately aggregated PHP and PPP, the PHP aggregation is not reversed on standing at 4°C for more than 12 hours. If however the lower molecular weight PHP and PPP are mixed at low temperature as solutions, then the turbidity profile of Figure 6E curve b is obtained and on immediately lowering the temperature all aggregates redissolve. This indicates that there is significant PHP-PPP association on raising the temperature. When higher molecular weight

PHP (n= 100) and PPP (n = 200) are combined as solutions at low temperature, the turbidity profile of Figure 6F is obtained. This mixed aggregation does not redissolve on standing at 4°C for several days. This means that there also can occur sufficient PHP-PHP interaction in the presence of PPP to achieve irreversibility i.e. interlocking. With these mixed results, it will be of interest to see the properties of a -irradiation crosslinked mixture of PHP + PPP at a ratio of 1:2, respectively, with the mixture being dissolved at 0°C and with shearing by means of the pestle and cryotube carried out in an ice bath prior to crosslinking by γ-irradiation.

Stress Strain Studies

With 20 MRAD γ-irradiation of polypentapeptide carried out in the same manner as with the hexapeptide containing samples, the elastic modulus of the crosslinked polypentapeptide, X^{20}-PPP, was in the range of 2 to 3 x 10^5 dynes/cm^2 with the measurement taking place in water at 40°C. A stress strain curve is shown in Figure 7A. This reference value provides an opportunity to look for possible enhancement of elastic modulus due to inclusion of hexapeptide in various ways. As a reference for the cellophane like polyhexapeptide, a sheet of PHP was cast from a high concentration in trifluoroethanol. The elastic modulus of the strip shown in Figure 5A was found to be 7 x 10^9 dynes/cm^2. This stress strain curve is shown in Figure 7B. A representative sample of cosequential polyhexapentapeptide, crosslinked by 20 MRAD -irradiation, i.e. X^{20}-P(HP)P, gave an initial value of 2.2 x 10^5 dynes/cm^2. On heating in water at 65°C for 12 hours and re-examining at 40°C in water, the elastic modulus had increased to 4.2 x 10^5 dynes/cm^2. The sample was then kept at 20°C in water for 5 hours and then heated at 80°C for 17 hours. The resulting elastic modulus at 40°C in water had decreased to 3.3 x 10^5 dynes.cm^2. These curves are given in Figure 7C. Averaged values for three samples were 1.9 x 10^5 dynes/cm^2, 4.2 x 10^5 dynes/cm^2 and 3.4 x 10^5 dynes/cm^2; these are for the initial values, the value after heat treatment at 65°C and the value after heating at 80°C, respectively. The rationale for the choice of temperatures for the heat treatments derives from the turbidity profiles in Figure 6A for the oligohexapeptides. A temperature of 65°C resulted in irreversible association of oligohexapeptides with n = 4. It would appear from the near doubling of elastic modulus on heating at 65°C that some association of hexapeptide units did occur which functioned like additional crosslinks. Irreversible association of oligohexapeptides with n = 3 did not occur until 80°C but association of P(HP)P was completely reversed on going below 25°C (see results of aggregation studies above). On this basis the X^{20}-P(HP)P, that had been heated at 65°C, was allowed to swell at 20°C for 5 hours. In analogy to the reversible aggregation of P(HP)P, this is expected to reverse all hexapeptide hexapeptide interactions. The samples were then heated at 80°C to see if additional putative (hexapeptide)$_n$-(hexapeptide)$_n$ associations might occur. This treatment caused a decrease in elastic modulus from what had been obtained after 65°C treatment. While association of repeating hexapeptides with n = 3 could possibly have been achieved, it is known that holding the temperature of PPP coacervate at 80°C results in a very slow destructuring of the β-turns in the polypentapeptide (25). This and even the heat destructuring of the conformations of hexapeptide repeats within the cosequential polypentapeptide are possible explanations of the decrease in elastic modulus resulting from the 80°C treatment.

A particularly interesting result is due to the elastomeric matrix obtained on mixing polyhexapeptide with polypentapeptide in a 1:2 ratio. These polymers were codissolved in a minimal amount of water (about 60% by weight) at 4°C in the cryotube. The pestle was then inserted causing the viscoelastic solution to flow into the channel turned in the pestle. The pestle was rotated with the sample in an ice bath for several hours and the sample was then placed in dry ice until γ-irradiation at 20 MRAD. The product can be referred to as X^{20}-(PHP+PPP). The resulting band of material is seen in Figure 5D and its stress strain curve is given in Figure 7D. The elastic modulus at between 50 to 60% extension was 2.6×10^6 dynes/cm^2. This value is essentially identical to that of natural fibrous elastin (14,16) which similarly exhibits a lower initial slope. As seen in Figure 7D, there is a marked reproducible hysteresis which is a result of the slow rate of retraction of the sample. The rate of change in length was approximately 1 mm/min.

DISCUSSION

The importance of elastic tissue in man can be demonstrated by the following observations. As noted in the introduction, it is presence of

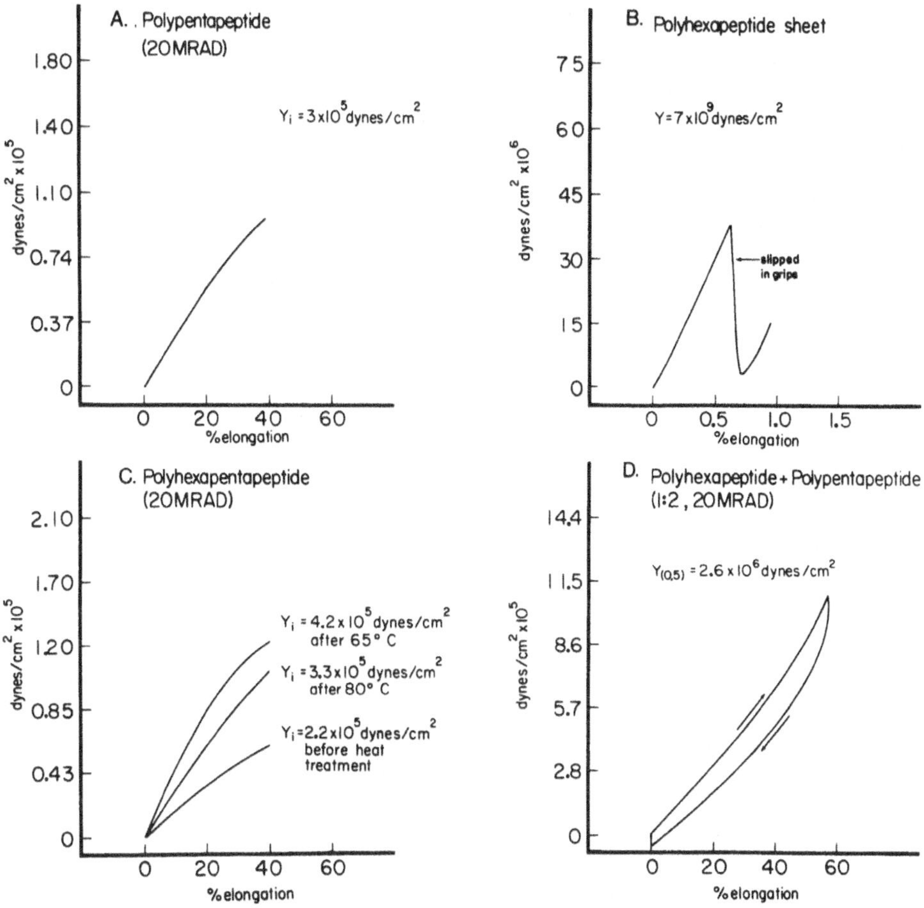

Figure 7. Stress strain characterization of (A) the polypentapeptide crosslinked by 20 MRAD of γ-irradiation, X^{20}-PPP, (B) the polyhexapeptide sheet cast from trifluoroethanol, (C) the cosequential polyhexapentapeptide, X^{20}-P(HP)P, and (D) the 1:2 mixture of polyhexapeptide plus polypentapeptide, X^{20}-(PHP+PPP).

twice as much elastin as collagen in the aortic arch and thoracic aorta that provides the compliance necessary to accomodate the blood volume resulting from left ventricular contraction without excessive pressure surge; it is the high elasticity of ligaments, being some 90% elastic fiber by weight, that allows repeated flexing joint movements; it is the elastic fibers of skin that, as joints bend, allow the skin to stretch and to recover without folding like an accordion; and it is the absence of functional elastic fibers in scar tissue that results in the often debilitating rigidity of scars. It seems apparent that a favorable situation for prostheses would be to match the elasticity (compliance) of the tissue it replaces or with which it interfaces. If this is the case, then in the area of biomedical polymers there is much to learn from the biological elastic fiber, that is, from the nature of protein and polypeptide elasticity.

In regard to the elasticity of the most striking primary structural feature of elastin, the polypentapeptide, the proposed structural motif and librational entropy mechanism were outlined in the Introduction. It is further noted that the heat elicited association of polyhexapeptide with polyhexapeptide in water was irreversible in the practical sense and that there is about half as much polyhexapeptide as polypentapeptide. With n just greater than five for the repeating hexapeptide in the precursor protein, the repeating hexapeptide sequence has been suggested as assisting in the chain alignment process prior to crosslinking by formation of interlocking hydrophobic ridges on association of polyhexapeptide in one chain with polyhexapeptide in a second chain. The derived polyhexapeptide conformation which gave rise to the concept of interlocking hydrophobic ridges is shown in Figure 8. Associations which in the synthetic polyhexapeptide can result in formation of cellophane like sheets would result in relatively stiff domains in the elastic fiber not unlike the hard segments of segmented polyurethanes and, of course, the polypentapeptide stretches would be analogous to the soft segments. Thus, it is not unlikely that biology has already utilized the concept of hard and soft segments in the development of its elastomeric fibers. It is that aspect which this report considers in a primary way. Specifically can polyhexapeptide be introduced into dominantly polypentapeptide matrices to modulate mechanical properties, for example, to increase elastic moduli?

The preliminary studies reported here demonstrate quite clearly that polyhexapeptide-polyhexapeptide interactions occur within a dominantly polypentapeptide matrix. This is demonstrated by an increased elastic modulus which is consistent with a hard segment role in the synthetic elastomeric polypeptide material. The results are also consistent with the previously proposed structural role in the elastic fiber where the repeating hexapeptides with n = 5 were suggested to interlock by means of interdigitating hydrophobic ridges (37). The results now point to the preparation of polypentapeptide containing predetermined runs of repeating hexapeptide with the general formula $[(VAPGVG)_l(VPGVG)_m]_n$. It would appear that l should be 3 or greater whereas the values for m could be in the range of 10 to 15. For the most satisfactory results when using direct γ-irradiation for crosslinking, n would be of a magnitude such that a single cosequential polypeptide would have as many as 1000 residues.

Interestingly the hexapeptide, which has been implicated in this report as having a useful structural role of interlocking chains, has been previously shown to be chemotactic toward fibroblasts (47). The chemotaxis toward fibroblasts appears to be specific for those fibroblasts which have differentiated for elastin synthesis. This suggests a scenario of fiber rupture, and partial enzymatic degradation effecting

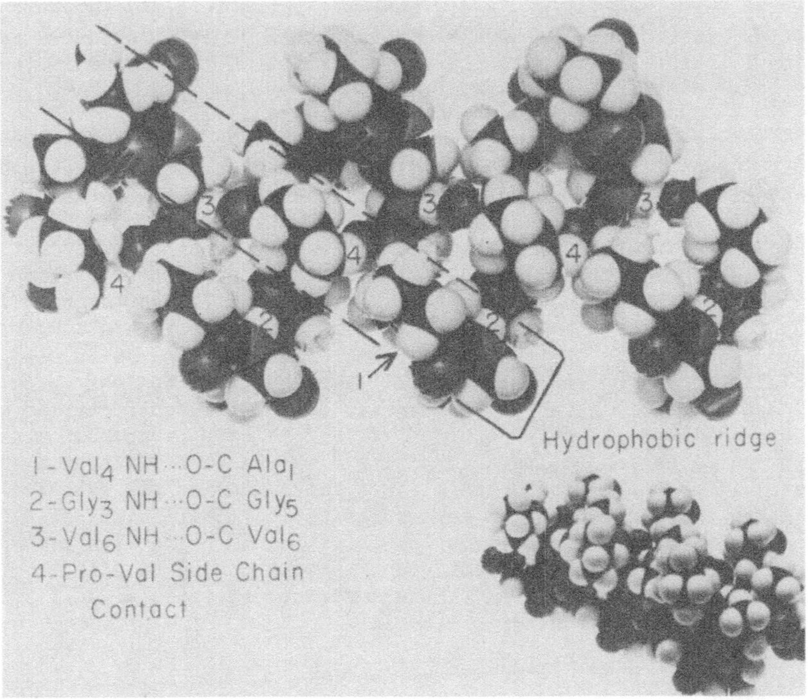

Figure 8. Proposed β-spiral conformation shown by space filling models of the polyhexapeptide of elastin. Three intramolecular hydrogen bonds are noted as well as an intramolecular hydrophobic contact. The inset at lower right is from the perspective of the hydrophobic ridges one of which is outlined in the upper view. The proposal is that as the temperature is raised in water the molecules optimize intermolecular hydrophobic interactions by means of an interlocking of these hydrophobic ridges. The view is that on lowering the temperature there is not sufficient kinetic energy to dislodge the interlocked ridges. Once developed these hexapeptide-hexapeptide intermolecular interactions form domains functionally much like the hard segments of segmental polyurethanes. These domains are effectively crosslinks that, once formed, increase the elastic modulus of the elastomeric polypeptide biomaterial.

release of hexapeptides, followed by migration of fibroblasts to the site of damage and repair of the damaged elastic fiber. A related aspect of the biomedical use of these elastomeric polypeptide biomaterials is that polypentapeptide containing an occasional lysine in place of Val^4 can function as substrate for the extracellular crosslinking enzyme, lysyl oxidase, and that the action of lysyl oxidase can result in the formation of natural crosslinks (48). Accordingly it should be possible to prepare the elastomeric polypeptide biomaterial which would elicit significant tissue responses. The presence of hexapeptide within the synthetic biomaterial could, on slow enzyme degradation, result in hexapeptide release which would induce migration of elastin synthesizing fibroblasts to the lysine containing biomaterial, and the newly synthesized elastin, released by the fibroblasts, could become covalently crosslinked to the synthetic biomaterial by means of the action of lysyl oxidase on the occasional lysine residues in the polypentapeptide and on the lysines in biosynthetic elastin.

Preliminary biocompatibility tests have been conducted in rabbits on the polypentapeptide and on γ-irradiation crosslinked polypentapeptide. The material was found to be biocompatible and to be biodegradable (49). Many issues and perspectives arise. One question would be whether the presence of hexapeptide and lysine could result in an elastin coating on the synthetic biomaterial which would provide a protective coating that would prevent further enzymatic degradation. But is slow degradation a problem or under certain circumstances might it be desirable? For vascular prostheses, if the synthetic biomaterial were to be a temporary scaffolding on which new vascular wall were built, then the prosthesis would need to retain structural integrity only until restructuring of the wall were achieved. In this regard S.-T. Li, at this symposium, presented results on a Dacron/collagen composite 6 mm vascular graft as a canine implant. After nine months there was a restructuring of vascular wall with neointimal formation and revascularization (50). With elastomeric polypeptide biomaterials there is the potential to begin with a compliant vascular prosthesis which, with, appropriate doping to elicit desired tissue responses, could result in a restructured vascular wall with no identifiable foreign prosthetic material.

ABBREVIATIONS

Boc, tert-butyloxcarbonyl; Bzl, benzyl; DMSO, dimethylsulfoxide; EDCI, 1-(3-dimethylaminopropyl)-3-ethylcarbodiimide; HOBt, 1-hydroxybenzotriazole; IBCF, isobutylchloroformate; NMM, N-methylmorpholine; ONp, p-nitrophenylester; TFA, trifluoracetic acid; A, alanine; G, glycine; V, valine; P, proline; PPP, polypentapeptide $(VPGVG)_n$; PHP, polyhexapeptide $(VAPGVG)_n$ and P(HP)P, cosequential polyhexapentapeptide.

ACKNOWLEDGEMENT

The authors gratefully acknowledge Vickie King for assisting in running the temperature profiles of aggregation, Tina Trapane and Ruth B. McMichens for the CMR spectra and for assisting in preparing the samples for crosslinking, and Dr. John Cooper and Richard Knight of the Nuclear Science Center of Auburn University for carrying out the γ-irradiation procedure. This work was supported in part by NIH grant HL 29578.

REFERENCES

1. S. M. Patridge, Adv. Prot. Chem., 17, 227-302 (1962).
2. L. B. Sandberg, W. R. Gray and C. Franzblau, Adv. Exp. Med. Biol., 79, 259-261 (1977) (general Discussion on Turnover and Elastolysis).
3. E. G. Cleary and M. Moont, Adv. Exp. Med. Biol., 79, 477-490 (1977).
4. L. B. Sandberg, N. T. Soskel and J. B. Leslie, N. Engl. J. Med., 304, 566-579 (1981).
5. L. B. Sandberg, private communication.
6. J. A. Foster, E. Bruenger, W. R. Gray, and L. B. Sandberg, J. Biol. Chem., 248, 2876-2879 (1973).
7. G. Cicila, M. May, N. Ornstein-Goldstein, Z. Indik, S. Morrow, H. S. Yeh, J. Rosenbloom, C. Boyd, J. Rosenbloom and K. Yoon, Biochemistry, 24, 3075-3080 (1985).
8. J. Rosenbloom, private communication.

9. D. W. Urry, R. D. Harris and M. M. Long, J. Biomed. Mater. Res., 16, 11-16 (1982).
10. D. W. Urry and M. M. Long, CRC Crit. Rev. Biochem., 4, 1-45 (1976).
11. R. S. Rapaka, K. Okamoto and D. W. Urry, Int. J. Pept. Protein Res., 11, 109-127 (1978).
12. J. R. Bell, R. C. Boohan, J. H. Jones and R. M. Moore, Int. J. Pept. Prot. Res., 7, 227-234 (1975).
13. J. R. Bell, R. C. Boohan, J. H. Jones and R. M. Moore, Int. J. Peptide Protein Res., 6, 155-156 (1974).
14. D. W. Urry, K. Okamoto, R. D. Harris, C. F. Hendrix and M. M. Long, Biochemistry, 15, 4083-4089 (1976).
15. D. W. Urry, R. D. Harris and M. M. Long, Biomater. Med. Dev. Art. Org., 9, 181-194 (1981).
16. D. W. Urry, J. Protein Chem., 3, 403-436 (1984).
17. R. Henze and D. W. Urry, J. Am. Chem. Soc., 107, 2991-2993 (1985).
18. D. W. Urry, R. Henze, P. Redington, M. M. Long and K. U. Prasad, Biochem. Biophys. Res. Commun., 128, 1000-1006 (1985).
19. D. W. Urry and M. M. Long, in: "Elastin and Elastic Tissue", L. B. Sandberg, W. R. Gray and C. Franzblau, eds., Plenum Press, New York, New York, Adv. Exp. Med. Biol., 79, 685-714 (1977).
20. D. W. Urry, T. L. Trapane, M. Iqbal, C. M. Venkatachalam and K. U. Prasad, Biochemistry, 24, 5182-5189 (1985).
21. J. R. Lyerla and D. A. Torchia, Biochemistry, 14, 5175-5183 (1975).
22. C. A. J. Hoeve and P. J. Flory, Biopolymers, 13, 677-686 (1974).
23. D. W. Urry, S. A. Wood, R. D. Harris and K. U. Prasad, in: "Polymers as Biomaterials", S. W. Shalaby, T. Horrbett, A. S. Hoffman and B. Ratner, eds., Plenum Publishing Corporation, New York, NY pp 17-32, 1985.
24. D. W. Urry, in: "Biomolecular Stereodynamics III", R. H. Sarma and M. H. Sarma, eds., Adenine Press, Guilderland, NY, (in press).
25. D. W. Urry, T. L. Trapane and K. U. Prasad, Biopolymers, 24, 2345-2356 (1985).
26. D. W. Urry, M. M. Long, T. Ohnishi and M. Jacobs, Biochem. Biophys. Res. Commun., 61, 1427-1433 (1974).
27. D. W. Urry, in:"New Comprehensive Biochemistry, Modern Physical Methods in Biochemistry", A. Neuberger, L. L. M. VanDeenen, eds., Elsevier Biomedical Press, Amsterdam, The Netherlands, pp 277-348 (1985).
28. This Laboratory, unpublished data.
29. D. W. Urry, T. L. Trapane, S. A. Wood, R. D. Harris, J. T. Walker and K. U. Prasad, Int. J. Pept. Protein Res., 23, 425-434 (1984).
30. W. J. Cook, H. M. Einspahr, T. L. Trapane, D. W. Urry and C. E. Bugg, J. Am. Chem. Soc., 102, 5502-5505 (1980).
31. D. W. Urry, T. L. Trapane, H. Sugano and K. U. Prasad, J. Am. Chem. Soc., 103, 2080-2089 (1981).
32. C. M. Venkatachalam, M. A. Khaled, H. Sugano and D. W. Urry, J. Am. Chem. Soc., 103, 2372-2379 (1981).
33. C. M. Venkatachalam and D. W. Urry, Macromolecules, 14, 1225-1229 (1981).
34. D. W. Urry, C. M. Venkatachalam, M. M. Long and K. U. Prasad, in: "Conformation in Biology", R. Srinivasan and R. H. Sarma, eds., G. N. Ramachandran Fetschrift Volume, Adenine Press, USA, pp 11-27, 1982.
35. D. W. Urry, T. L. Trapane, M. M. Long and K. U. Prasad, J. Chem. Soc., Faraday Trans. I, 79, 853-868 (1983).
36. D. W. Urry, T. L. Trapane, S. A. Wood, J. T. Walker, R. D. Harris and K. U. Prasad, Int. J. Pept. Prot. Res., 22, 164-175 (1983).
37. D. W. Urry, Perspect. Biol. Med., 21, 265-295 (1978).
38. E. W. Merrill, in: "Comtemporary Biomaterials", J. W. Boretos and M. Eden, eds., Noyes Publications, New Jersey, pp 27-45 (1984).
39. H. Ulrich, H. W. Bank and G. C. Colovos, in: "Synthetic Biomedical Polymers: Concepts and Applications", M. Szycher and W. J. Robinson,

eds., Technomic Publishing Co., Inc., Westport Connecticut, pp 29-38 (1980).
40. D. W. Urry, T. Ohnishi, M. M. Long and L. W. Mitchell, Int. J. Pept. Protein Res., 7, 367-378 (1975).
41. J. C. Sheehan, J. Preston and P. A. Cruickshank, J. Am. Chem. Soc., 87, 2492-2493 (1965).
42. D. W. Urry and K. U. Prasad, in: "Biocompatibilty of Natural Tissues and Their Synthetic Analogues", D. F. Williams, ed., CRC Press, Inc., Boca Raton, Florida, pp89-116 (1985).
43. J. R. Vaughan, Jr., and R. L. Osato, J. Am. Chem. Soc., 74, 676-678 (1952).
44. K. U. Prasad, M. A. Iqbal and D. W. Urry, Int. J. Pept. and Protein Res., 25, 408-413 (1985).
45. M. Bodanszky, Nature, 175, 685 (1955).
46. Th. Wieland, B. Heinke and J. Vogeler, Ann.,655, 189-194 (1962).
47. R. M. Senior, G. L. Griffin, R. P. Mecham, D. S. Wrenn, K. U. Prasad and D. W. Urry, J. Cell Biol., 99, 870-874 (1984).
48. H. M. Kagan, L. Tseng, P. C. Trackman, K. Okamoto, R. S. Rapaka and D. W. Urry, J. Biol. Chem., 255, 3656-3659 (1980).
49. S. A. Wood, J. E. Lemons, K. U. Prasad and D. W. Urry, J. Biomed. Mater. Res., (in press).
50. S.-T. Li, Polymeric Mater.: Sci. Eng., 53, 30 (1985).

POLYMER MODIFICATION BY RADIATION GRAFTING IN THE PRESENCE OF SALT ADDITIVES AND ITS APPLICATION IN ENZYME IMMOBILIZATION

J. L. Garnett[a], S. V. Jankiewicz[a], M. A. Long[a] and D. F. Sangster[b]

(a) School of Chemistry, The University of New South Wales NSW, Australia 2232
(b) CSIRO, Division of Chemical Physics, Sutherland NSW, Australia 2232

A novel radiation grafting method for modifying polymers of use in biomedical work is discussed. The procedure involves the use of particular inorganic salts as additives for accelerating the radiation grafting process. Styrene is used as typical monomer in these studies with cellulose, the polyolefins and PVC as representative backbone polymers. Two radiation grafting techniques are examined, namely pre-irradiation and simultaneous methods. The parameters affecting the observed grafting enhancement in the presence of these salt additives are treated. The results are compared with analogous data previously obtained with other additives, including acids. A new model for the mechanism of enhancement in radiation grafting in the presence of acids and/or salts is proposed. A typical biomedical application of the salt grafting technique for enzyme immobilization is briefly discussed.

INTRODUCTION

Radiation grafting is a valuable one-step process for producing unique copolymers of use in biomedical applications (1-4). In such work, low doses of radiation to achieve grafting are ideal since, under these conditions, competing homopolymerization is minimized and radiation degradation of the backbone polymer is readily controlled. The utilization of additives to enhance radiation grafting is thus important, since in the presence of such materials, lower total doses of radiation are required to achieve a particular percentage graft. Over recent years groups of additives have been discovered which enhance the radiation copolymerization of a wide range of monomers to a variety of backbone polymers. During mechanistic studies of these additive effects, a new class of compounds has been found which possess unique accelerating properties in radiation grafting reactions (5).

It is the purpose of this present work to review briefly the earlier studies with additives and to compare their use with these new compounds in typical radiation grafting reactions. Because in biomedical applications, grafting by both preirradiation and simultaneous techniques can be valuable, examples will be cited of the use of these additives in both types of grafting processes. The effect of these additives on the properties of the resulting radiation graft copolymers in a typical biomedical application, namely enzyme immobilization, will also be briefly discussed.

The new class of additives recently discovered are particular inorganic salts. For the current experiments lithium perchlorate and nitrate are used as representative salts because of their suitable solubility characteristics in the grafting solutions. Styrene has been utilized as representative monomer with the polyolefins, PVC and cellulose as typical backbone polymers.

EXPERIMENTAL

Grafting Procedures

The techniques of grafting utilized were the following modifications of previously adopted procedures (1,6). For the simultaneous method, grafting experiments were performed in stoppered pyrex tubes (15 x 2.5 cm) containing styrene/solvent solutions (20 ml) at 20 $\pm 1^{\circ}C$. Backbone polymer films (40 x 20 x 0.12 mm) or strips of cellulose of the appropriate size were fully immersed in the monomer solutions and the tubes irradiated in a 1200 Ci cobalt-60 source. At the completion of the grafting, the films were removed from the solution, solvent washed and soxhlet extracted for 72 hours. With the acid solutions, especially for the cellulose runs, the films and paper strips were washed with methanol: dioxan (1:1) prior to soxhlet treatment. After removal of homopolymer, grafted polymers were dried to constant weight at 65%RH. For the powder grafting runs, polypropylene (200-400) and PVC (100μ) were reacted in stoppered pyrex bottles thus enabling the monomer solutions to be stirred continuously during irradiation. Grafted powders were collected in cellulose thimbles immediately after irradiation, the copolymer washed with appropiate cold solvent, extracted in a modified soxhlet for 72 hours, dried in air to remove most of the solvent and then in a ventilated oven at $45^{\circ}C$. For the preirradiation grafting technique (7), cellulose samples were dried in an oven at $80^{\circ}C$ for 24 hours before sealing under high vacuum in a breakseal apparatus for irradiation. This process was necessary to give reproducible data since cellulose contains a finite equilibrium moisture content which must be kept constant during irradiation. Following irradiation, the cellulose was reacted in evacuated vessels with the monomer grafting solution in the dark at $20^{\circ}C$. After reaction, copolymerized celluloses were treated as for the simultaneous method.

Enzyme Immobilization Procedure

For the attachment of the enzyme trypsin, a nitrostyrene copolymer was prepared as intermediate using two methods. Firstly, p-nitrostyrene was directly grafted to the backbone polymer. In the second method, styrene was grafted to the trunk polymer and the resulting styrene copolymer nitrated with a mixture of concentrated nitric and sulfuric acids at $35\text{-}60^{\circ}C$ to give a nitrostyrene copolymer. The copolymer produced from the nitration reaction or by the direct grafting of p-nitrostyrene was reduced with stannous chloride and hydrochloric acid in dimethyformamide, or, with the PVC based copolymers, in absolute ethanol. The resulting polyaminostyrene graft copolymer, suspended in dimethylformamide, was treated at $0^{\circ}C$ successively with carbon disulfide, triethylamine and ethylchloroformate to give the isothiocyanato derivative. The number of isothiocyanate groups in the copolymer was determined by a modification of the method of Karten and Ma (8). The covalent attachment of the trypsin to the copolymer-isothiocyanate derivative was achieved by suspending the support in a solution of trypsin (0.2%) in a bicarbonate buffer (0.05 M, pH 9.6) and shaking overnight at 5°C (9). The amount of protein covalently bound to the support was determined by acid hydrolysis (10) followed by spectrophotometric analysis of the amino acids in the hydrolysate (11,12). The esterase activity of the bound trypsin was determined by a spectrophotometric method using N-α-benzoyl-L-arginine ethyl ester, a synthetic low molecular weight substrate (13).

RESULTS AND DISCUSSION

In previous grafting studies (1,6,7), the effect of solvents on the radiation process has been discussed. Those solvents which wet and swell the backbone polymer are particularly attractive for giving high grafting yields. With cellulose the low molecular weight alcohols such as methanol are suitable for this purpose. Methanol both wets and swells cellulose thus enabling monomers like styrene to penetrate the trunk polymer and graft. With the synthetic backbone polymers, polypropylene and PVC, solvents such as methanol do not swell the polymer, however, in this instance the monomer, styrene, swells the polymer, thus styrene in methanol remains a satisfactory system for efficient radiation grafting, even for the synthetic backbone polymers. Hence, overall, styrene in methanol constitutes a useful general purpose grafting system and is used as model for the studies reported in this paper.

Effect of Acid in Radiation Grafting

Acid is a useful additive for enhancing the radiation grafting of a range of monomers to a wide variety of backbone polymers. Typical of the behavior expected of acid in such reactions are the data shown in Table 1 where styrene in methanol is grafted to cellulose, polypropylene and PVC in the presence of acid. With acid and cellulose, enhancement in grafting yield occurs at monomer concentrations of up to 50%, whereas with polypropylene and PVC, at all monomer concentrations studied, a significant increase in yield is observed with the inclusion of sulfuric acid. The

Table I. Acid effect in radiation grafting styrene to cellulose, polypropylene (PPE) and PVC (a).

Styrene in Methanol (% V/V)	Cellulose		PPE Graft (%)		PVC	
	N.A.	H^+	N.A.	H^+	N.A.	H^+
20	8	24	29	218	0	10
30	17	33	94	150	13	34
40	22	31	50	85	65	70
50	21	27	37	55	6	25
60	28	25	36	45	3	10

(a) Irradiated to total dose of 2.0×10^6 rad at dose rates of 4.0×10^4 rad/hr (cellulose), 3.0×10^5 (PPE) and 1.0×10^5 (PVC); PPE (doubly oriented film, 0.06 mm); PVC (film, 0.20 mm).

results in Table I also show that the acid enhancement in yield is generally larger with the synthetic polymers than with cellulose. The synthetic materials can tolerate relatively high acid concentrations whereas the naturally occuring polymers, like cellulose, are readily degraded at acidities above 0.2 Molar. Thus the amount of acid capable of being used with backbone polymers like cellulose is limited and hence the magnitude of the enhancement in such systems is restricted to low acidities.

Addition of acid also has a further effect in cellulose grafting. Inclusion of this additive leads to a peak in copolymerization at 30% monomer concentration. This peak is referred to as the Trommsdorff effect and is important in grafting reactions (1). Not only does the Trommsdorff effect lead to a maximum in grafting yield but the chain length of the grafted polymer is longer under the experimental conditions of the Trommsdorff effect. Thus in a preparative context it is important to know the experimental conditions under which a Trommsdorff peak may be observed. When polypropylene and PVC are used as backbone polymers (Table 1), a gel effect is already present in the neutral solutions without acid. Inclusion of acid in these monomer solutions leads not only to an increase in the intensity of the Trommsdorff peak but also to an enhancement in grafting yields with all other solutions. Inclusion of acid can thus induce a Trommsdorff peak in a grafting system or alternatively, if a Trommsdorff peak is already present, addition of acid can enhance the intensity of that peak.

Mechanistically the role of acid in enhancing radiation grafting involves a number of competing processes. With cellulose it has been suggested that acid at the concentration used does not markedly effect the precipitation of the grafted chains or the swelling of the backbone polymer. Instead it had been proposed (1,14) that the acid enhancement with this backbone polymer can be essentially attributed to a radiation chemistry phenomenon involving increased G (H) yields and thermalized electrons (Equations 1-3, CellOH = cellulose).

$$CH_3OH + H^+ \rightarrow CH_3OH_2^+ \qquad \text{(Equation 1)}$$

$$CH_3OH_2^+ + e \rightarrow CH_3OH + H \qquad \text{(Equation 2)}$$

$$CellOH + H^+ \rightarrow CellOH_2^+ \qquad \text{(Equation 3)}$$

These processes can lead to increased numbers of grafting sites by abstration reactions with the trunk polymer (Equation 4).

$$CellOH + H \rightarrow CellO\cdot + H_2 \qquad \text{(Equation 4)}$$

More recent work on the acid effect (15) indicates that acid leads to an increase in styrene-solvent intermediates (MS·) in the grafting solution, such species resulting in more grafting sites (Equation 5). This latter mechanism is particularly relevent to polypropylene since this backbone polymer does not possess hydroxyl groups and therefore would not be able to participate in hydroxylation reactions similar to cellulose. In these acid catalysed reactions PVC is unique since, during irradiation of the background polymer, HCl is generated in situ and this can act as an acid additive and thus accelerate grafting further.

$$MS\cdot + CellOH \rightarrow MSH + CellO\cdot \qquad \text{(Equation 5)}$$

Effect of Polyfunctional Monomers as Additives in Radiation Grafting

In addition to acid, a new series of additives, which are polyfunctional monomers such as divinylbenzene (DVB) and trimethylolpropane triacrylate (TMPTA), have been found to accelerate the above grafting reactions when used in concentrations of up to 1% (v/v). When these polyfunctional monomers are added to the monomer solution the results show that radiation grafting of styrene in methanol to the polyolefins is enhanced to a comparable degree to that of inclusion of acid (Table II). The data in Table II indicate that inclusion of DVB in the monomer solution leads to enhancement in grafting to polyethylene at all styrene concentrations studied. With polypropylene an analogous observation can be made with the exception of the 30% monomer solution. Similar to the acid result, inclusion of DVB leads to an increase in grafting yield at 50% styrene which is the monomer concentration corresponding to the presence of the Trommsdorff peak.

With propylene the presence of DVB leads to a shift in the monomer concentration of the observation of the Trommsdorff peak from 30% styrene in methanol to 40%. When DVB is replaced by TMPTA a similar enhancement in grafting yield is observed with polyethylene (Table II). The polyfunctionality of DVB and TMPTA is obviously important in these reactions because the number of grafting sites in the system is potentially enhanced by the increase in unsaturation due to the presence of the type of additive. Thus with polyfunctional monomers, such as DVB in additive amounts, branching of the grafted polystyrene can occur when one end of the DVB, immobilized during grafting, is bonded to the growing chain. The other end is unsaturated and free to initiate a new chain growth by a scavenging reaction. The new branched polystyrene may eventually terminate or cross-link by reacting with an adjacent polystyrene chain or immobilized DVB radical. Grafting is thus enhanced mainly through the branching of the grafted polystyrene chain. The results from the addition of TMPTA can be explained in an analagous fashion. The data in Table II are reported for grafting to the polyolefins, however, additional evi-

Table II. Effect of polyfunctional monomers as additives in radiation grafting of styrene to polyolefins (a).

Styrene in Methanol (%v/v)	Polyethylene (PE)	Graft (%)		Polypropylene (PPE)	
	N.A.	DVB	TMPTA	N.A.	DVB
20	14	15	--	37	38
30	37	41	39	132	92
40	76	74	73	75	132
50	109	136	137	--	--
60	89	121	--	56	101
80	68	74	--	35	55

(a) Irradiated to 2.4×10^5 rads at a dose rate of 4.1×10^4 rad/hr; DVB (Divinylbenzene) and TMPTA (trimethylolpropanetriacrylate) at 1% (v/v); PE (low density film 0.12 mm); PPE (doubly oriented film, 0.06 mm).

dence is now available to indicate that the use of polyfunctional monomers as additives for enhancing radiation grafting is also relevent to other backbone polymer systems and thus is a general phenomenon in radiation grafting overall.

Synergistic Effect of Acid and Polyfunctional Monomers in Radiation Grafting

If the mechanisms of the effect of acid and the polyfunctional monomers in enhancing radiation grafting reactions are different as suggested in the preceding paragraph, then inclusion of both additives in the same grafting solution should theoretically lead to a synergistic effect of the two additives. The data in Table III for the radiation grafting of styrene in methanol to low density polyethylene film confirm this suggestion, because the effect of DVB is very much enhanced in the presence of acid. Not only is there a significant enhancement in grafting yield at all monomer concentrations studied in the presence of both additives, there is also a very large increase in yield at the monomer concentration corresponding to the Trommsdorff peak. The other significant feature of the data in Table III are the results for grafting efficiency which reflect the degree to which homopolymer participates in the grafting process. This is important in a preparative context because high grafting efficiency reflects low homopolymerization and efficient use of monomer in copolymerization. At low homopolymer levels purification of graft copolymer is simplified since the amounts of occluded homopolymer are small. The effect of the additives on the grafting efficiency is significant because using sulfuric acid alone has only a marginal effect on grafting efficiency, whereas inclusion of both DVB and acid in the grafting solution leads to a very appreciable increase in grafting efficiency at all monomer concentrations studied, particularly

Table III. Synergistic effect of polyfunctional monomers and acid in radiation grafting styrene to polyethylene (a).

Styrene in Methanol	Graft (%)				Grafting Efficiency (%)		
	N.A.	H_2SO_4	DVB	DVB + H_2SO_4 (b)	N.A.	H_2SO_4	DVB + H_2SO_4 (b)
(%v/v)		(0.2M)	(1 % v/v)			(0.2M)	
20	14	19	15	27	42	33	50
30	37	51	41	58	44	47	57
40	76	61	74	119	57	58	68
50	109	134	136	188	56	58	69
60	89	119	121	156	--	--	--
70	89	73	--	112	36	31	42
80	68	62	74	69	--	--	--

(a) Irradiated to 2.4×10^5 rads at dose rate of 4.1×10^4 rad/hr.
(b) DVB (1 % v/v) + H_2SO_4 (0.2M).

those concentrations corresponding to the Trommsdorff peak. Inclusion of the two additives thus leads not only to an increase in overall grafting yields, it leads to a significant reduction in competing homopolymerization and thus is beneficial to the grafting system.

Mechanistically both acid and polyfunctional monomers participate as indicated in the preceding section. Thus inclusion of acid increases the concentration of H· and MS· species (Equations 2 and 5) in the grafting solution. This increase in radical concentration when combined with increased unsaturation in the grafted chains due to polyfunctional monomers leads to branched polystyrene chains of higher molecular weight and hence higher graft.

Acid Effect in Preirradiation Grafting

The preirradiation method for grafting involves initial irradiation of the backbone polymer followed by subsequent reaction with the monomer. By contrast the mutual method, which has already been previously discussed in the earlier tables, involves simultaneous irradiation of cellulose in contact with monomers. This latter technique thus reduces the possibility of bond rupture during irradiation. In order to achieve comparable grafting yields to the simultaneous method, the preirradiation technique requires higher radiation doses with the subsequent risk of potential degradation in the finished copolymer. However, for simplicity, especially in a preparative context, for high volume processing, the preirradiation technique is a very convenient procedure for achieving grafting. The fact that acid as an additive in the grafting solution can also very significantly enhance grafting yields in the preirradiation process is very important (Table IV). The increase in yield with acid is even very much greater than the corresponding enhancement observed previously in Table I for the simultaneous grafting procedure. This acid effect in the preirradiation technique is particularly valuable for grafting to cellulose in the low monomer concentration solutions where yields are very low in neutral solutions and, for synthetic purposes, inclusion of acid in these solutions is very beneficial. If the total dose used in the preirradiation technique is doubled from 0.5 Megarads to 1.0 Megarads there is a corresponding increase in grafting yields which is small in the neutral solutions at low monomer concentration, but large in the corresponding acid solutions (Table IV). The inclusion of acid in the monomer solutions in the preirradiation technique also induces a Trommsdorff peak at 20% monomer concentration whereas no equivalent peak is observed at this concentration in the corresponding neutral solutions. This observation presumably reflects mechanistically the role of the radiation-induced trapped radicals as initiators for grafting in the cellulose and the fact that monomer must diffuse into the cellulose to the radical site to react.

Effect of Inorganic Salts as Additives in Both Preirradiation and Simultaneous Methods for Grafting to Cellulose

The data in Table V shows that certain inorganic salts, in particular lithium perchlorate, when used as additives in the grafting solution, enhance the grafting of styrene in methanol to cellulose using the preirradiation method. The significant feature of the data in Table V is that lithium perchlorate is much more efficient than acid in increasing grafting yields. At 30% styrene in methanol, the presence of lithium perchlorate leads to a very large Trommsdorff peak in the grafting system. By contrast, with acid, no such large gel peak is observed and therefore the lithium perchlorate additive effect is of extreme value in a preparative context for grafting by the preirradiation method.

Table IV. Acid enhancement in preirradiation grafting of styrene to cellulose at different doses (a)

Styrene in Methanol	Graft (%)			
	5.0×10^5 Rad		1.0×10^6 Rad	
(%v/v)	N.A.	H_2SO_4(0.1M)	N.A.	H_2SO_4 (0.1M)
10	3	34	4	19
15	5	42	6	58
20	6	54	8	91
25	6	42	11	82
30	6	36	18	78
35	10	40	22	56
40	51	36	84	45
45	46	27	76	36

(a) Dose of 3.3×10^4 rad/hr.

In Table VI the analogous result for the use of lithium perchlorate as an additive in the radiation grafting of styrene in methanol to cellulose by the simultaneous procedure is reported. As before the inorganic salts are even more efficient than mineral acids in enhancing the grafting yield under certain experimental conditions. A range of inorganic salts have been examined in these studies and certain lithium salts of which lithium perchlorate is representative are particularly effective. The data in Table VI show that, at lower monomer concentrations, inclusion of lithium perchlorate leads to a significant enhancement in grafting yields. The enhancement occurs even at the Trommsdorff peak corresponding to the 25% styrene in methanol solution. This lithium perchlorate result with cellulose is important, because not only does the yield of copolymer increase in the presence of this additive when compared with acid, but lithium perchlorate is chemically neutral and does not attack the cellulose. Sulfuric acid, however, degrades cellulose and therefore can only be used sparingly in these experiments. When the samples in Table VI are acid digested and the graft copolymer isolated, gel permeation chromatography analysis of the product indicates that the copolymers from the lower styrene concentration solutions prepared in the presence of both additives possess higher number average molecular weights ($\bar{M}_n$) than the copolymers isolated from the corresponding additive free grafting solutions (Table VII). Furthermore the samples from those solutions in Table VI which yield enhanced grafting with both acid and salt correspond to the copolymers in Table VII possessing the higher $\bar{M}_n$ values. Thus overall, when acid and lithium perchlorate are compared as additives in the simultaneous grafting of styrene, in methanol, to cellulose similar trends in enhancement are observed with both additives suggesting that common mechanistic pathways are involved in both systems.

Effect of Inorganic Salts on Radiation Grafting to Polyethylene

When the backbone polymer in these reactions is changed from cellulose to low density polyethylene (Table VIII) the salt enhancement effect in grafting only occurs at the lower monomer concentrations

Table V. Comparison of acids with lithium perchlorate as additives for enhancing the preirradiation grafting of styrene to cellulose (a)

Styrene in Methanol (% v/v)	Graft (%) Additive Free	H_2SO_4 (0.1M)	$LiClO_4$ (0.2M)
10	3	34	1
15	5	42	3
20	6	54	58
25	6	42	180
30	6	36	242
35	10	40	229
40	51	36	147
45	46	27	98

(a) Irradiation at 3.3×10^4 rad/hr to total dose of 5.0×10^5 rad followed by reaction for 72 hr in the dark at 20°C.

Table VI. Comparison of acid with lithium perchlorate as additives for enhancing the radiation grafting styrene in methanol to cellulose by the simultaneous procedure (a).

Styrene (%v/v)	Graft (%) No Additive	H_2SO_4 (0.1M)	$LiClO_4$ (0.2M)
15	32	34	54
20	66	120	155
25	106	153	159
30	112	95	96
40	82	52	72

(a) Irradiation in air at a dose rate of 3.3×10^4 rad/hr to a total dose of 5.0×10^5 rad at 20°C using Whatman 41 filter paper.

Table VII. Comparison of acid with lithium perchlorate on number average molecular weights ($\bar{M}_n$) of copolymer from radiation grafting styrene in methanol to cellulose by the simultaneous procedure (a).

Styrene in Methanol (% v/v)	$\bar{M}_n$ (x 10^5) No Additive	H_2SO_4 (0.1M)	$LiClO_4$ (0.2M)
15	3.5	4.8	3.7
20	5.0	9.8	8.0
25	7.7	10.0	10.0
30	12.0	8.6	9.0
40	12.0	9.5	12.0

(a) Irradiation conditions as in Table I.

whereas, with acid, increased grafting yields are observed in all styrene solutions. The nature of the salt used for the enhancement effect is also important because it must not only be soluble in the grafting solution, it must also contribute significantly to the overall polarity of the system. In this respect the results in Table VIII, involving a comparison of lithium perchlorate and nitrate as additives, show that both salts enhance grafting in the lower monomer concentrations, however, lithium nitrate is less efficient than the perchlorate under the radiation conditions used. The effect of structure of inorganic salt on the grafting enhancement is further accentuated by the data in Table IX where a range of acetates have been used as salt additives and no increase in grafting yields of styrene in methanol to polyethylene film is observed at any of the monomer concentrations studied. The overall polarity of the system and the effect of the salts on this polarity is obviously critical in any grafting enhancement with these additives.

Mechanism of Acid and Salt Effects in Radiation Grafting

From the earlier studies of the acid effect in the radiation grafting of styrene (Table I) the enhancement due to this additive was attributed essentially to two predominant factors, namely the radiolytic yield of hydrogen atoms and also the extent to which grafting polymer (polystyrene) was solubilized in the bulk solution. Solvent structure was also considered to be important, because, in addition to possessing the appropriate wetting and swelling properties, those solvents with high radiolytic yields of hydrogen atoms, such as methanol and the lower molecular weight alcohols, were considered to be efficient reagents for grafting. This proton model (Equations 1-5) for treating these acid effects is consistent with the known increase in yield of hydrogen atoms from irradiated methanol upon the addition of acid (16,17), however, the scheme fails to explain several observations including (i) the occurrence of enhancement only at certain acid and monomer concentrations and (ii) the efficacy of H_2SO_4 and HNO_3 over other acids (1). In addition, it is now known that (1) the presence of acid induces large increases in yield

Table VIII. Comparison of lithium perchlorate with lithium nitrate as additives in radiation grafting styrene in methanol to polyethylene by the simultaneous procedure (a).

Styrene	Graft (%)			
(% v/v)	No Additive	H_2SO_4 (0.1M)	$LiClO_4$ (0.2M)	$LiNO_3$ (0.2M)
15	34 (31)	(32)	42 (44)	28
20	60 (64)	(70)	77 (81)	68
25	78 (103)	(148)	140 (192)	122
30	142 (187)	(240)	160 (196)	102
40	117 (150)	(157)	92 (114)	80

(a) Irradiation in air at a dose rate of 3.3×10^4 rad/hr to a total dose of 1.8×10^5 rad at 20°C with low density polyethylene film (0.12 mm). Data in brackets at 2.0×10^5 rad.

Table IX. Effect of inorganic acetate salts on radiation grafting of styrene in methanol to polyethylene film (a).

Styrene	Graft (%)			
(%v/v)	No Additive	Li^+	Na^+	NH_4^+
15	26	4	5	1
20	46	33	19	1
25	94	64	54	19
30	166	78	68	97
35	159	97	71	124
40	132	86	63	114

(a) Irradiation in air at a dose rate of 3.3×10^4 rad/hr to total dose of 1.8×10^5 rad at 20°C using salt concentrations of 0.2 M with low density polyethylene film (0.12 mm).

in the preirradiation technique (Table IV) and (2) solutions of metal salts such as $LiClO_4$ produce grafting patterns similar to acid (Tables V-VIII). These two later results suggest that radiolytically produced hydrogen atoms are not essential to the operation of graft enhancement. In addition, the differences in grafting reactivities between cellulose and polyethylene clearly indicate that polarity of backbone polymer is important in these processes.

From the above results and more extensive evidence to be published elsewhere (18), a new model is proposed to explain the effects of acids and metal salts such as $LiClO_4$ in enhancing radiation grafting. In any grafting system at any time, there is an equilibrium concentration of monomer absorbed within the grafting region of the backbone polymer. This grafting region may be continually changing as grafting proceeds. Thus in grafting styrene to cellulose, during the initial part of the reaction, the grafting region will be essentially cellulosic in nature, however, as reaction proceeds, the grafting region will become more styrenated. The degree to which monomer will be absorbed by this grafting region will therefore depend on the chemical structure of the region at the specific time of grafting. Further experimental evidence now available indicates that increased partitioning of monomer occurs in the graft region when ionic solutes are dissolved in the bulk grafting solution. Thus higher concentrations of monomer are available for grafting at a particular backbone polymer site in the presence of these salt additives. The extent of this improved monomer partitioning depends on the polarity of monomer, substrate and solvent and also on the concentration of ionic solute. Metal salts such as $LiClO_4$ are more efficient than acids in enhancing grafting due to the overall monomer partitioning effect. The type of salt used and its effectiveness will depend upon its contribution to the total polarity of the system. Overall, the mechanism of the enhancement process is complicated and other processes such as radiolytically produced hydrogen atoms will make some contribution, however, present evidence suggests that such processes will not be involved in the predominant reaction pathway. Finally, in very recent work (19), it has been shown that both acid and $LiClO_4$ enhance copolymerization yields in a similar manner in the analogous grafting system initiated by UV irradiation. These UV data (19) are also consistent with the above model proposed for gamma irradiation systems, thus the new model based on partitioning of monomer would appear to be of general applicability. This partitioning concept explains in physical terms how increased concentrations of monomer can be found at grafting sites, however, in order to complete the grafting process, the actual chemical steps involving bond rupture and reformation in grafting need to be defined. An appropiate mechanism for these subsequent chemical steps involves the charge-transfer concept previously discussed (7,20) and now generally accepted.

Application of Additive Effect in Grafting to Enzyme Immobilization

In practical context, the discovery of the new salt additives has an additional advantage in radiation grafting to modify polymer structures. Acid additives could not previously be used with acid sensitive backbone polymers, whereas salts can now be used with such materials to achieve grafting enhancement. This development is of advantage in enzyme immobilization work where a range of backbone polymers is used, including acid sensitive materials. For enzyme immobilization processes, grafting is used to prepare suitable supports, however for this application a convenient form of the backbone polymer is a powder or as beads. Two representative polymer powders have been utilized in the current experiments, namely polypropylene which is relatively inert, and PVC, which is polar. In addition, studies have also been carried out on polystyrene beads which have been cross-linked with 20%DVB using chemical procedures. Two different monomers have been used for this work. Firstly, p-nitrostyrene has been grafted to all three polymers and the resulting nitrostyrene copolymers converted to the corresponding isothiocyanato derivatives to which enzymes have been attached (8-13). Secondly, styrene has also been grafted to polypropylene powder and the resulting copolymer nitrated then converted to the isothiocyanato derivatives as in the preceding example (8-13). In the grafting of styrene to polypropylene powder inclusion of mineral acid in the vessel, if stirring is not used, leads to an enhancement in grafting at monomer concentrations up to 40% (Table X). At 50% styrene in methanol there is a reduction in efficiency in the presence of acid which is reversed again at 60% styrene in methanol. If DVB is added to this acid system, again without stirring during irradiation, an additional enhancement in grafting is observed at all monomer concentrations examined except for 50% styrene in methanol (Table X). The other interesting aspect of the data in Table X is that in all solutions studied, with and without additive, the grafting yield increases virtually in a linear fashion with increasing concentration of styrene, this observation being of particular value in a preparative context. When the same system is stirred during irradiation, the overall grafting yields at all monomer concentrations studied in additive free solutions are lower than when the solutions are not stirred (Table XI). When mineral acid is added to these stirred solutions there is an increase in grafting yield at all monomer concentrations studied, however, overall the yields again, except for the 20% monomer concentration, are generally lower than for the non-stirred solutions. The significant feature of these styrene grafting data is that at relatively low total doses (of the order of 0.2 megarads) appreciable styrene can be grafted to polypropylene powder under conditions where the radiation dose is not high and therefore degradation of the backbone polymer is minimized.

When p-nitrostyrene is used as a monomer the radiation doses required to graft this material to polypropylene powder are significantly higher than when styrene is used (Table XII. The nitro group in the styrene is deactivating in radiation grafting because the grafting yields with styrene are significantly higher than with p-nitrostyrene yet the dose used with the former monomer is one order of magnitude lower. Inclusion

Table X. Effect of acid and DVB on radiation grafting of styrene to PPE powder without stirring (a)

Styrene	Graft (%)		
(% v/v)	Additive Free	H_2SO_4 (0.2M)	DVB (1 % v/v) + H_2SO_4 (0.2M)
20	4	2	16
30	21	22	34
40	36	52	55
50	108	99	105
60	156	172	180

(a) Total dose 2.4×10^5 rad at 4.1×10^4 rad/hr; polypropylene powder not stirred during irradiation.

of sulfuric acid in the p-nitrostyrene methanol solution leads to a significant graft enhancement at the only monomer concentration studied, namely 7%. This observation is particularly important because doses of almost 4 megarads are required to achieve 10% graft with this particular monomer and any enhancement at all in the presence of additive would be very beneficial at this high radiation dose. A similar situation occurs when p-nitrostyrene is grafted to polystyrene beads and PVC powder (Table XIII) where radiation doses of at least 3 megarads are required to obtain finite yields for subsequent immobilization studies. Inclusion of sulfuric acid in the solutions leads to a very significant enhancement in grafting and is thus beneficial. If DVB is used as additive, marginal enhancement is observed with p-nitrostyrene. The reasons for the poor reactivity of p-nitrostyrene are interesting since based on the Hammett value ($-NO_2$, $\sigma = 0.78$) it would be expected that p-nitrostyrene would be more reactive than indicated by the data in Tables XII and XIII. However, it is well known that in polymerization reactions of substituted styrenes, steric effects and resonance stabilization of the propagating radical can have a strong influence on the resulting rates of reaction. For p-nitrostyrene the molecule is essentially planar (21) and thus steric hindrance should not play a major role in its polymerization rate. However, aromatic nitro compounds are known to inhibit strongly thermal or radical initiated polymerization of styrene and related monomers. The mechanism of this inhibition is complex, however, in part it can be attributed to the stability of the aromatic nitro radicals formed during these polymerizations. The stability of a free radical is influenced by the ability of the group attached to the radical center to delocalize the unpaired electron. Such effects diminish the tendency of the radical to undergo subsequent reactions by decreasing the spin density at the radical center. The propagating radical (III, Equation 6) contributes strongly to the delocalization of the unpaired electron generated during graft-

Table XI. Effect of acid and DVB on radiation grafting of styrene to PPE powder with stirring (a).

Styrene	Graft (%)				
(%v/v)	Additive Free	H_2SO_4 (0.1M)	DVB[b] + A	H_2SO_4 B	(0.2M) C
15	9	10	--	--	--
20	15	17	--	--	--
30	23	29	71	67	37
35	30	35	127	113	86
40	42	46	--	--	--
45	50	57	--	--	--

(a) Dose rate of 4.1×10^4 rad/hr to total dose of 2.3×10^5 rad polypropylene powder stirred during irradiation.

(b) DVB concentrations:

A = 0.5 % v/v
B = 2.0 % v/v
C = 8.1 % v/v

Table XII. Effect of acid on simultaneous radiation grafting of p-nitrostyrene to polypropylene powder in methanol (a).

p-Nitrostyrene (% v/v)	Graft (%) No Additive	Graft (%) H_2SO_4 (0.1M)
7	10	14
14	22	--
28	33	--

(a) Dose rate of 4. x 10^4 rad/hr to total dose of 3.8 x 10^5 rad.

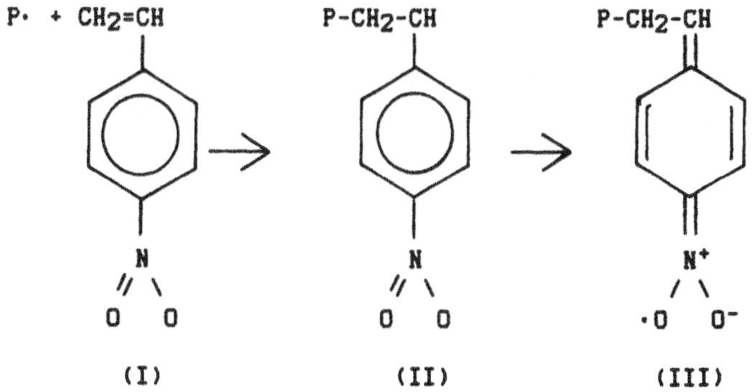

(Equation 6)

Table XIII. Effect of acid and DVB on radiation grafting of p-nitrostyrene to polystyrene beads and PVC powder (a).

Trunk Polymer	No Additive	Graft (%) H_2SO_4 (0.1M)	DVB (1 % v/v)
PS[b]	1.5	6.3	1.8
PVC	7.3	16.7	7.6

(a) Monomer 30% (w/v) in dioxan/methanol (3.2, v/v); dose-rate of 2.6 x 10^4 rad/hr to total dose of 3 x 10^6 rad.
(b) Polystyrene beads cross-linked with 20% DVB (250-850 μ).

ing of p-nitrostyrene, leading to a decreased chance of further reaction with monomer and thus a lower grafting yield. The data in Tables XII and XIII are consistant with this theory and indicate why additive effects in grafting of p-nitrostyrene particularly are very important for minimizing the total dose in these immobilization studies.

For the application of the above radiation graft copolymers to enzyme immobilization, trypsin was used as model together with two types of copolymers, one prepared by the direct grafting of p-nitrostyrene and the other by nitration of grafted styrene. In each case reduction of the nitro group in the copolymer to the amine and conversion to the isothiocyanate was carried out by chemical means. Enzyme was bound to the support by covalent bonding between the isothiocyanate group and an amine group in the enzyme forming a thiourea linkage. The amount of enzyme bound to each support is shown in Table XIV. There is a reasonable correlation between the extent of grafting, the number of isothiocyanate groups introduced and the amount of enzyme bound. The activity of the bound enzyme was determined using a low molecular weight synthetic substrate, N- α-benzoyl-L-arginine ethyl ester. The activity (Table XIV) is expressed in several forms, namely μmoles of substrate hydrolysed per minute per gram of support, μmoles of substrate hydrolysed per minute per milligram of immobilized enzyme and the percentage that has been retained by the immobilized enzyme when compared to the activity of the enzyme free in solution.

Two important aspects of the results in Table XIV are obvious. Firstly, there appears to be little difference in activity for copolymers prepared by grafting p-nitrostyrene and for copolymers chemically nitrated after grafting styrene. This result is not unexpected as the possible side reactions during nitration such as multiple nitration of

Table XIV. Activity of immobilized enzyme on radiation graft copolymer.

System (a)	mEqCNS/ g Support	mqEnz/ g Support	μmoles/min/ g Support	μmoles/min/ mg Enz	% Retained Activity
Free Trypsin	--	--	--	18.0	100
10%NS-PP	0.33	2.1	2.4	0.9	5
15%NS-PP	0.56	2.9	3.1	0.9	5
19%NS-PP	--	4.9	2.2	0.4	2
22%NS-PP	1.00	3.8	2.0	0.5	3
33%NS-PP	1.46	10.4	8.5	0.8	4
13%S-PP	0.92	1.8	2.2	1.0	5
6%NS-PS20	--	0.1	2.4	5.9	33
17%NS-PVC	--	0.4	2.8	3.1	17

(a) NS-PP signifies direct nitrostyrene graft to polypropylene; S-PP signifies direct styrene graft to polypropylene, followed by chemical nitration; NS-PVC signifies direct nitrostyrene graft to polystyrene beads (Table XIII); NS-PVC signifies direct nitrostyrene graft to PVC powder. All nitrostyrene copolymers were converted to isothiocyanato derivatives for enzyme attachment.

the styrene ring would not have important consequences on the subsequent reactions. The second factor to emerge from the data in Table XIV is that the activity is not directly proportional to the amount of protein bound. In fact the copolymer showing the highest activity retention contained the lowest amount of enzyme. This is observed through both p-nitrostyrene and styrene grafted copolymers and with the three trunk polymers. The obvious conclusion from this result is that if too many isothiocyanate groups are made available for binding to enzymes multiple linkage occurs with resultant deformation of the enzyme. As the steric conformation of an enzyme is as important for substrate binding and reaction as is the amino acid sequence at the reactive site, this deformation would lead to loss of activity in highly substituted copolymers. Overall the current data are encouraging because the results obtained with low grafts on polystyrene, polypropylene and PVC involving retained activity figures of up to 10% are consistent with other covalent attachment techniques (22,23).

<u>Significance of Additive Effects in Grafting on Enzyme Immobilization</u>

The present additive studies in grafting are important for enzyme immobilization applications. The discovery of a range of metal salts that can be used for this purpose, in addition to acid, is valuable since the scope of the enhancement technique is widened. Thus grafting systems which may have been sensitive to acid, can now be treated with metal salts to achieve the same type of enhanced reactivity. Although the efficiency of the metal salts in increasing grafting yields is higher than for acid, some grafting systems may have solubility problems with the salts, so, in a preparative context, it is fortunate that acids are available as an alternative and can be used to overcome such solubility difficulties. In the present work, where p-nitrostyrene has been grafted to polypropylene and PVC powders for enzyme immobilization, relatively high radiation doses are required to achieve reasonable grafting since the nitro group in styrene is deactivating in these reactions. Inclusion of acid, increases the grafting yields significantly and thus is beneficial for grafting to backbone polymers, such as polypropylene and PVC, which are somewhat radiation sensitive, particularly PVC which tends to dehydrochlorinate at medium levels of radiation. Overall the radiation grafting techniques described in this paper offer a number of advantages for enzyme immobilization, namely (1) flexibility in choice of trunk polymer and (2) availability of a one-step method for surface attachment of enzyme such that minimum steric hinderence and diffusional problems occur when the actual reagents react with the immobilized enzyme. In the future an obvious extension of the process is to use faster polymerization techniques such as radiation rapid cure (RRC) where polymerization is achieved in a fraction of a second. Preliminary studies (24) with this faster process have been performed and the results are sufficiently encouraging to justify further enzyme immobilization work in this area.

ACKNOWLEDGEMENTS

The authors thank the Australian Institute of Nuclear Science and Engineering, the Australian Research Grants Committee and the International Atomic Energy Agency for the support of this research.

REFERENCES

1. J. L. Garnett, Radiation Phys. Chem., <u>14</u>, 79 (1979)
2. B. D. Ratner and A. S. Hoffman, J. Appl. Polym. Sci., <u>18</u>, 3183 (1974)
3. V. Stannett, "Graft Copolymerization of Lignocellulose Fibers", Am.

Chem. Soc. Symp. Ser. No. 187, D.N.-S Hon, Ed., Am. Chem. Soc. Washington, D.C., 1982, p. 3.
4. J. C. Arthur, Jr., "Graft Copolymerization of Lignocellulose Fibers", Am. Chem. Soc. Symp. Ser. No. 187, D. N. -S. Hon, Ed., Am. Chem. Soc., Washington, D. C., 1982, p. 21.
5. J. L. Garnett, S. V. Jankiewicz, R. Levot and D. F. Sangster, Radiation Phys. Chem., in press.
6. C. H. Ang, J. L. Garnett, R. Levot and M. A. Long, J. Appl. Polym. Sci., 27, 4893 (1982).
7. S. Dilli and J. L. Garnett, Aust. J. Chem., 24, 981 (1971).
8. B. S. Karten and T. S. Ma, Microchem. J., 3, 507 (1959).
9. L. Goldstein, Biochem. Biophys. Acta, 315, 1 (1973).
10. S. Blackburn, Amino Acid Determination, Arnold, London, 1968.
11. J. P. Greenstein and M. Winitz, Chemistry of the Amino Acids, Wiley, New York, 1961.
12. E. W. Yemm and E. C. Cocking, Analyst, 80, 209 (1955).
13. H. Bergmeyer, Methods of Enzymatic Analyses, Academic Press, New York, 1963.
14. W. J. Chappas and J. Silverman, Radiat. Phys. Chem., 14, 487 (1979).
15. C. H. Ang, J. L. Garnett, S. V. Jankiewicz and D. F. Sangster, "Graft Copolymerization of Lignocellulose Fibers", Am. Chem. Soc. Symp. Ser. No. 187, D. N. -S. Hon, Ed., Am. Chem. Soc., Washington, D. C., 1982, p.141.
16. J. H. Baxendale and F. W. Mellows, J. Am. Chem. Soc., 83, 4720 (1962).
17. A. Ekstrom and J. L. Garnett, J. Phys. Chem., 70, 324 (1966).
18. J. L. Garnett, S. V. Jankiewicz, R. Levot, M. A. Long and D. F. Sangster, Radiat. Phys. Chem., in press.
19. J. L. Garnett, S. V. Jankiewicz and D. F. Sangster, unpublished work.
20 S. Dilli and J. L. Garnett, Aust. J. Chem., 21, 397 (1968).
21. M. Borzaghi, A. Gamba, G. Morosi and M. Simonetta, J. Phys. Chem., 78, 49 (1974).
22. D. L. Regan and P. F. Greenfield, Chem. Aust., 45, 317 (1978).
23. S. Venkataraman, T. A. Horbett and A. S. Hoffman, J. Biomed. Mater. Res. Symp., 8, 111 (1977).
24. S. Bett and J. L. Garnett, unpublished work.

USE OF ORGANIC COSOLVENTS FOR ENHANCED LEACHING OF HEMA/MAA COPOLYMERS

M. R. Van De Mark[a], N. D. Lian[b], and E. C. Eckstein[c]

(a) Department of Chemistry
(b) Department of Biomedical Engineering
University of Miami, Coral Gables, FL 33124
(c) Chemistry Department
University of Missouri-Rolla, Rolla, MO 65401

Leaching processes that use solvents to shrink and swell hydrogels by ionizing constituents of the gel and/or by solvating the hydrophobic regions with organic cosolvents were explored as a means to cleanse implantable hydrogels. Gels of poly(2-hydroxyethyl methacrylate/methacrylic acid) (pHEMA/MAA) were chosen because of their common use as biomaterials; a composition having 2.5 mole % MAA and a crosslinker of tetraethylene glycol dimethacrylate (TEGDMA) at 0.15 mole % was used. Leaching solutions with pH 10 and 3, which lie above and below the dissociation point of the MAA moieties on the gel network, were used either alone or as 50% volume mixtures with ethanol or isopropanol. Execution of three shrink/swell cycles shows that as much as an 4% of additional material was lost from the gel compared to simple leaching in pH 6 water baths. Sequential exposure to acidic and basic aqueous alcohol solutions was shown to be a faster, but equivalent method of leaching as compared with sequential exposure to acid and base. A method for controlled swelling and shrinking was developed to allow delicate or irregularly shaped gel structures to be leached without damage due to stress associated with rapid swelling/shrinking.

INTRODUCTION

Leaching is a crucial step in production of polymeric biomaterials since it removes initiator, accelerator, solvents, unreacted monomer, oligomer and other byproducts remaining after polymerization. Such small molecules are generally acknowledged to be a major reason for biological reaction to implanted polymers (1). We show here that the ionized groups common to some classes of hydrogels and the composition of the leaching solvents can be manipulated to enhance greatly the leaching process over common methods wherein the leaching water bath is periodically changed. The method uses organic cosolvents and inorganic salts to control the swelling or deswelling during the leaching process. Our interest in leaching processes stemmed from the diverse times, ranging from one year

(2) to three days (3), that were allowed in prior work for the gels to equilibrate or be purified.

A gel with ionizable groups will generally exhibit two distinct levels of swollen volume in a high dielectric media, the larger one reflecting the extra force on the gel network caused by ionization of the groups on the gel. For example, gels composed of 2.5 % methacrylic acid (MAA), (0.15 mol%) tetraethylene glycol dimethacrylate (TEGDMA) and the remainder of 2-hydroxyethyl methacrylate (HEMA) have water contents of about 40% when swollen in acidic physiological saline and 60% when swollen in basic physiological saline (4). In basic solutions of lower ionic strength there is less counter shielding of the ionized species in the gel network and accordingly, both the force among the ions on the gel and the swollen gel volume are larger in such solutions. Organic cosolvents, such as ethanol or isopropanol, also increase the swollen volume; values of solvent content of over 95% are easily reached in basic water-ethanol solutions (5). The value of organic cosolvents is tied to their ability to solvate the hydrophobic regions of the gel and hence speed the process of network rearrangement that occurs during swelling.

Swollen volumes that exceed that of the unionized state are important in the leaching process because there is more bulk water (solvent) within the gel in such states. Yasuda et al. have shown that the permeability of the gel is increased by this water (6). Diffusional transport of aqueous solutes in such gels is primarily enhanced because the extra water increases the free volume of the system. For very large swollen volumes there is also increased permeability since sieve mechanisms become less restrictive. The principal concept is that in highly swollen gels molecular impurities are less likely to be ensnared in the water associated with the gel and instead, more closely approximate their behavior in simple solutions. Of particular interest is that some impurities, e.g. oligomer, can take a "free chain" form in the solutions that induce maximal gel swelling. Such species then would have sufficient freedom within the swollen gel network so they can move out of the gel by a process similar to diffusion of solvated species. With more constricted network sizes, the repitational movements of the hydrated oligomer would be the primary means for movement of such molecules (7); as these movements are typified by diffusivities of polymer strands, they are impractical for leaching. Yet over the longer period while the device is indwelling or implanted, such transport will be sufficient to deliver oligomer to the tissues interface, where it could cause irritation. Perhaps the most important point is that a dilutional reduction of impurities can be achieved by transferring the shrunken gel to pure media, expanding it in that media, and then shrinking it again. The change of swollen volume of gel discs in acidic and basic 50% solutions (volume/volume) of alcohol and water provides a sense of the potential of dilutional leaching. Values of swollen volume (cm^3 solvent/g. polymer) for gels with 2.5 mol% MAA in ethanol/water and isopropanol/water solutions respectively 9.3 and 5.5 for acidic solutions and 18.3 for both basic alcohol solutions (5). The key variables for such dilutional leaching are the organic cosolvent, solution pH and the ionic strength.

As not all organic solvents swell pHEMA gels (5) and as many of the solvents are themselves quite toxic, the choice of a solvent is important. Gels of pHEMA swell in water because of the hydrophilic nature of the alkyl side group. To induce additional swelling by opening the gel network, it is desirable to use an organic molecule that solvates the methacrylic backbone. Facts relevent to the selection of the organic species come from theories of swelling of polymers, and in particular, from the concept of changing the affinity of the polymer for the solvent relative to itself (8). Practicality demands that the organic solvent be miscible with water over a wide range of proportions and that the solvent be inexpensive in pure form. Alcohols are an obvious choice since the organic portion can solvate the backbone and the hydroxyl group can provide water miscibility and interaction with the hydrophilic side group of the HEMA. Ethanol is particularly attractive because it is a sterilizing agent at high concentrations and is biodegradable.

Certain protocols for manipulation of the swollen volume by use of the ionizable groups and organic cosolvents can produce undesirable large changes of gel volume for very slight changes of the leaching solvent. Following concepts elucidated by Tanaka et al. (9), the swollen volume is set by the interplay of rubber elasticity, polymer-polymer affinity and electrostatic interaction among the charged groups. The repulsion of like-charged ionic groups fixed to the polymeric network leads to larger swelling than occurs for unionized polymer in the same polymer mixture. It is known that for certain small changes of environmental condition, relatively large (approximately step-like) changes of swollen volume can occur because the balance among the events setting the swollen volume is drastically altered (9). Our work has shown that these changes of size can be rapid and lead to gel damage. For large gel structures, such changes of swollen volume may crack the gel as outer regions shrink or swell more rapidly than inner regions. Counter shielding of the ionized groups offers a convenient method of manipulating the amount of swelling that occurs upon ionization of the polymer, generally by allowing equilibrium to be established throughout the material before the level of counter shielding is changed. We have previously shown that the size of MAA/HEMA gels in pH 10 NaCl solutions is well behaved and can be logarithmically related to the water content with the relationship:

$$\text{Water Content (in\%)} = -23 \log C + 36.$$

where C is the concentration of NaCl in moles/liter and is valid between 0.01 and 1 (10).

The amounts of material that can be removed from typical HEMA/MAA copolymers, the special concerns needed to use such leaching techniques with delicate structures, and an apparatus for partially automating the leaching process are discussed below. These techniques are especially suited for thin structures with long-term exposure to tissue or implantation, such as contact lens or membrane for controlled drug release or containment of cells or tissue, such as pancreatic islets. Of special interest is that the leaching process enhances the speed and reproducibility of subsequent shrink/swell behavior. These aspects of gel behavior are crucial for drug delivery systems that contain ionizable gel membranes and enzyme systems, wherein substrates, such as glucose, are degraded to transiently alter the local pH and thus, the membrane permeability.

EXPERIMENTAL

Materials

The monomer, HEMA, was obtained from Alcolac Chemicals, Inc. (Baltimore, MD) under the trade name SIPOMER CL-100 and purified by the method discussed in reference 4. Analysis of the water, cyclohexane and carbon tetrachloride by HPLC should be performed prior to their use in the purification scheme. Carbon tetrachloride in particular can contain a substantial impurity that interferes with the quantitative analysis of HEMA and its contaminants, ethylene glycol dimethacrylate (EGDMA) and

MAA. The comonomer and crosslinker, MAA and TEGDMA, were purchased from Polysciences, Inc. (Warminster, PA); the inhibitor in each material was removed prior to use by passing it through a column packed with DEHIBIT-100, also from Polysciences. All water was distilled, deionized and boiled for 20 minutes to dispel carbon dioxide and to sterilize it. Isopropyl alcohol was distilled from bulk over magnesium turnings with a catalytic amount of mercuric chloride. Ethanol solutions were made with freshly opened bottles of absolute alcohol (Pharmco, Philadelphia, PA). All other reagents were ACS grade and were used as received.

Fabrication

Rods of pHEMA/MAA copolymer were prepared by polymerizing the monomeric solution inside 18.5 mm diameter polypropylene syringes (11). The monomers were in the following concentration: 6.21 g (97.35 mole %), MAA 0.100 g (2.5 mole %) and TEGDMA 0.023 g (0.15 mole %). Two milliliters of ethylene glycol were used as a cosolvent with an aqueous initiation system consisting of 1.0 ml ammonium peroxysulfate (0.044 M) and 1.0 ml sodium metabisulfite (0.443 M). Ingredients in these proportions produce gels with an initial solids content of 60%. All solutions were thoroughly degassed with nitrogen. Once the initiators were added to the syringe, the mixture was degassed for five additional minutes and the plunger of the syringe inserted. The sryinge was inverted, the nitrogen expelled, and the tip sealed. The sryinge was kept at ambient temperature (26°C) and was pressurized at 100 pounds per square inch for 24 hours to prevent nitrogen bubble formation within the gel.

The polymerized hydrogel rod was partially removed from the syringe, the syringe barrel mounted in the chuck of a lathe and the exposed gel cut with a razor-edged knife into disks 2 mm thick. The disks were then placed into pH 6 water and soaked for 7 days with daily exchanges of water. Since no buffers were used in the study, the pH was monitored twice each day. In this initial standardization procedure, the pH 6 condition was well maintained.

The pH 6 leached disks were then weighed, dried in vacuum at 1 mm Hg for 12 hours, then dried at 50 um Hg for 12-14 hours at 56°C in a drying pistol. Separate tests, which involved drying the gels for longer periods or with higher driving forces to remove the solvent, showed that this procedure produced a constant weight. Weights shown in Table I as dry weights at the start were obtained by this process.

Leaching Protocol

The disks from six independently prepared rods were used in this study to obtain data representative of the possible differences in molecular detail that occur in different polymerizations. The dry disks and 100 ml of a leaching medium were placed in jars with tightly sealing teflon lined caps. The leaching media, identified by letters for convenience in Table I, were:

A. Water at pH 3 maintained with the addition of HCL.

B. Water at pH 10 maintained with the addition of NaOH.

C. A solution of 50% ethanol : 50% water (volume:volume) with the water being at pH 10.

D. A solution of 50% isopropanol : 50% water (volume:volume) with the water being at pH 10.

Table I. Data and Analysis of the Leaching Protocol.

Order of Media 123456	Time* of Exposure.	Average Dry Wt. Start	Average Dry Wt. Final	% Loss	Standard Deviation
AAAAAA	24 hr.	0.1194g	0.1187g	0.63	0.32
A-----	6 days	0.1176g	0.1167g	0.76	0.28
B-----	6 days	0.1169g	0.1160g	0.79	0.24
BBBBBB	24 hr.	0.1189g	0.1180g	0.83	0.28
BABABA	24 hr.	0.1176g	0.1163g	1.11	0.29
BABABA	equil.	0.1189g	0.1151g	3.26	0.57
CACACA	24 hr.	0.1177g	0.1134g	3.66	0.32
DADADA	24 hr.	0.1119g	0.1071g	4.24	0.63

* Time of exposure is the time between changing of the media in sample jars.

The number and duration of leaching cycles for each medium are given in Table I. The gels were kept at room temperature for the entire leaching protocol. The pH of the solutions was regularly checked and when necessary, adjustments were made with HCL or NaOH. When cycling was complete, the gels were shrunk in pH 3 water and dried as described above. By shrinking the gels in pH 3 water prior to drying them, inclusion of sodium ions was avoided; this aspect of the protocol provided a consistent basis for comparison of treatments.

Gradient Leaching

Three gel disks, prepared in the fashion described above, were leached through a gradient procedure. The gels were suspended in a cell that was stirred magnetically and that had a maximum volume of 150 ml because of a drain part. The initial solvent in the cell was 1.0 molar sodium chloride solution at pH 10. A peristaltic pump added 5 ml aliquots of dilute to the cell at 30 minute intervals. The dilute added to the solvent in the cell during the first 48 hours was 0.1 molar sodium chloride solution at pH 10. The dilute for the next 48 hours was pH 10 water and that for the final 48 hours was 50% ethanol: 50% pH 10 water (volume:volume). At this point in the protocol the gels reached their maximum swollen volume and the dilute was charged to begin shrinking them. The first dilute was pH 10 water for 24 hours, the next 0.9% NaCl at pH 10 for 24 hours, and finally pH 3 water for 48 hours. Three stages achieved shrinkage by first diluting the organic swelling agent, then by reducing the ionic force among ions on the gel through increased counter shielding of the ion by the salt, and finally by neutralization of the gel. The gels were dried as discussed above.

To test the gradient leaching procedure on a delicate shape, a gel of complex shape was cast inside a polypropylene syringe, using the pressurized polymerization technique. A silicone rubber insert was placed in the syringe to provide the desired outer shape of the gel structure. To provide the inner shape, a core was made of teflon and stainless steel hypodermic tubing; the core fit inside the silicone insert in a way that allowed for the through flow of monomeric mixture. The HEMA-based gel was formulated with 2.5 mole% MAA, 0.15 mole% TEGDMA. Figure 1 shows the

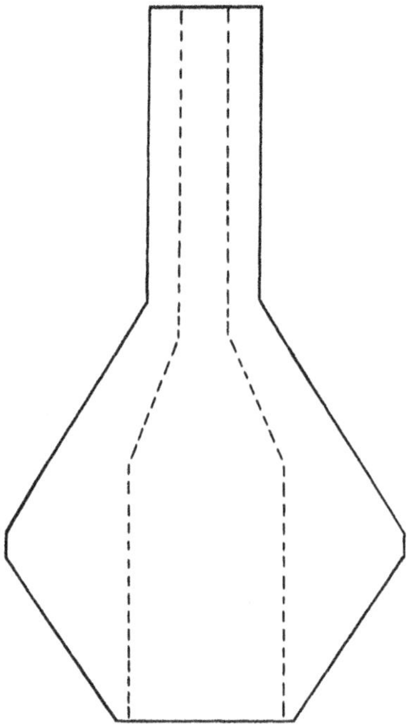

Figure 1. Cross-section of the nonuniform gel body leached by the gradient technique. The dashed lines illustrate the interior space in the axisymmetric object.

shape of the gel; the largest dimension is about 1 cm when shrunken in pH 3 water.

RESULTS AND DISCUSSION

The results of the leaching protocol are displayed in Table I. The data are presented in an order wherein the leaching protocol becomes more efficient, i.e., the protocol that removes the least material is presented first, etc. Note that the material lost in these tests is above and beyond that lost during an initial one week leaching period with daily exchanges of pH 6 water. Inspection of the disks after the initial standardization procedure showed that it had not totally purified them; the disks had a white color which disappeared upon use of the protocols described below. It is noteworthy that this standardization procedure exceeded a published method for gel purification (3). Data in the table show that simple soaking or soaking with exchange of the medium for identical medium, whether, pH 3 or pH 10, produce an equal effect. (Note: these results are for 6 days). In the acidic solvent there is about 40% water content for these gels. It is worth remembering that this water forms the five molecules of hydration associated with each monomeric unit (12). As this water is highly associated with the polymer, it does not form "pores" that allow enhanced diffusion of the impurities. Changing the swelling media from acid to base every twenty four hours leaches slightly more material than simply soaking the gel disks in identical solutions. Such a process does not allow these disks time to equilibrate completely to the swollen volume associated with the different pH states. It is likely that the surface regions of the disks are leached in a shrink/swell process similar to that of the last three entries of the table. There was visual evidence for this viewpoint as the gels developed clear edges but maintained opaque cores.

Of greatest interest is that the swell/shrink process associated with the last three entries produces significantly larger weight losses. A t-test shows that the mean weight loss of any one of the three last entries in Table I differs from that of any of the first entries with a confidence level of over 99.5%. In each of the processes for the last three entries, the gel disks reach a state of maximum swelling well before the media are exchanged. The overall time required for the three exchanges of acid and base was three weeks, which was substantially longer than the six day period required for the ethanol/water solutions. For practical reasons, it would appear that few would want to use simple exchange of acidic and basic water. Qualitatively, the disks leached with alcohol solvents have a greater clarity than the gels leached inpure aqueous solvents. During the protocols for the three last entries of the table, there was ample opportunity for the transport processes discussed in the introduction to move impurities into the bulk-like solution in the gel and then to expel it during the shrinking process. Clearly, the shrink/swell approach described here is a strong method for ridding gels of impurities.

Gel disks that have been leached by any of the last three methods using the shrink/swell process for at least three cycles show a more rapid response to changed environments. After the pH 6 standardization protocol a gel disk that is 2 mm thick requires from 6to 10 days to become clear, swollen and flat during its first exposure to pH 10 (solution B). After the first two stages of shrink/swell leaching, the same acid equilibrated disk will require only 1 to 2 days to reach final dimensions and clarity in base, thus clearly demonstrating the enhancement of speed of transformation between states of different swollen volume.

When the gel disks are removed from a solution that induces large swelling to one of the causes shrinkage or vice versa, the disk goes through a highly warped state and generally looses its optical clarity. Simple geometries of small size withstand the stresses induced by the change of environment without obvious damage. More complex shapes or thicker specimens can crack or split. To avoid these problems, the method of gradient leaching was investigated which used the concepts on swollen volume and environment which are treated in the introduction. A programmed change of salinity is easily achieved by periodically adding solutions of lesser salinity to a stirred vessel containing a more concentrated solution that can overflow. Disks treated in this manner remain flat and clear throughout the entire shrink/swell process. For a group of three disks, there was a mean loss of 2.63% associated with the gradient leaching process described above. This weight loss occured over the first cycle, when the danger of swelling induced damage is greatest. The gradient leaching process was also successfully applied to a nipple with the shape depicted in Figure 1. To provide a scale for this gel object, we note that in acidic solutions (40% water content), the tubular projection has a wall thickness of about 0.5 mm and an outer diameter of 2 mm. The major diameter of the bi-cone is 1.2 cm.

In conclusion, the use of swell/shrink protocols for leaching of HEMA/MAA copolymers offers the possibility for relatively quick removal of materials remaining after polymerization. Gels of this generic chemical composition will swell in the body since the physiological pH is sufficient to ionize the MAA moieties partially. The extent of ionization and subsequent swelling will be dependent upon salitiy, the amount of MAA incorporated and the crosslinked density. The potential for swelling makes it important that impurities be removed before implantation. Organic cosolvents, such as ethanol, offer a way to speed the process of leaching the copolymer greatly and, indeed, seem to be the preferred system because of the great degree of clarity with gels treated by these processes. The pH of the baths should be routinely checked during the leaching process lest the residual base or acid in the gel alter their pH. Use of such methods should be considered for items ranging from contact lens to membranes for drug delivery systems.

REFERENCES

1. R. I. Leininger, CRC Critical Rev. in Bioengineering, 333-381 (1972).
2. M. Refojo and H. Yasuda, J. Applied Poly. Sci., 92, 425 (1965).
3. R. A. Abrahams and S. H. Ronel, Polymer Prepints, 16 (2), 535-540 (1975).
4. L. Pinchuk, E. C. Eckstein and M. R. Van De Mark, J. Applied Pol. Sci., 29, 1749 (1984).
5. N. Lian, "Purity protocol, leaching, and swelling behavior of pHEMA/MAA hydrogels in water, electrolytes, and organic solvents", MS thesis, University of Miami, Coral Gables, FL.
6. H. Yasuda, A. Peterlin, C. K. Colton, K. A. Smith and E. W. Merrill, Die Makromolekulare Chemie., 126, 177-186 (1969).
7. P. G. deGennes, "Scaling Concepts in Polymer Physics", Cornell University Press, Ithaca, NY, 1979.
8. A. F. M. Barton, "Handbook of Solubility Parameters and Other Cohesion Parameters", CRC Press, Boca Raton, FL (1983).
9. T. Tanaka, D. Fillmore, S-T Sun, I. Nishio, G. Swislow and A. Shah, Phys. Rev. Let., 45, 1636-1639 (1980).
10. E. C. Eckstein, L. Pinchuk, M. R. Van De Mark and E. G. Barroso, Trans. 10th Ann. Meeting of Soc for Biomaterials, 7, 14 (1984).
11. L. Pinchuk and E. C. Eckstein, J. Biomed. Mater. Res., 15, 183-189 (1981).
12. N. Lian, M. R. Van De Mark and E. C. Eckstein, Polym. Prepints, 26, 168 (1985).

EFFICIENT REMOVAL OF EXCESS INORGANIC PHOSPHATE BY HEMOPERFUSION

N. S. Mason[a], E. C. Eckstein[b], T. C. Finley, S. A. Gagneux[c], M. P. Gold, B. D. Ward[c] and R. E. Sparks[a]

(a) Chemical Engineering Department
Washington University, St. Louis, MO 63130
(b) Department of Biomedical Engineering
University of Miami, Coral Gables, FL 33124
(c) Hospital Supply Research Division
Sandoz-Wander, Salt Lake City, UT 84106

In uremia, the inability of the diseased kidney to clear phosphate causes an increase in blood phosphate concentration leading to disorders associated with secondary hyperparathyroidism. Hemodialysis does not efficiently remove phosphate because of the low concentration gradient across the dialysis membrane. When the phosphate concentration is out of control it may be beneficial to add a phosphorus removal cartridge to the hemodialysis circuit. A phosphate sorbing blood perfusion cartridge was designed and tested _in vitro_ and in sheep. Phosphate sorbing fibers and beads were made by extrusion of a suspension of an aluminum gel in a viscous solution of cellulose acetate into an aqueous gelling bath. Typically, 64 gm of dry fibers packed into 4.4 cm diam. x 5.5 cm long cartridges removed 800-1000 mg phosphorus from blood in 5 hours.

INTRODUCTION

Patients with kidney diseases, especially those who have been on hemodialysis for a long time, often have symptoms that are associated with chronically elevated concentrations of inorganic phosphate in blood. The symptoms may include, skeletal pain, changes in bone mineralization, spontaneous fractures and calcium phosphate-like deposits in soft tissues and blood vessels (1). Excess of phosphate in serum above the normal (3-5 mg/100) causes a decrease in calcium ion concentration which in turn causes the parathyroid to secrete an excess amount of parathyroid hormone (PTH). High concentrations of PTH plus the inability of the diseased kidney to hydroxylate 25-hydroxycholecalciferol to the active forms of vitamin D, are the principal biochemical factors underlying most of these symptoms.

Hemodialysis is less effective in removing phosphate than other solutes because the concentration driving force is low relative to the amounts to be transferred. Physicians control the serum phosphate concentration by restricting phosphorus in the diet to 0.8 to 1 gm/day and by prescribing large doses of oral aluminum hydroxide or aluminum carbonate antacids (4-5 gm/day.

Despite these measures, in some patients serum phosphate may remain out of control. In this case one might propose to adsorb the phosphate ion directly from the blood stream during hemodialysis. This paper describes an initial approach to develop a cartridge which would remove inorganic phosphate and would be considered in series with the hemodialyzer. The goal was to assemble a cartridge that could remove 1000 mg phosphorus from blood in 6 hours at a blood flow rate of 200 ml/min. It would have a blood priming volume of 45 ml or less and a pressure drop of 40 torr or less. This device would remove about 1/2 of the phosphorus ingested by the patient between dialysis periods, the remainder would leave via the GI tract.

EXPERIMENTAL DETAILS

Fabrication of the Packing for the Cartridge

The packing to adsorb phosphate was based on an aluminum hydroxide sorbent available as finely divided (mean size = 15 μm) spray dried spheres. This sorbent has low toxicity and is more specific for phosphate than anion exchange resins. It has been evaluated for oral use at the Washington University School of Medicine (2). Figure 1 shows the adsorption isotherms for phosphate on these materials (3). The binder was mixed with a solution of cellulose acetate in acetone and formamide. This thick slurry was extruded through dies into a water gelling bath.

The purpose of the cellulose acetate, (CA), was to prevent blood from contacting the aluminum hydroxide sorbent surface directly. Cellulose acetate was known from previous work to be a good film former and to be at least equal to cellophane (used in hemodialyzers) in blood compatibility. Formamide draws water into CA causing it to become highly swollen, thus decreasing the resistance to mass transfer.

Either beads or fibers were produced. To produce the beads, the

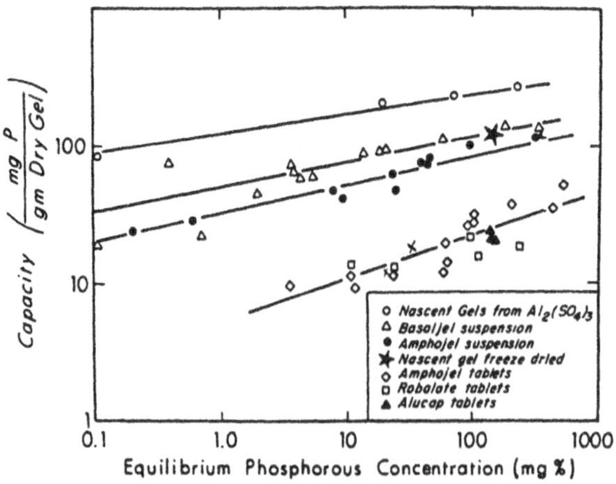

Figure 1. <u>in vitro</u> phosphorous binding capacity of liquid and dry gels as a function of the equilibrium phosphorous concentration.

extruding cylindrical fibers were chopped by means of a wire passing in front of the die. When fibers were produced they were stored as coils. Residual solvent was removed by extensive washing with deionized water. The packing was soaked in a 30 vol % solution of glycerol in water and dried in air. Most of the preprototype cartridges were packed using heat-shrink tubing and incorporating a plastic funnel at each end.

The following manufacturing variables were investigated: beads vs. fibers, size of packing, ratio of binder to CA, addition of ethanol, and drying/sterilization cycles. The effect of these variables on the performance of the packing was measured by batch adsorption tests. The packing was contacted with phosphate buffer at 37°C and placed in a shaker. The initial phosphorus concentration was 10 mg/100 ml. Most runs were of 6 hours duration. Initial and final phosphate concentration were run on an autoanalyzer using a colorimetric method based on the reduction of phosphomolybdate to produce an intensely blue color (5).

Cartridge Design

The results of batch sorption tests were used to generate an empirical prediction equation using a statistical package (6). The following equation accounts for 82.4 % of the spread from the mean of all the experimental points:

$$Y = 249.9 - (5.43\ X1) + (6.45\ X1X2) + (1.81\ X1X4) + (10.7\ X1X5) - (25.80\ X2) - (18.22\ X2) - (18.22\ X3) \qquad (\pm\ 23.5)$$

Where Y is the weight of beads required to remove 1 gm of phosphorus,
X1 is the bead diameter (mm)
X2 is the ratio of binder to CA,
X3 is the volume % formamide divided by 10,
X4 is the concentration of CA (gm/100 ml),
X5 represents the effect of drying the beads (0 not dried, 1 dried).

(The concentration of additive was found not to improve the prediction).

Figure 2 shows that as the sorbent content increases, larger beads may be used to absorb the same amount of phosphate. The rate of diffusion of phosphate into the bead determines how much is taken up in 6 hours, the diffusion path is shorter when the sorbent concentration is higher.

The curvature observed with the 150 gm bead line (_._._._.) is caused by an experimental artifact. During the sorption test when the beads are shaken in buffer, the larger beads which do not contain as much sorbent, are softer and deform more readily. After sufficient flexing they develop cracks and finally break up entirely. This increases the area for diffusion of phosphate and decreases the path length, hence a lower concentration of gel appears to be adequate.

In a packed column this would not happen. Hence for cartridge design, the line should follow about the same curvature as the other bead weight lines. Tests in a packed column through which buffer was pumped showed that typically 10 % more beads were required to pick up the same amount of phosphate than in the shaker tests, but due to their labor-saving convenience the shaker tests were preferred.

The pressure-drop through the cartridge is also an important consideration. The well-known Ergun equation (7), was used to compute the pressure drop shown as a dashed line on Figure 2.

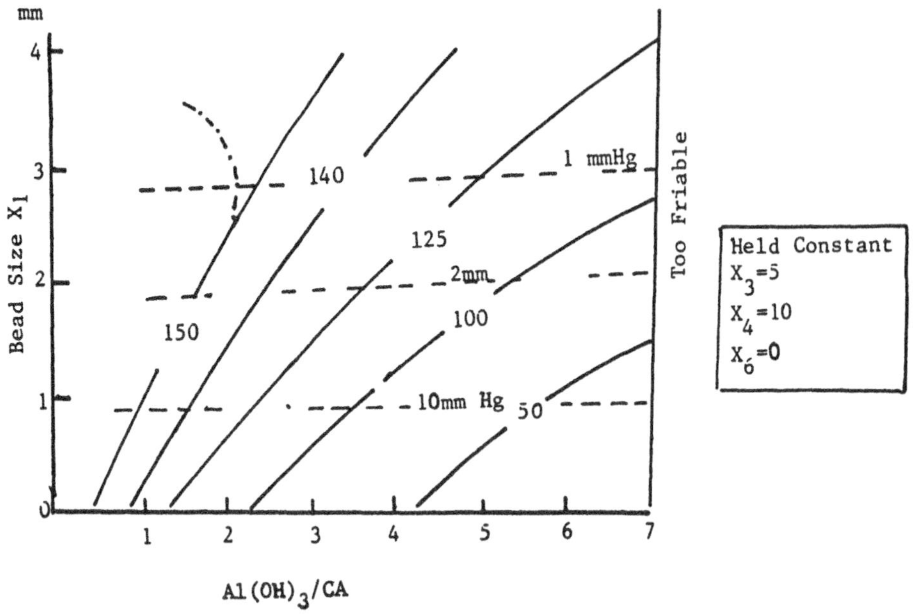

Figure 2. Bead weight required to remove 1 gm P in 6 hours.

Tests of Flow Distribution

Because stagnant regions in such a cartridge could cause blood clots, two types of flow distribution tests were carried out. In the first test, while buffer was pumped through the cartridge, a small volume of a solution of crystal violet dye was injected just upstream of the cartridge. The cartridge was then disassembled and it was observed that the dye was distributed among essentially 75 to 100 % of the fibers despite a mean residence time of 6 seconds. The few white fibers were single fibers and not packets.

The second test was also a dye pulse test but it used a pulse of sodium chloride solution which is not adsorbed by the packing and is easy to detect. Figure 3 is a sketch of the experimental set-up. Residence time distribution was measured after a pulse of 0.5 ml of 1.3 % NaCl was injected. Mixing was studied with the procedure of Levenspiel (8) which assumes a vessel with negligible entrance exit effect and D = 0 at inlet and outlet. Table I shows the results obtained.

Dispersion numbers, D/vL, greater than 0.2 are considered large. Despite the lower mean residence time for the fibers (because of the

Table I. Results of Tracer Response Study.

	mean residence time (min) corrected	uncorrected	$\sigma_t^{2'}$	σ^2	D/vL
Beads	1.76 ± 0.14	1.96 ± 0.12	1.65 ± 0.47	0.43	0.3
Fibers	1.02 ± 0.08	1.21 ± 0.07	0.86	0.59	0.53

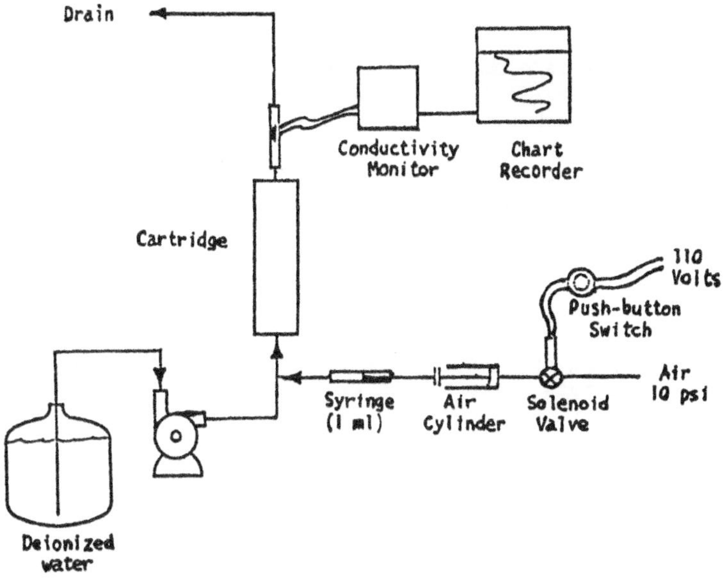

Figure 3. Apparatus for pulse test measurement.

lower priming volume) the dispersion was greater than with beads. However, the entrance geometry was not the same for the two packings and may have contributed to the greater dispersion number for the fibers.

<u>Control of Erosion of Fine Particles</u>

Even though NIH or FDA have not formulated standards for particle emissions from devices such emission must be low (no more than a few μg per treatment) (9). Effluent from the device was pumped through a Nucleopore ultrafilter and the membrane is weighed on a Cahn microbalance. Table II shows some early results. To decrease the generation of fines further, two approaches may be considered. The fibers may be sheathed by coextruding pure polymer or the binder may be pumped as a slurry into capillary tubes.

Table II. Release of Fines from Beads and Fibers.

	wt. fines*	N
1.7 mm beads 5 parts binder/1 part CA	170 ± 80	8
0.5 mm fibers 3 parts binder/1 part CA	27 ± 15	8
No packing	10	

* Fines in μg generated in 2 hours by 100 gm packing

Perfusion Tests

In vitro tests were carried out as follows: citrated porcine blood, to which was added sufficient phosphate to give an inorganic phosphorus concentration in plasma of 10mg/100 ml, was pumped through the cartridge by means of a roller pump at the rate of 200 ml/min. Periodically samples were removed from the inlet and outlet and analyzed for inorganic phosphorus and for total phosphorus. After a 6 hour run the packing itself was dissolved in perchloric acid and analyzed for phosphate.

In vivo experiments were carried out with anesthetized sheep and were run with and without added phosphate. These sheep were used to evaluate blood oxygenators at the same time and the phosphate binder cartridge was usually connected in parallel with a blood oxygenator. Calibrated roller pumps insured the rate through each section.

RESULTS

Sorption of Inorganic Phosphate from Blood

Figure 4 shows the changes in phosphorus concentrations as porcine blood is circulated over large beads (4.3 mm) in a cartridge. In this partical test, 770 mg P was taken up by 157 gm of beads in 6 hours. The initial concentration of inorganic phosphorus was adjusted to 2.5 times the normal (human) plasma concentration. To pick up 1 gm of phosphorus would require 202 gm of beads. From our buffer tests only 159 gm of beads were predicted. Therefore 27 % more sorbent is required to achieve the same performance in blood. It may be noted that while the total phosphate decreased with time, inorganic phosphate concentration actually increased during the runs with live sheep.

Figure 5 shows the *in vivo* performance of the large (4.3 mm). Table

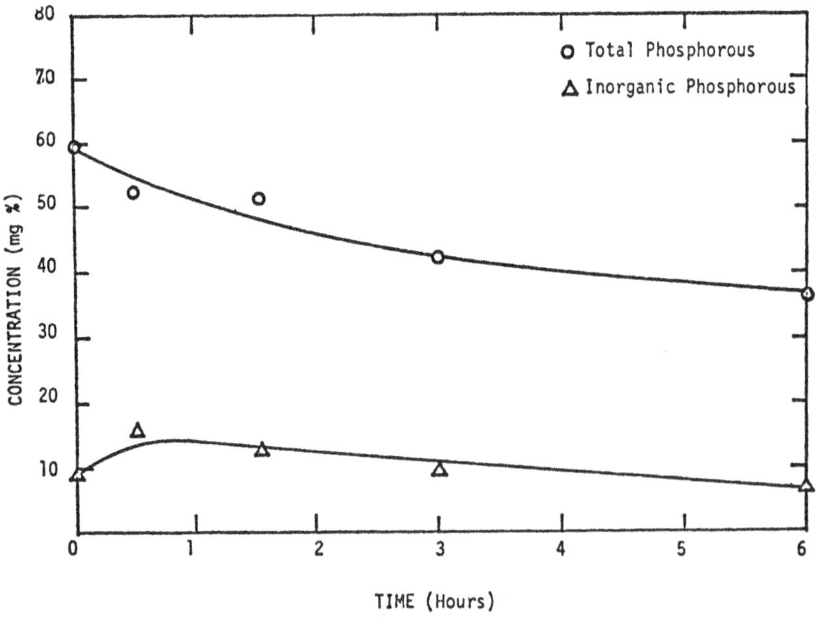

Figure 4. Inorganic and total phosphorus in entering blood.

III compares the performance data of small and large beads. The pressure drop was 3-7 times higher than calculated. Sometimes this was caused by small clots which formed near the inlet. In vivo evaluation of cartridges packed with 0.8 mm fibers was also carried out. This is summarized in Table IV. The fact that the inlet concentration of inorganic phosphorus increases even while phosphate was being removed was unexpected. It may indicate that organic P is transformed to inorganic P as the inorganic P is removed. The causes of this effect and its implications are not well understood.

<u>Changes in Calcium and Magnesium Concentration During Perfusion</u>

A net amount of calcium is removed from plasma and blood (Figure 6) during the first 1-2 hours of contact. After this time the concentration stabilizes. The change occuring with blood is less pronounced than in

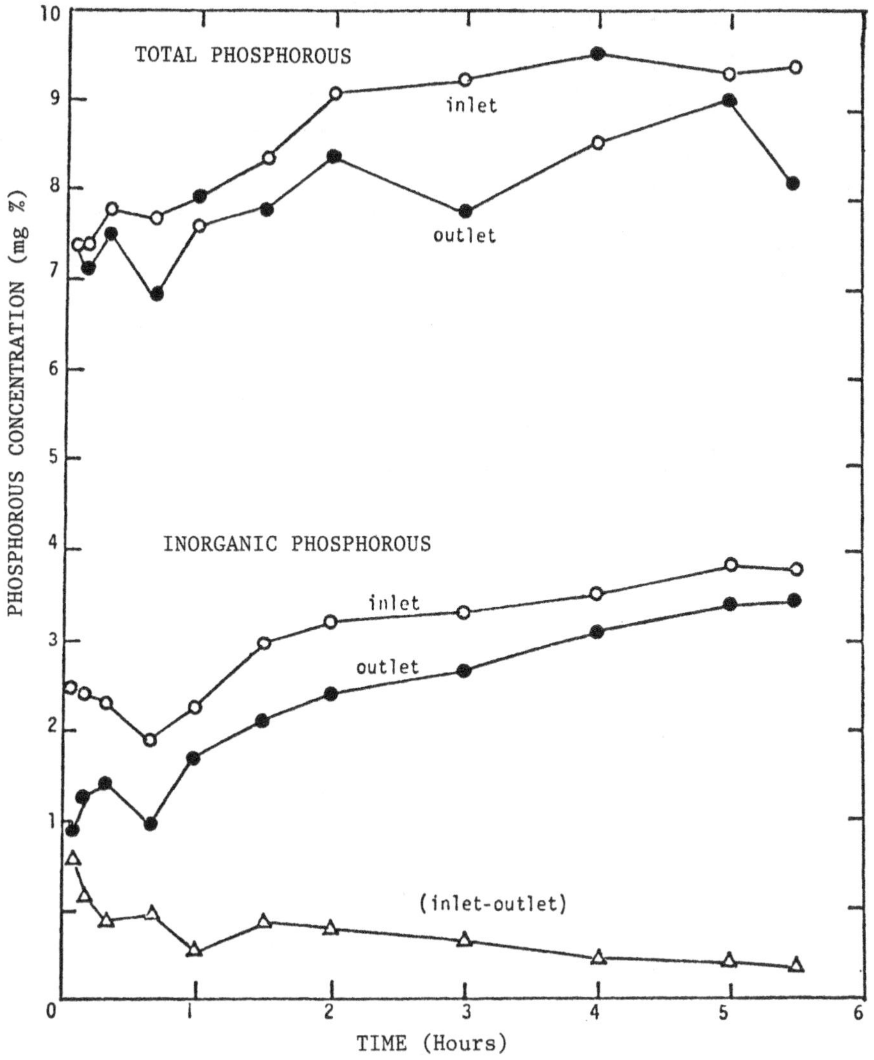

Bead Diameter = 4.3 mm
Weight of Beads = 167 gm
Animal: Sheep

Figure 5. Phosphorus blood levels during in vivo perfusion.

Table III. Summary *In vivo* performance.

Bead Diameter mm	Wt Beads g	Test Time hr	Mean Conc. inorg P mg/100ml	P absorbed		Pressure Drop	
				blood analysis mg	packing analysis mg	calcd. mm Hg	measured mmHg
4.3	157	5.5	4.1	437	433	1	7.0
1.7	167	4.7	6.0	1200	1300	9.0	25

Table IV. Sorption Data on Fibers.

Cartridge No.	Weight of fibers (wet) (gm)	Perfusion time hrs.	Phosphorus removed fr. (mg)	Phosphorus recovered on fibers (mg)
1	52.3	1	76	219
2	183	2	569	683
3	184	1.5	427	688
4	188	5.0	1075	1245

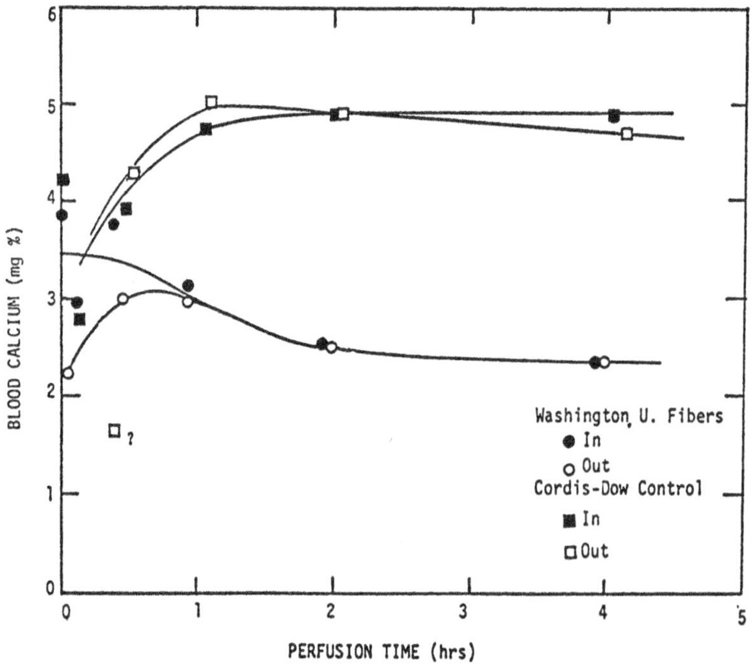

Figure 6. Whole blood calcium changes during perfusion over fibers.

plasma. Most of the calcium is present in plasma red blood cells contain little, but calcium is present in several bound and free forms (10) and these forms are readily interconverted.

When the packing was contacted with a simple calcium salt such as calcium acetate, no calcium was taken up. From plasma, calcium could be taken up as a complex ion, e.g., especially $CaPO_4$, because a small amount of inorganic phosphate was added in the test to set the initial concentration of P. The ratio of Ca/P taken up was highly variable from test to test.

Magnesium concentration decreased in blood and plasma in contact with the cartridge for the first half hour of perfusion. It remained unchanged after that time. The ratio of Mg/P taken up was 0.23 to 0.077.

CONCLUSION

While it is relatively simple to meet the design goals for sorption of phosphate, it may be concluded that the biochemical complexity of a fluid such as blood causes many subtle and not so subtle effects, which must be carefully evaluated before one could plan for patient tests.

REFERENCES

1. E. Slatopolsky, S. Cagler, J. P. Pennell, D. Tagant, J. Canterbury, E. Reiss and N. S. Bricker, "On the Pathogenesis of Hyperparathyroidism in Chronic Renal Disease", J. Clin. Invest., 50, 492 (1971).
2. E. Rutherford, S. King, B. Perry, J. Blondin, T. Finley, N. S. Mason, R. E. Sparks and E. Slatopolsky, "Use of a New Phosphate Binder in Chronic Renal Insufficiency", Kidney International. 17, 528-534 (1980).
3. E. Rutherford, A. Mercado, K. Hruska, N. S. Mason, R. E. Sparks, S. Klahr and E. Slatopolsky, "An Evaluation of a Phosphorus Binding Agent", Trans. Amer. Soc. Artif. Int. Organs, XIX, 446 (1973).
4. M. P. Gold, "Aluminum Hydroxide Gels for Phosphate Control in Uremia", Master's Thesis, Washington University, St. Louis, 1974.
5. G. Gomori, "A Modification of the Colorimetric Phosphorus Determination", J. Lab. Clin. Sci., 27, 955-959 (1942).
6. J. O. Kim and F. J. Kohout, in: "Statistical Package for the Social Sciences", N. H. Nie, ed. McGraw Hill, New York 1975.
7. R. B. Baird, W. E. Stewart and E. N. Lightfoot, "Transport Phenomena", John Wiley and Sons, New York, 1960, p. 200.
8. O. Levenspiel, "Chemical Reaction Engineering", John Wiley and Sons, New York, 1962, p. 243-267.
9. D. L. Coleman, in: "Guidelines for Phsicochemical Characterization of Biomaterials", NIH Publication No. 80-2186, Sept. 1980, p. 91-93.
10. J. T. Irving, "Calcium and Phosphorus Metabolism", John Wiley and Sons, New York, 1973, p. 54.

CONTRIBUTORS

Roland Barbucci (259),
C.R.I.S.M.A. University of Siena,
Le Scotte - Nuovo Poloclinico,
53100 Siena (Italy)

G. C. Battistone (111),
U. S. Army Institute of Dental Research,
Walter Reed Army Medical Center,
Washington, DC 20307-5300

Manuela Benvenuti (259),
C.R.I.S.M.A. University of Siena,
Le Scotte - Nuovo Policlinico,
53100 Siena (Italy)

W. A. Berthelson (159),
Medtronic, Inc.,
Fridley, MN 55432

Stephen D. Bruck, Ph.D. (125),
Biomedical Technology Risk Assessment Group,
Stephen D. Bruck Associates, Inc.,
7315 Wisconsin Avenue, Suite 727-E,
Bethesda, Maryland 20814 USA

J. A. Cameron (101),
Institute of Materials Science,
University of Connecticut,
Storrs, Connecticut 06268

H. Carr, Jr. (111),
U. S. Army Insitute of Dental Research,
Walter Reed Army Medical Center,
Washington, DC 20307-5300

Charles E. Carraher, Jr. (311,325),
Florida Atlantic University,
Department of Chemistry,
Boca Raton, Florida 33431

Robert A. Casper (75),
Southern Research Institute,
2000 Ninth Ave., South,
Birmingham, AL 35255-5305

Cynthia H. Cholakis (253),
Department of Chemical Engineering and Applied Chemistry,
University of Toronto,
Toronto, Ontario, M5S 1A4

C. C. Chu (185),
Department of Textile and Apparel,
Martha Van Rensselaer Hall,
Cornell University,
Ithaca, NY 14853-4401

M. W. Davis (147,159),
Medtronic, Inc.,
Frindley, MN 55432

Bruce Dunn (117),
Engineering Department,
University of California,
Los Angeles, CA

Richard L. Dunn (75),
Southern Research Institute,
2000 Ninth Ave., South,
Birmingham, AL 35255-5305

E. C. Eckstein (373,381),
Department of Biomedical Engineering,
University of Miami,
Coral Gables, FL 33124

Peter G. Edelman (101),
Insitute of Materials Science,
University of Connecticut,
Storrs, Connecticut 06268

Charles L. Farris (55),
Gulf South Research Institute,
P.O. Box 26518,
New Orleans, LA 70186-6518

Paolo Ferruti (259),
C.R.I.S.M.A. University of Siena,
Le Scotte - Nuovo Policlinico,
53100Siena (Italy)

T. C. Finley (381),
Chemical Engineering Department,
Washington University,
St. Louis, MO 63130

S. A. Gagneux (381),
Chemical Engineering Department,
Washington University,
St. Louis, MO 63130

J. L. Garnett (355),
School of Chemistry,
The University of New South Wales,
NSW, Australia 2232

Charles G. Gebelein (277),
Department of Chemistry,
Youngstown State University,
Youngstown, Ohio 44555

Paul H. Gebert (31,55),
Gulf South Research Institute,
P.O. Box 26518,
New Orleans, LA 70186-6518

Lawrence Gettleman (31,55),
Gulf South Research Institute,
P.O. Box 26518,
New Orleans, LA 70186-6518

David Giron (311,325),
Department of Microbiology and Immunology,
Wright State University,
Dayton, Ohio 45435

Jean Goggins (215),
Meadox Medicals, Inc.,
Oakland, NJ 07436

M. P. Gold (381),
Hospital Supply Research Division,
Sandoz-Wander,
Salt Lake City, Utah 84106

David Grainger (229),
University of Utah,
Department of Pharmaceutics,
Salt Lake City, Utah 84112

J. R. Griffith (63),
Chemistry Division,
Naval Research Laboratory, Code 6120,
Washington, DC 20375-5000

Mutaz B. Habal (17,41,117),
Human Resources Institute,
University of South Florida,
The Tampa Bay Craniofacial Center,
Tampa, FL

R. D. Harris (335),
Laboratory of Molecular Biophysics,
The University of Alabama at Birmingham,
P.O. Box 311/University Station,
Birmingham, Alabama 35294

A. D. Haubold (131),
CarboMedics, Inc.,
1300-D East Anderson Lane,
Austin, TX 78752

Takatoshi Hayashi (285),
Department of Agricultural Biochemistry,
Tottori University,
Tottori 680, Japan

Hitoshi Higashijima (285),
Department of Agricultural Biochemistry,
Tottori University,
Tottori 680, Japan

Shigehiro Hirano (285),
Department of Agricultural Biochemistry,
Tottori University,
Tottori 680, Japan

Anne Hiltner (215),
Departments of Biomedical Engineering and Macromolecular Science,
Case Western Reserve University,
Cleveland, Ohio 44106

Samuel J. Huang (101),
Insitute of Materials Science,
University of Connecticut,
Storrs, Connecticut 06268

T. Q. Hung (131),
CarboMedics, Inc.,
1300-D East Anderson Lane,
Austin, TX 78752

A. C. Ibay (111),
U.S. Army Institute of Dental Research,
Walter Reed Medical Center,
Washington, DC 20307-5300

Wan F. Ip (253),
Department of Chemical Engineering and Applied Chemistry,
University of Toronto,
Toronto, Ontario, M5S 1A4

B. Jadhav (87),
Johnson & Johnson Products, Inc.,
Orthopaedic Research,
U.S. Route #1 South,
North Brunswick, NJ 08902

S. V. Jankiewicz (355),
School of Chemistry,
The University of New South Wales,
NSW, Australia 2232

Richard D. Jones (215),
Division of Surgical Research,
St. Luke's Hospital,
Cleveland, Ohio 44104

George Joseph (249),
Division of Biosurface Technology,
BMT Wing, Sree Chitra Tirunal Institute for Medical
Sciences and Technology,
Trivandrum 695012, India

Atsushi Kato (299),
Faculty of Pharmaceutical Sciences,
Science University of Tokyo,
Shinjuku-ku, Tokyo 162, Japan

Benjamin S. Kelley (75),
Southern Research Institute,
2000 Ninth Ave., South,
Birmingham, AL 35255-5305

Sung Wan Kim (229),
University of Utah,
Department of Pharmaceutics,
Salt Lake City, Utah 84112

Junko Kinugawa (285),
Department of Agricultural Biochemistry,
Tottori University,
Tottori 680, Japan

Tamotsu Kondo (299),
Faculty of Pharmaceutical Sciences,
Science University of Tokyo,
Shinjuku-ku, Tokyo 162, Japan

F. Kummer (87),
Hospital for Joint Diseases,
Orthopaedic Institute,
301 E 17th Street,
New York, NY 10003

Donald L. Leake (41,117),
Dental Research Institute,
University of California,
Los Angles, CA

Ralph J. LeBoeuf, Jr. (55),
Gulf South Research Institute,
P.O. Box 26518,
New Orleans, LA 70186-6518

L. E. Lecaroz (185),
Department of Textiles and Apparel,
Martha Van Rensselaer Hall,
Cornell University,
Ithaca, NY 14853-4401

W. B. Lehman (87),
Hospital for Joint Diseases,
Orthopaedic Institute,
301 E. 17th Street,
New York, NY 10003

Shu-Tung Li (171),
Meadox Medicals, Inc.,
Oakland, NJ 07436

N. D. Lian (373),
Department of Biomedical Engineering,
University of Miami,
Coral Gables, FL 33124

Raymond Linville (325),
Air Force Wright Aeronautical Laboratories (AFWAL/MLPJ)
Wright-Patterson AFB, Ohio 45433

M. A. Long (355),
School of Chemistry,
The University of New South Wales,
NSW, Australia 2232

M. M. Long (335),
Laboratory of Molecular Biophysics,
The University of Alabama at Birmingham,
P.O. Box 311/University Station,
Birmingham, Alabama 35294

Isabel Lopez (311,325),
Department of Chemistry,
Wright State University,
Dayton, Ohio 45435

Annie Y. C. Ma (11),
Institute of Biomedical Engineering,
University of Toronto,
Toronto, Ontario, Canada M5S 1A8

N. S. Mason (381),
Chemical Engineering Department,
Washington University,
St, Louis, MO 63130

R. A. Miller (111),
U.S. Army Institute of Dental Research,
Walter Reed Army Medical Center,
Washington, DC 20307-5300

Daniel Murphy (277),
Department of Chemistry,
Youngstown State University,
Youngstown, Ohio 44555

Mila Nocentini (259),
C.R.I.S.M.A. University of Siena,
Le Scotte - Nuovo Policlinico,
53100 Siena (Italy)

Yasuharu Noishiki (285),
Department of Agricultural Biochemistry,
Tottori University,
Tottori 680, Japan

T. V. Norregaard (1),
Neurosurgical Service,
Massachusetts General Hospital,
Boston, MA 02114

E. Nyilas (131),
CarboMedics, Inc.,
1300-D East Anderson Lane,
Austin, TX 78752

Teruo Okano (229),
University of Utah,
Department of Pharmaceutics,
Salt Lake City, Utah 84112

J. G. O'Rear (63),
Chemistry Division Laboratory, Code 6120,
Washington, DC 20375-5000

D. P. Orgill (1),
Fibers and Polymers Laboratories,
Massachusetts Institute of Technology,
Cambridge, MA 02139

Z. Oser (69),
Research Division,
Johnson & Johnson Products Inc.,
501 George St.,
New Brunswick, NJ 08903

R. E. Phillips (131),
Intermedics. Inc.,
P.O. Box 617,
Freeport, TX 77541

K. U. Prasad (335),
Laboratory of Molecular Biophysics,
The University of Alabama at Birmingham,
P.O. Box 311/University Station,
Birmingham, Alabama 35294

H. Ralph Rawls (31,55),
Gulf South Research Insitute,
P.O. Box 26518,
New Orleans, LA 70186-6518

M. W. Rohovsky (87),
Johnson & Johnson Products, Inc.,
Orthopaedic Research,
U.S. Route #1 South,
North Brunswick, NJ 08902

D. F. Sangster (355),
CSIRO, Division of Chemical Physics,
Sutherland,
NSW, Australia 2232

W. C. Schoene (1),
Department of Pathology (Neuropathology),
Brigham and Women's Hospital,
Boston, MA 02114

Michael V. Sefton (253),
Department of Chemical Engineering and Applied Chemistry,
University of Toronto,
Toronto, Ontario, M5S 1A4

C. P. Sharma (249),
Division of Biosurface Technology,
BMT Wing, Sree Chitra Tirunal Institute for Medical
Sciences and Technology,
Trivandrum 695012, India

J. Silver (1),
Department of Developmental Genetics and Anatomy,
School of Medicine,
Case Western Reserve University,
Cleveland, Ohio 44106

R. E. Sparks (381),
Chemical Engineering Department,
Washington University,
St. Louis, MO 63130

K. B. Stokes (147,159),
Medtronic, Inc.,
Fridley, MN 55432

A. Strongwater (87),
Hospital for Joint Diseases,
Orthopaedic Institute,
301 E 17th Street,
New York, NY 10003

H. Sugano (335),
Laboratory of Molecular Biophysics,
The University of Alabama at Birmingham,
P.O. Box 311/University Station,
Birmingham, Alabama 35294

R. J. Thoma (131),
CarboMedics, Inc.,
1300-D East Anderson Lane,
Austin, TX 78752

D. C. Tunc (87),
Johnson & Johnson Products Inc.,
Orthopaedic Research,
U.S. Route #1 South,
North Brunswick, NJ 08902

D. W. Urry (335),
Laboratory of Molecular Biophysics,
The University of Alabama at Birmingham,
P.O. Box 311/University Station,
Birmingham, Alabama 35294

M. R. Van De Mark (373),
Department of Chemistry,
University of Miami,
Coral Gables, FL 33124

Janet Vargo (31,55),
Gulf South Research Insitute,
P.O. Box 26518,
New Orleans, LA 70186-6518

Paul Y. Wang (11),
Insitute of Biomedical Engineering,
University of Toronto,
Toronto, Ontario, Canada M5S 1A8

B. D. Ward (381),
Hospital Supply Research Division,
Sandoz-Wander,
Salt Lake City, Utah 84106

I. V. Yannas (1),
Fibers and Polymers Laboratories,
Massachusetts Institute of Technology,
Cambridge, MA 02139

N. T. Zervas (1),
Neurosurgical Service,
Massachusetts General Hospital,
Boston, MA 02114

W. Zingg (253),
Department of Chemical Engineering and Applied Chemistry,
University of Toronto,
Toronto, Ontario, M5S 1A4
Division of Surgical Research Hospital for Sick Children,
Toronto, Ontario, M5G 1X8

INDEX

GPSR Compliance

The European Union's (EU) General Product Safety Regulation (GPSR) is a set of rules that requires consumer products to be safe and our obligations to ensure this.

If you have any concerns about our products, you can contact us on ProductSafety@springernature.com

In case Publisher is established outside the EU, the EU authorized representative is:

Springer Nature Customer Service Center GmbH
Europaplatz 3
69115 Heidelberg, Germany

Batch number: 10755782

Printed by Printforce, the Netherlands